547 H537m
High resolution
chromatography : a practical
approach

 S0-ERN-442

WITHDRAWN

High Resolution Chromatography

The Practical Approach Series

SERIES EDITOR

B. D. HAMES
Department of Biochemistry and Molecular Biology
University of Leeds, Leeds LS2 9JT, UK

See also the Practical Approach web site at **http://www.oup.co.uk/PAS**
★ **indicates new and forthcoming titles**

Affinity Chromatography
★ Affinity Separations
Anaerobic Microbiology
Animal Cell Culture (2nd edition)
Animal Virus Pathogenesis
Antibodies I and II
Antibody Engineering
★ Antisense Technology
Applied Microbial Physiology
Basic Cell Culture
Behavioural Neuroscience
Bioenergetics
Biological Data Analysis
Biomechanics—Materials
Biomechanics—Structures and Systems
Biosensors
Carbohydrate Analysis (2nd edition)
Cell-Cell Interactions
The Cell Cycle
Cell Growth and Apoptosis
★ Cell Separation

Cellular Calcium
Cellular Interactions in Development
Cellular Neurobiology
★ Chromatin
★ Chromosome Structural Analysis
Clinical Immunology
Complement
★ Crystallization of Nucleic Acids and Proteins (2nd edition)
Cytokines (2nd edition)
The Cytoskeleton
Diagnostic Molecular Pathology I and II
DNA and Protein Sequence Analysis
DNA Cloning 1: Core Techniques (2nd edition)
DNA Cloning 2: Expression Systems (2nd edition)
DNA Cloning 3: Complex Genomes (2nd edition)
DNA Cloning 4: Mammalian Systems (2nd edition)

- ★ Drosophila (2nd edition)
- Electron Microscopy in Biology
- Electron Microscopy in Molecular Biology
- Electrophysiology
- Enzyme Assays
- Epithelial Cell Culture
- Essential Developmental Biology
- Essential Molecular Biology I and I
- ★ Eukaryotic DNA Replication
- Experimental Neuroanatomy
- Extracellular Matrix
- Flow Cytometry (2nd edition)
- Free Radicals
- Gas Chromatography
- Gel Electrophoresis of Nucleic Acids (2nd edition)
- ★ Gel Electrophoresis of Proteins (3rd edition)
- Gene Probes 1 and 2
- Gene Targeting
- Gene Transcription
- ★ Genome Mapping
- Glycobiology
- ★ Growth Factors and Receptors
- Haemopoiesis
- ★ High Resolution Chromotography
- Histocompatibility Testing
- HIV Volumes 1 and 2
- ★ HPLC of Macromolecules (2nd edition)
- Human Cytogenetics I and II (2nd edition)
- Human Genetic Disease Analysis
- ★ Immobilized Biomolecules in Analysis
- Immunochemistry 1
- Immunochemistry 2
- Immunocytochemistry
- ★ In Situ Hybridization (2nd edition)
- Iodinated Density Gradient Media
- Ion Channels
- ★ Light Microscopy (2nd edition)
- Lipid Modification of Proteins
- Lipoprotein Analysis
- Liposomes
- Mammalian Cell Biotechnology
- Medical Parasitology
- Medical Virology
- MHC Volumes 1 and 2
- ★ Molecular Genetic Analysis of Populations (2nd edition)
- Molecular Genetics of Yeast
- Molecular Imaging in Neuroscience
- Molecular Neurobiology
- Molecular Plant Pathology I and II
- Molecular Virology
- Monitoring Neuronal Activity
- Mutagenicity Testing
- ★ Mutation Detection
- Neural Cell Culture
- Neural Transplantation
- Neurochemistry (2nd edition)

Neuronal Cell Lines
NMR of Biological Macromolecules
Non-isotopic Methods in Molecular Biology
Nucleic Acid Hybridisation
Oligonucleotides and Analogues
Oligonucleotide Synthesis
PCR 1
PCR 2
★ PCR 3: PCR In Situ Hybridization
Peptide Antigens
Photosynthesis: Energy Transduction
Plant Cell Biology
Plant Cell Culture (2nd edition)
Plant Molecular Biology
Plasmids (2nd edition)
Platelets
Postimplantation Mammalian Embryos
Preparative Centrifugation
Protein Blotting

★ Protein Expression Vol 1
★ Protein Expression Vol 2
Protein Engineering
Protein Function (2nd edition)
Protein Phosphorylation
Protein Purification Applications
Protein Purification Methods
Protein Sequencing
Protein Structure (2nd edition)
Protein Structure Prediction
Protein Targeting
Proteolytic Enzymes
Pulsed Field Gel Electrophoresis
RNA Processing I and II
★ RNA-Protein Interactions
Signalling by Inositides
Subcellular Fractionation
Signal Transduction
★ Transcription Factors (2nd edition)
Tumour Immunobiology

High Resolution Chromatography
A Practical Approach

Edited by
PAUL MILLNER
School of Biochemistry and Molecular Biology,
University of Leeds, Leeds

OXFORD
UNIVERSITY PRESS

Great Clarendon Street, Oxford OX2 6DP

Oxford University Press is a department of the University of Oxford
and furthers the University's aim of excellence in research, scholarship,
and education by publishing worldwide in
Oxford New York
Athens Auckland Bangkok Bogotá Buenos Aires Calcutta
Cape Town Chennai Dar es Salaam Delhi Florence Hong Kong Istanbul
Karachi Kuala Lumpur Madrid Melbourne Mexico City Mumbai
Nairobi Paris São Paulo Singapore Taipei Tokyo Toronto Warsaw
and associated companies in Berlin Ibadan

Oxford is a registered trade mark of Oxford University Press

Published in the United States
by Oxford University Press Inc., New York

© Oxford University Press 1999

All rights reserved. No part of this publication may be reproduced,
stored in a retrieval system, or transmitted, in any form or by any means,
without the prior permission in writing of Oxford University Press.
Within the UK, exceptions are allowed in respect of any fair dealing for the
purpose of research or private study, or criticism or review, as permitted
under the Copyright, Designs and Patents Act, 1988, or in the case
of reprographic reproduction in accordance with the terms of licenses
issued by the Copyright Licensing Agency. Enquiries concerning
reproduction outside those terms and in other countries should be
sent to the Rights Department, Oxford University Press,
at the address above.

This book is sold subject to the condition that it shall not, by way
of trade or otherwise, be lent, re-sold, hired out, or otherwise circulated
without the publisher's prior consent in any form of binding or cover
other than that in which it is published and without a similar condition
including this condition being imposed on the subsequent purchaser

Users of books in the Practical Approach Series are advised that prudent
laboratory safety procedures should be followed at all times. Oxford
University Press makes no representation, express or implied, in respect of
the accuracy of the material set forth in books in this series and cannot
accept any legal responsibility or liability for any errors or omissions
that may be made.

A catalogue record for this book is available from the British Library

Library of Congress Cataloging in Publication Data
(Data available)

ISBN 0-19-963649-4 (Hbk)
0-19-963648-6 (Pbk)

Typeset by Footnote Graphics,
Warminster, Wilts
Printed in Great Britain by Information Press, Ltd,
Eynsham, Oxon.

Preface

The traditional biochemical track, from protein purification and sequencing through to gene cloning, has largely been subverted by a panoply of molecular genetic techniques, augmented in the last few years by the massive output from the many systematic sequencing programmes. However, the growing ease of gene discovery has, in many cases, meant a methodological shift from protein isolation for discovery to protein isolation (often of recombinant material) for characterization and determination of function. Increasingly the proteins which are to be resolved are low abundance proteins such as regulatory proteins, present in low concentrations within the cell, in some cases even after overexpression of the recombinant protein. This situation has required the development of new chromatographic approaches, and the optimization of existing techniques, in order to give the highest resolution and yield reasonable quantities of protein for study.

The first four chapters of this volume address some of the instrumental aspects of chromatography. Often, attention to small details within the optimization of separation protocols can yield significant improvements in resolution and yield. This is particularly the case with 'non-specific' separations based on the physico-chemical aspects of the macromolecules to be resolved, such as charge, size, and hydrophobicity (Chapter 1) and which often represent the initial clean-up step, prior to the application of affinity, or other specific separations. In addition, the development of new and increasingly sophisticated detector technologies (Chapter 3) now allows much more information, concerning the separated macromolecules emerging from a chromatographic column, to be determined. In some cases, specialized chromatographic 'niches', such as microbore chromatography, which allow the resolution of extremely small amounts of sample to be performed, require special attention to chromatographic parameters that could normally be taken for granted in laboratory scale preparations. The application of microscale separation procedures is discussed in detail in Chapter 2 by Yang. Finally, the technology of capillary electrophoresis and allied techniques is covered by Hu and Martin (Chapter 4). Although not strictly chromatographic, capillary electrophoretic techniques are often used in conjunction with high resolution chromatographic procedures and represent a powerful suite of analytical procedures, applicable to a wide range of biomolecules.

The remainder of this volume covers a range of chromatographic procedures whose common feature is that they are based on the interaction of a specific ligand with its target protein or other macromolecule. Some of the contributions to this volume are focused on specific groups of molecules, for example complex oligosaccharides and glycosylated proteins (Chapter 5), nucleotide-binding proteins (Chapter 6), proteins that bind free and chelated metal ions

Preface

(Chapter 10), and DNA-binding proteins (Chapter 11). Others are more generally applicable to a wide range of macromolecules and describe the use of peptides (Chapter 7), inhibitors (Chapter 8), and antibodies (Chapter 9) as affinity ligands, although the latter approaches still rely on specific interaction between immobilized ligand and target macromolecule. In one or two cases, similar protocols are provided in different chapters. Whilst these could have been edited with brevity in mind, it was felt that that it was preferable to give the choice of the slightly different style of protocol offered by different contributing authors.

There are doubtless many other groups of ligands whose chromatographic use could be described. For example, many of the more recent expression vectors permit the overexpression of proteins which bear a specific affinity tag. Appropriate chromatographic matrices are then available to purify the tagged recombinant protein with high selectivity. As the use of these matrices are already well covered by the manufacturer's instructions they are accordingly not included. However, it is hoped that this volume is reasonably comprehensive and will provide the necessary information in most experimental situations to enable development of rapid and effective purification protocols.

Leeds   PAUL MILLNER
September 1998

Contents

List of contributors   xvii
Abbreviations   xix

Section A. Techniques and equipment

1. Separation by charge, size, and hydrophobicity   1

Paul A. Millner

1. Introduction   1

2. Choice of separation method   1
    Gel filtration   1
    Ion exchange chromatography   5
    Hydrophobic interaction chromatography (HIC)   9
    Affinity chromatography   11

3. Choice of hardware   11
    Pre-packed or self-packed columns   11
    Fluidics   12
    Detection   13
    Fraction collection   14

4. Packing columns   14

5. Running the column   16
    Equilibration   16
    Loading the sample   16
    Elution   18
    Column sanitation   20

References   21

2. Practice and benefits of microcolumn HPLC   23

Frank J. Yang

1. Introduction   23

2. Practice of microcolumn HPLC   25
    Preparation and routine care of microcolumns   25
    Injection techniques   29
    Detector considerations   32
    Pumping system considerations   33

Flow splitting	35
New generation microbore column pumping systems	37
3. Conclusion	44
References	46

3. Detection devices

Raymond P. W. Scott

1. Introduction	49
Detector specifications	49
2. The classification of detectors	53
Bulk property and solute property detectors	54
Mass-sensitive and concentration-sensitive detectors	54
Specific and non-specific detectors	54
3. Detecting devices for high efficiency columns	55
4. Light absorption detectors	57
The fixed wavelength UV detector	58
The multi-wavelength UV detector	61
5. Fluorescence detectors	63
The single wavelength fluorescent detector	63
The multi-wavelength fluorescence detector	66
6. Electronic detectors	66
The electrical conductivity detector	66
The electrochemical detector	68
The multi-electrode array detector	70
7. Light scattering detectors	72
References	75

4. Capillary electrophoresis of peptides and proteins

Bi-Huang Hu and Lenore M. Martin

1. Introduction	77
2. The capillary	78
Choosing a capillary diameter and length	78
Modifying the capillary wall	79
3. Buffer	90
Buffer selection	90
Preparing the CE buffer	92

49

77

x

Contents

4. Sample preparation	93
Preparation of samples for CZE, MEKC, and cIEF	93
Preparation of samples for CGE	94
Sample concentration	94
Desalting biological samples	97
5. Modes used in CE	97
Capillary zone electrophoresis (CZE)	98
cITP	99
Micellar electrokinetic capillary chromatography (MEKC)	100
Capillary isoelectric focusing (cIEF)	100
Capillary gel electrophoresis (CGE)	102
6. CE detection strategies	105
UV detection	105
Fluorescence detection	106
CE-mass spectrometry	107
7. Applications of CE to peptide and protein research	108
Evaluation of sample purity	108
Protein mapping	108
Binding constant determinations	109
Peptide and protein molecular weight determinations	111
Micropreparative CE	112
8. CE of peptides and proteins: future prospects	113
References	114

Section B. Selected affinity approaches

5. Lectins as affinity probes   119

Philip S. Sheldon

1. Properties and uses of lectins—a historical perspective	119
2. The types of oligosaccharides found in glycoproteins	120
N-linked oligosaccharides	120
O-linked oligosaccharides	124
3. Carbohydrate recognition and lectin specificity	125
Abrus precatorius	125
Agaricus bisporus	125
Aleuria aurantia	125
Allomyrina dichotoma	125
Amaranthus caudatus	125
Anguilla anguilla	125
Arachis hypogea	128
Artocarpus integrifolia	128
Bauhinia pururea	128

Canavalia einsiformis	128
Codium fragile	128
Datura stramonium	129
Dolichos biflorus	129
Erythrina species	129
Galanthus nivalis	129
Glycine max	129
Griffonia (Bandeiraea) simplicifolia	129
Helix pomatia	130
Lens culinaris	130
Limax flavus	130
Lotus tetragonolobus	130
Lycopersicon esculentum	130
Maackia amurensis	130
Phaseolus limensis	131
Phaseolus vulgaris	131
Phytolacca americana	131
Pisum sativum	131
Ricinus communis	132
Sambucus nigra	132
Solanum tuberosum	132
Triticum vulgare	132
Ulex europaeus	133
Vicia faba	133
Vicia villosa	133
Wisteria floribunda	133

4. Use of lectins: practical aspects   133
    Immobilization of lectins   135
    Purification of glycoproteins by lectin affinity chromatography   136

5. Lectin affinity chromatography of oligosaccharides, glycosylasparagines, and glycopeptides   138
    Cleavage of glycans from glycoproteins   138
    Purification of oligosaccharides, glycosylasparagines, and glycopeptides by serial lectin chromatography   142

References   146

6. Nucleotide– and dye–ligand chromatography   153

Bo Mattiasson and Igor Yu. Galaev

1. Introduction   153

2. Nucleotide–ligand chromatography   153
    Nucleotides coupled via base   156
    Nucleotides coupled via sugar   160
    Nucleotides coupled via phosphate   161
    Protein purification strategy   165

Contents

3. Dye–ligand chromatography	165
Dye coupling techniques	165
Purity, leaching, and toxicity of triazine dyes	170
Designer dyes	171
Polymer shielding	173
Routine purifications	175
References	186

7. Synthetic peptides as affinity ligands 191

Gary J. Reynolds and Paul A. Millner

1. Introduction	191
Peptides (production, purification, and handling)	191
Immobilization chemistries	192
Choice of matrix	196
2. Coupling peptides via amine groups directly to the matrix	196
Cyanogen bromide (CNBr) activation	196
Cyanuric chloride (2,4,6,-trichloro-1,3,5-triazine)	198
Glutaraldehyde	202
N,N'-carbonyldiimidazole (CDI)	203
N-hydroxysuccinimide (NHS) esters	205
3. Coupling peptides via sulfydryl groups	206
Reduction of disulfide bonds	207
Determination of free sulfhydryl groups using DTNB	209
Activation	209
Coupling cysteinyl peptides to iodoacetyl or maleimide activated gels	211
4. Coupling peptides and introducing spacer arms using carbodiimides	212
5. Chromatography on peptide affinity gels	213
Designing the peptide ligand	213
Choosing the elution conditions	214
References	214

8. Drugs and inhibitors as affinity ligands 217

Nigel M. Hooper

1. Introduction	217
2. Choice of ligand	217
Specificity	218
Maintenance of binding affinity when coupled	218

  Affinity and reversibility of binding   219
  Stability   219

 3. Choice of matrix   219

 4. Choice of spacer arm   220

 5. Coupling method   222

 6. Determination of ligand coupling efficiency   225

 7. Affinity chromatography   226

 8. Method of elution   227

 9. Affinity chromatography with biotinylated ligands   228

 10. Regeneration of the affinity matrix   230

Acknowledgements   231

References   231

9. Immunoaffinity chromatography   233

Paul Cutler

 1. Introduction   233

 2. Antibody selection   236

 3. Applications of immunoaffinity separations   239

 4. Matrices   240

 5. Activation and immobilization   242
  Cyanogen bromide   244
  Immobilization with *N*-hydroxysuccinimide   247
  Avidin immobilization of biotinylated immunoglobulins   250
  Site-directed immobilization with protein A/protein G   250
  Site-directed immobilization with hydrazide   251

 6. Determining coupling efficiency   253

 7. Equipment and operation   254
  Immunoaffinity purification   254
  Sample preparation   255
  Binding   255
  Flow rates   256
  Pre-elution washing   256
  Elution   256
  Matrix stability and ligand leakage   259
  Analysis of purified proteins   259

References   259

Contents

10. Strategies and methods for purification of Ca^{2+}-binding proteins   263

Sandra L. Fitzpatrick and David M. Waisman

1. Introduction   263
2. Purification of the annexin Ca^{2+}-binding proteins   265
3. The heparin column   274
4. The Chelex-100 competitive Ca^{2+}-binding assay   277

References   281

11. Purification of DNA-binding proteins   283

Hiroshi Handa, Yuki Yamaguchi, and Tadashi Wada

1. General procedure for DNA-binding proteins   283
2. Preparation of nuclear extracts   283
   - Cell culture   284
   - Preparation of nuclear extracts   284
3. Heparin–Sepharose   286
   - Protein fractionation by heparin–Sepharose   286
   - Batchwise method for determination of salt concentration of elution buffer   289
4. DNA affinity chromatography for purification of sequence-specific DNA-binding proteins   290
   - Design and preparation of oligonucleotides   291
   - Preparation of DNA oligomers   292
   - Preparation of DNA affinity resins   293
   - DNA affinity Sepharose   294
   - DNA affinity latex beads   296
   - Troubleshooting   299
   - Comments   301

References   301

Appendix   303

Index   309

Contributors

PAUL CUTLER
Analytical Sciences, SmithKline Beecham Pharmaceuticals, New Frontiers Science Park-North, Coldharbour Road, The Pinnacles, Harlow CM19 5AW, UK.

SANDRA L. FITZPATRICK
Cell Regulation Research Group, Department of Medical Biochemistry, University of Calgary, 3330 Hospital Drive NW, Calgary, Alberta T2N 4N1, Canada.

IGOR YU. GALAEV
Department of Biotechnology, Center for Chemistry and Chemical Engineering, Lund University, PO Box 124, S-221 00, Lund, Sweden.

HIROSHI HANDA
Faculty of Bioscience and Biotechnology, Tokyo Institute of Technology, 4259 Nagatsuta-cho, Modori-ku, Yokohama 226–8501, Japan.

NIGEL M. HOOPER
School of Biochemistry and Molecular Biology, University of Leeds, Leeds LS2 9JT, UK.

BI-HUANG HU
The Department of Biomedical Sciences, College of Pharmacy, The University of Rhode Island, Kingston, Rhode Island 02881–0809, USA.

LENORE M. MARTIN
The Department of Biomedical Sciences, College of Pharmacy, The University of Rhode Island, Kingston, Rhode Island 02881–0809, USA.

BO MATTIASSON
Department of Biotechnology, Center for Chemistry and Chemical Engineering, Lund University, PO Box 124, S-221 00, Lund, Sweden.

PAUL A. MILLNER
School of Biochemistry and Molecular Biology, University of Leeds, Leeds LS2 9JT, UK.

GARY J. REYNOLDS
School of Biochemistry and Molecular Biology, University of Leeds, Leeds LS2 9JT, UK.

RAYMOND P. W. SCOTT
Georgetown University, Washington DC, USA.

Contributors

PHILIP S. SHELDON
Horticultural Research International, Wellesbourne, Warwick CV35 9EF, UK.

TADASHI WADA
Faculty of Bioscience and Biotechnology, Tokyo Institute of Technology, 4259 Nagatsuta-cho, Modori-ku, Yokohama 226–8501, Japan.

DAVID M. WAISMAN
Cell Regulation Research Group, Department of Medical Biochemistry, University of Calgary, 3330 Hospital Drive NW, Calgary, Alberta T2N 4N1, Canada.

YUKI YAMAGUCHI
Faculty of Bioscience and Biotechnology, Tokyo Institute of Technology, 4259 Nagatsuta-cho, Modori-ku, Yokohama 226–8501, Japan.

FRANK J. YANG
Micro-Tech Scientific Inc., 140 South Wolf Road, Sunnyvale, California 94086, USA.

Abbreviations

ACE	affinity capillary electrophoresis
BSA	bovine serum albumin
CDI	N,N'-carbonyldiimidazole
CE	capillary electrophoresis
CE/MS	combination of capillary electrophoresis and mass spectrometry
CGE	capillary gel electrophoresis
Ches	2-(cyclohexylamino)ethanesulfonic acid
cIEF	capillary isoelectric focusing
cITP	capillary isotachophoresis
CM	carboxymethyl
CMC	critical micelle concentration
CNBr	cyanogen bromide
Con A	concanavalin A
CTAC	cetyltrimethylammonium chloride
CZE	capillary zone electrophoresis
DEAE	diethylaminoethyl
DIFP	diisopropylfluorophosphate
DMF	dimethylformamide
DMS	disuccinimidyl succinate
DMSO	dimethyl sulfoxide
DSC	N,N'-disuccinimidyl carbonate
DTAB	dodecyltrimethylammonium bromide
DTE	dithioerythritol
DTNB	5,5′-dithiobis(2-nitrobenzoic acid)
DTT	dithiothreitol
EDTA	ethylenediaminetetraacetic acid
ELISA	enzyme-linked immunosorbent assay
EOF	electro-osmotic flow
ESI-MS	electrospray ionization mass spectrometry
FAB-MS	fast atom bombard mass spectrometry
FACS	fluorescence-activated cell sorting
HAP	hydroxylapatite
HEC	hydroxyethylcellulose
Hepes	N-2-hydroxyethylpiperazine-N-2-ethanesulfonic acid
HIC	hydrophobic interaction chromatography
HPIAC	high performance immunoaffinity chromatography
HPLC	high performance liquid chromatography
HPMC	hydroxypropylmethylcellulose
HTAB	hexadecyltrimethylammonium bromide
KPi	potassium phosphate buffer (KH_2PO_4 and K_2HPO_4)

Abbreviations

mAb	monoclonal antibody
MAPS	3-methacryloxypropyltrimethoxysilane
MEKC	micellar electrokinetic capillary chromatography
MES	2-(*N*-morpholino)ethanesulfonic acid
MS/MS	tandem mass spectrometry
NE	nuclear extract(s)
NHS	*N*-hydroxysuccinimide
PAGE	polyacrylamide gel electrophoresis
PBS	phosphate-buffered saline
PBSA	PBS containing 0.1% (w/v) sodium azide
PCV	packed cell volume
PDMS	plasma desorption mass spectrometry
PEI	poly(ethylene imine)
PEO	poly(ethylene oxide)
pI	isoelectric point
PMSF	phenylmethylsulfonyl fluoride
RP-HPLC	reversed-phase high performance liquid chromatography
SDS	sodium dodecyl sulfate
SDS–PAGE	polyacrylamide gel electrophoresis in the presence of SDS
St-GMA	styrene-glycidyl methacrylate
STI	soybean trypsin inhibitor
TEMED	*N,N,N′,N′*-tetramethylethylenediamine
Tes	*N*-tris(hydroxymethyl)methyl-2-aminoethanesulfonic acid
Tricine	*N*-tris(hydroxymethyl)methylglycine
UV	ultraviolet

Section A
Techniques and equipment

1

Separation by charge, size, and hydrophobicity

PAUL A. MILLNER

1. Introduction

The material in this chapter, which covers perhaps the most generic forms of chromatography, is written very much from a pragmatic standpoint and results from experience of chromatography of every sort. Where appropriate, links have been drawn with later chapters, that cover more specific chromatographic approaches, although many of the procedures and precautions within this chapter are generic to all chromatographic procedures. General maxims for chromatographic work are to do the fewest chromatography runs possible and as quickly as possible. These aims stem not only from a desire for speed, but often from constraints imposed by the viability of the sample or target protein. The protocols presented are basic chromatographic procedures which cover the most common situations encountered.

2. Choice of separation method

The most common initial chromatographic step is ion exchange chromatography, principally due to the inherently high capacity of most ion exchange media. However, HPLC ion exchange media (e.g. Pharmacia Mono Q, Mono S) are often capable of impressive resolution and are sometimes used as a final stage. Since proteins are eluted from ion exchange media in salt containing buffers, gel filtration or hydrophobic interaction chromatography are then often used as subsequent steps prior to a final affinity step. A typical preparation procedure for GTP-binding proteins (1), which employs successive separations on DEAE–Sephacel (ion exchange), Ultragel AcA34 (gel filtration), heptyl-amine agarose (hydrophobic interaction), and hydroxylapatite (affinity) illustrates this point.

2.1 Gel filtration

This is often referred to as gel permeation chromatography, which is perhaps a more accurate description. A good general description of gel filtration is provide in refs 2 and 3. Media for this type of chromatography are usually small beads which contain an open meshwork of cross-linked polysaccharides, e.g. Sephadex (Pharmacia). In some cases the beads are composed of polymerized acrylamide or other polymers, or of silica, e.g. BioGel (Bio-Rad), many HPLC gel filtration media. A number of factors will affect the gel permeation media that is chosen for a particular application.

2.1.1 Exclusion limit

The main factor to consider in gel permeation chromatography is the exclusion limit, also called the M_r cut-off. Gel permeation relies on the fact that within a particular sized mesh smaller molecules will diffuse within the beads rather than in the eluent stream for a larger proportion of time than will larger molecules. This is depicted in *Figure 1*. Accordingly, larger molecules will travel down a gel filtration column faster, and elute sooner, than smaller molecules. For any particular gel matrix there will be an apparent molecular weight beyond which molecules are excluded, i.e. cannot penetrate in the beads' mesh, and elute from the column without ever entering the beads. This molecular weight is the exclusion limit and the volume at which molecules of greater molecular weight emerge is the void volume (V_o). For any particular column this will be the smallest volume any component can emerge in and it represents the space between the beads. Typically, V_o is about 30–50% of the total column volume. More precisely, separation in gel permeation is by Stokes radius. The best way of visualizing this quantity is to think of two proteins of the same M_r but different shapes, one being spherical and the other

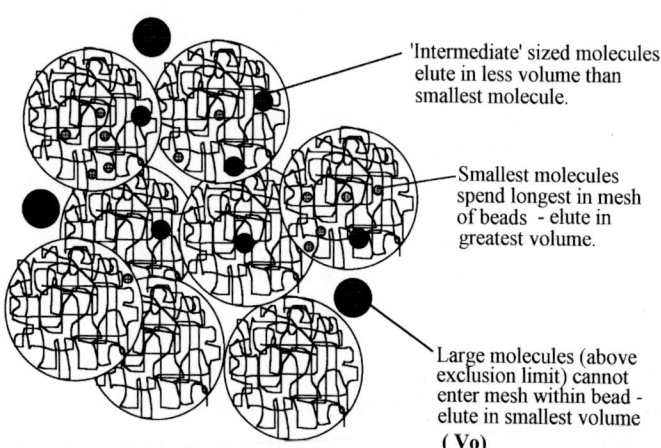

Figure 1. Principle of gel filtration chromatography. Molecules of a greater size than the exclusion volume elute in the void volume (V_o), whilst those below the exclusion limit emerge at an elution volume (V_e) characteristic of their molecular size. The ratio V_e/V_o is dependent on the gel filtration medium but *independent* of column size.

elongated (like a bread roll). The elongated protein will sweep out a much larger diameter (imaginary) sphere than the spherical protein and will elute at an apparently higher M_r (and lower volume). Determination of V_o is described in *Protocol 1*.

Protocol 1. Determination of V_o[a]

Equipment and reagents
- Packed column, e.g. 1 cm × 40 cm Sephadex G25 column
- UV/visible spectrophotometer or detector
- Eluting buffer: 25 mM Tris–HCl pH 7.5, 100 mM NaCl
- 1 mg/ml dextran blue[b] in eluting buffer

Method

1. Apply 100 μl[c] dextran blue solution to the column.[d] If a detector is used then mark the chart at this point.

2. Elute the column with elution buffer at a flow rate of 0.5 ml/min. If a detector is not used, collect 0.5 ml fractions and measure A_{280}.

3. Determine the volume at which the peak concentration of dextran blue emerges; this is V_o for that particular column. If measuring chart output from a detector, the volume to the centre of the peak should be measured.

[a] V_o is a useful quantity to know for any particular column, since molecules which elute in the void volume cannot have bound to the column matrix. Conversely molecules which should elute at V_o, but actually elute at greater volumes must be interacting with the column matrix.
[b] This is a blue dye linked to dextran; its average molecular weight is in excess of several million.
[c] The volume applied should be small, around 0.2–0.4% of the column volume.
[d] Sample application is most easily performed using a sample loop (see Section 5.2).

An additional important factor is to have sufficient counterions in the elution buffer to prevent interaction with the matrix; for example with Sephadex or Sepharose this occurs via interaction with polar hydroxy groups. Interaction effects are easily revealed by greater than expected elution volumes and/or a trailing edge to what should otherwise be a symmetrical peak. They are easily suppressed by inclusion of 100 mM NaCl or other appropriate salts in the eluting buffer.

2.1.2 Choice of exclusion limit

The grade of gel filtration medium chosen, i.e. exclusion limit, will depend on the purpose for which it is required, i.e. separation of a number of proteins on a size basis or 'desalting' (see below).

For separation of several proteins of differing sizes the grade of gel filtration medium should be chosen so that the target proteins cluster in the middle of the separation range. Normally, suppliers of gel media will indicate both the upper and lower M_r limits, for a particular grade of medium, between which useful separation of proteins will take place.

By choosing a grade of gel so that the target protein is in the middle of this range its separation from other proteins is maximized. A plot of log M_r versus the ratio V_e/V_o may also be used analytically, to determine the apparent M_r of a protein (3–5) or its shape (6). This can be extremely useful to study the oligomeric state of proteins in solution.

For 'desalting', a grade of gel should be chosen which has a M_r exclusion limit below the target protein(s). In many cases Sephadex G25 (M_r cut-off ≈ 5 kDa) is sufficient although if small proteins or peptides are to be desalted, Sephadex G10 (M_r cut-off > 700 Da) must be used. Desalting columns are very versatile in that they also provide a rapid, gentle means of exchanging the target protein(s) from one solution, i.e. with salt or detergent present, to another solution, i.e. without salt or with a different detergent. The general principle behind 'desalting' is that the column is equilibrated in the eventual solution (B) in which the target protein(s) are to end up. The sample (in solution A) is then applied and elution continued with eluent B. Since the protein(s) are beyond the exclusion limit it emerges in the void volume whilst the salts or detergent to be removed, being small, remain behind in the gel and elute at much greater volume.

2.1.3 Choice of gel bead size

Gel permeation matrices generally come in a range of bead sizes. These will vary from 50–100 μm for, e.g. Sephadex and other soft gels which can be self-packed, down to 10 μm or less for HPLC gel media which must usually be bought pre-packed. Sometimes the gel media will be described as 'coarse', 'medium', 'fine', etc. with the 'coarser' grades corresponding to larger bead sizes and higher permissible flow rates. In general, choose 'coarse' or 'medium' grades for desalting and 'fine' or even 'superfine' for separations on size basis. The effect of decreasing bead size is to make an increasingly larger surface area available for permeation and to consequently give better resolution, i.e. narrower peaks. The trade off is that smaller beads means using slower flow rates in soft gels, e.g. Sephadex or higher pressures in HPLC-type gels.

2.1.4 Sample size

Whilst it is the most widely applicable chromatographic method, gel permeation chromatography suffers from the disadvantage that the target protein(s) do not bind to the gel matrix and so there is nothing to counter diffusion and hence sample dilution. The only solution to this is to minimize the sample volume applied (7). As a rule of thumb, for separation by size of a number of proteins keep down to a sample size of 1–2% of total gel volume in the column, whilst for desalting applications, up to 10% of total gel volume may be applied. (Note: the actual optimum value to load will vary slightly with the

1: Separation by charge, size, and hydrophobicity

make and grade of gel filtration medium. Recommended sample loadings for specific gels can normally be found in the technical literature supplied.)

2.2 Ion exchange chromatography

Along with gel filtration, ion exchange chromatography is the other most commonly applicable chromatographic procedure. The principle requirement to permit ion exchange separations is that the target protein(s) is charged at the pH in which the chromatography is carried out: the protein(s) can either be anionic (negatively charged) or cationic (positively charged). At neutral pH many proteins are anions, although there are many important exceptions, e.g. DNA-binding proteins are usually strongly cationic. The charged protein is bound to the ion exchange medium which carries the opposite charge, unbound proteins washed away, and then the target protein(s) eluted from the medium by increasing concentration of a competing ion of the same charge as the target protein(s) (*Figure 2*), hence 'ion exchange chromatography'. Most commonly, in order to elute the target protein from the medium, a concentration gradient of NaCl or some other salt is supplied. At a critical ionic concentration which will depend on the charge on the protein, all of the target protein will de-bind and will be eluted in relatively small volume. Since

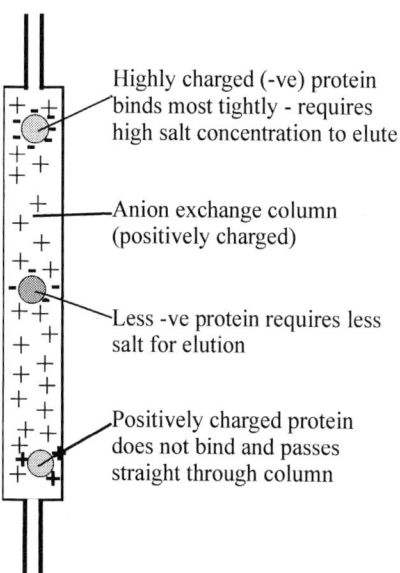

Figure 2. Principle of ion exchange chromatography. A protein above its isoelectric point (pI) will be anionic and will bind to a positively charged matrix. Proteins of increasingly negative charge at the chromatographic pH will bind tighter and require more competing anion (usually Cl^- or $CH_3CO_2^-$) to be released. The same principle applies to cationic proteins (i.e. below their pI), which require a cation (usually Na^+, K^+, or Li^+) for release from a negatively charged matrix.

the protein is actually bound by ionic interaction to the ion exchange medium, this form of chromatography can be very useful for concentrating the target protein(s) from dilute samples. Below the critical ionic (i.e. NaCl) concentration all of the target protein will remain bound. Once all of the sample has been loaded, potentially in a very large volume, the appropriate NaCl concentration can be applied to elute the target protein. Also, by judicious choice of pH for the separation, many unwanted proteins can be removed, by virtue of not binding to the medium. More detailed coverage of ion exchange chromatography can be found in refs 8–10.

2.2.1 Choice of ion exchange medium

Ion exchange media fall into two classes; strong ion exchangers and weak ion exchangers. (Note: the terms 'strong' and 'weak' exchanger do *not* refer to how tightly the proteins bind to the matrix.) The most common example of a strong anion exchanger is QAE (quarternary ammonium ethyl), whilst DEAE (diethylammoniumethyl) is a weak exchanger. Conversely, strong cation exchangers are the sulfonate (S) and sulfopropyl (SP) groups, whilst carboxymethyl (CM) is a weak exchanger (*Figure 3*). The principal difference between strong and weak exchangers is that the charge on the latter will vary with pH whilst for strong exchangers the charge carried is essentially independent of pH.

In most cases it is advantageous to choose a strong exchanger *but* in some

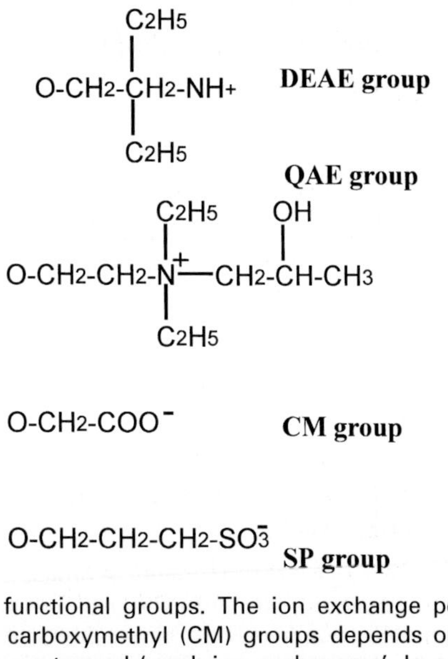

$$O-CH_2-\underset{\underset{C_2H_5}{|}}{\overset{\overset{C_2H_5}{|}}{C}}H_2-NH^+ \quad \textbf{DEAE group}$$

$$O-CH_2-CH_2-\underset{\underset{C_2H_5}{|}}{\overset{\overset{C_2H_5}{|}}{N^+}}-CH_2-\overset{\overset{OH}{|}}{C}H-CH_3 \quad \textbf{QAE group}$$

$$O-CH_2-COO^- \quad \textbf{CM group}$$

$$O-CH_2-CH_2-CH_2-SO_3^- \quad \textbf{SP group}$$

Figure 3. Ion exchange functional groups. The ion exchange performance of diethylaminoethyl (DEAE) and carboxymethyl (CM) groups depends on the ambient pH and these functional groups are termed 'weak ion exchangers'. In comparison the binding capacity of quaternary aminoethyl (QAE) and sulfopropyl (SP) groups does not change at high or low pHs.

1: Separation by charge, size, and hydrophobicity

cases the target protein binds very tightly and high salt concentrations, i.e. harsh conditions, are required for elution. In these cases a weak exchanger should be chosen. As with gel filtration media, ion exchange media come in a range of grades; Levison *et al.* (11) review the performance characteristics of 70 different ion exchange matrices. Choice of the gel chosen is less critical with perhaps gel bead size being the major determining factor governing the resolution that can be obtained. The choice of anion versus cation exchanger will depend on the charge carried by the target protein at the pH of the chromatographic step. Clearly this pH should be chosen so as to be as close as possible to that optimum for the target protein's activity and stability.

2.2.2 Choice of pH for the separation

If the isoelectric point (pI) of the target protein is known, it is an easy matter to choose the appropriate pH for the separation. For anion exchange the buffer should be about one pH unit above the pI whilst for cation exchange it should be one pH unit below pI. Secondly, in many cases the pI of the target protein is not known. In this case there are two possible ways to proceed. The first possibility is to perform an isoelectric focusing experiment (12)—this can be useful if the target protein is easily detectable, e.g. it is coloured or can be stained in some specific way, i.e. activity stain. The second, and much easier method, is to determine the approximate pI empirically. The method described in *Protocol 2* should suffice for most applications.

Protocol 2. Determination of approximate pI for ion exchange chromatography

Equipment and reagents
- Ion exchange resin, e.g. DEAE–Sepharose FastFlow
- 1.5 ml Eppendorf tubes
- Microcentrifuge
- Blood wheel, roller table, or rocking table
- 20 ml each of 50 mM Tris, 50 mM MES buffer adjusted to pH 5, 5.5, 6, 6.5, 7, 7.5, 8, 8.5, 9

Method

1. A small quantity (≈ 150 μl settled volume) of ion exchange gel is placed in several 1.5 ml Eppendorf tubes.

2. Equilibrate the aliquots of ion exchange gel with the different pH value Tris/MES buffers[a] by resuspending the ion exchange resin in the appropriate pH buffer, leaving it for 2–3 min, and then removing the buffer after a brief (3–5 sec pulse) centrifugation in a microcentrifuge. The equilibration step should be carried out three or four times before leaving the gel in 300 μl buffer.

3. Add a small amount[b] of the target protein or protein mixture, e.g. 50 μl containing the target protein to each aliquot of gel.

Protocol 2. Continued

4. Agitate the mixture, containing gel plus the target protein(s), gently[c] for 5 min before sedimenting by microcentrifuging for 3–5 sec.
5. Remove the supernatant and retain. Add 200 µl of fresh buffer at the same pH and repeat step 4.
6. Combine the supernatants and assay[d] for the target protein.
7. The first pH value at which the activity is bound is appropriate for chromatography. For an anion exchange resin, when the pH is above the pI of the target protein, the latter will be bound and its activity will disappear from the supernatant. (For a cation exchange resin the opposite is true.)

[a] MES (pK_a = 6.1) and Tris (pK_a = 8.1) mixtures enables the range from pH 5–9 to be buffered effectively. If is suspected that the pI of the target protein falls outside of this range then other buffer pairs can be used, e.g. citrate/Pipes (pH 4–8), MOPS/borate (pH 6–10).
[b] The main consideration is not to completely saturate the ion exchange capacity, or it can be difficult to distinguish the pH point at which the protein does not bind.
[c] End-over-end mixing is most efficient but the Eppendorf tubes can also be placed inside a bottle which is placed on a roller table, or laid down on a rocking table.
[d] Convenient assays could be enzyme activity, absorbance or fluorescence, radiolabelled protein, or immunologically detectable protein.

The same method as in *Protocol 2*, but with the pH fixed at the value determined to give protein binding, and buffers containing 0.2–2 M NaCl (or other appropriate salt) can be used to determine the salt concentration needed for elution of the target protein. If greater than 0.5 M salt is required for elution, the pH used should be lowered (for anion exchange), or raised (for cation exchange) by 0.2 units.

2.2.3 Choice of salt for elution

This is chiefly dictated by the sensitivity of the target protein to particular salt(s). If in doubt, start with NaCl; Na(CH$_3$CO$_2$) is also a possible, and gentle eluent for anion exchange. Other anions are possible, e.g. SO$_4^{2-}$, SCN$^-$, Br$^-$, but are somewhat harsher and at higher concentrations maybe chaotropic, i.e. may cause partial unfolding of the protein. If freeze-drying of the sample is subsequently required, NH$_4$(CH$_3$CO$_2$) and NH$_4$(HCO$_3$) are possible eluents depending on pH. Their advantage is that these NH$_4^+$ salts, along with NH$_4$Cl will lyophilize to produce NH$_3$ and the appropriate acid, i.e. acetic acid, HCl, leaving the sample on its own. If SDS–polyacrylamide gels are contemplated, avoid KCl, or K$^+$ salts, as eluents; K$^+$SDS$^-$ is quite insoluble at low temperatures and can sometimes be problematic. Finally, when applying the sample to the ion exchange matrix it is important that the solution containing the target protein is at least of similar composition and pH to that of the initial eluent. For example, if a protein sample is applied to an ion exchange matrix in a high salt concentration it will not bind to the column. Similarly, too low or high pH

1: Separation by charge, size, and hydrophobicity

may mean that the target protein carries insufficient charge (or the wrong charge) to enable binding to the matrix. (Note: the pH is simple to adjust by addition of the appropriate acid or base. Removal of salt can be easily accomplished by use of a desalting column; see Sections 2.1.2 and 2.1.4.)

2.2.4 Volumes/concentration gradients

If separating a complex mixture of proteins, a gradient of 0–1 M NaCl in the appropriate buffer is generally sufficient. The gradient can be delivered to the column in a variety of ways (see Section 3 on Hardware). Usually, it is easier to deliver a linear gradient to the ion exchange column (*Figure 4*). However, under some circumstances exponential gradients, or 'steep/shallow/steep' type gradients (*Figure 4*) can be useful where the target protein elutes close to contaminating proteins. As a rule of thumb, the gradient should be delivered over approximately 20 column volumes, e.g. for a 1 ml column use a 20 ml gradient.

2.3 Hydrophobic interaction chromatography (HIC)

HIC is closely related to reverse-phase chromatography (RPC). However RPC is utilized mainly for small molecules (13), including peptides, whilst HIC is more widely applicable to larger biomolecules such as proteins.

2.3.1 Principles of HIC

The basis of HIC is the interaction between hydrophobic parts of the target protein and a hydrophobic matrix (14, 15). Functional groups for HIC can be

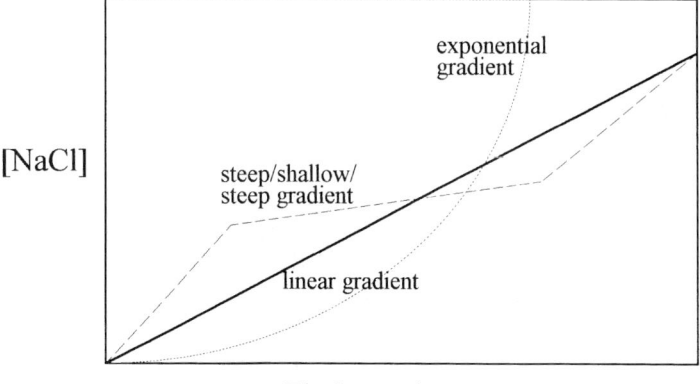

Figure 4. Typical eluent profiles for ion exchange chromatography. During method development, simple linear gradients of the competing anion (or cation) are usually used. Subsequent fine-tuning may lead to the use of more complex gradients. For example in the 'steep/shallow/steep' gradient shown, the NaCl concentrations would be chosen so that the target protein elutes in the middle of the shallow part.

-O-CH$_2$-CH$_2$-CH$_2$-CH$_3$ Butyl Group

 Phenyl Group

Figure 5. Functional groups for hydrophobic interaction chromatography (HIC). The butyl and phenyl groups shown represent the two alkyl and aryl ligands respectively that are most commonly used for HIC.

obtained in a range of hydrophobicities. *Figure 5* shows two common HIC groups, butyl, and the more hydrophobic phenyl. Since binding to the HIC gel is by hydrophobic forces, in order to elute the target protein conditions must be employed under which these forces can be reduced. To achieve this end it is usual to load the protein sample at high salt concentrations (which promotes hydrophobic interaction) and then elute by running a reverse salt concentration gradient, i.e. decreasing concentration.

2.3.2 Choice of hydrophobic group

One potential pitfall of HIC is that the target protein may bind so tightly as to be only released under harsh, denaturing conditions, e.g. by SDS. As a rule of thumb start with phenyl–Sepharose for soluble proteins whilst for more hydrophobic proteins, such as membrane proteins, butyl–Sepharose should be the initial choice. Rather than commit all of the available sample to HIC separation straight away, a similar exercise to that used for finding the pH optimum for ion exchange chromatography should be carried out (see Section 2.2.2). However, this time the sample should be incubated with the HIC gel in the presence of 2 M NaCl (or another appropriate salt) whilst de-binding of the target protein should be tested using decreasing NaCl concentrations, e.g. 1.5 M, 1 M, 0.5 M, and 0.2 M.

2.3.3 When to use HIC?

Since proteins elute with *decreasing* salt concentrations, HIC is often used as a separation step immediately after ion exchange chromatography. The advantage here is that the ion exchange fractions containing the target proteins, which have eluted in relatively 'high' salt, can often be applied directly to an HIC column without prior sample processing such as dialysis. In some cases it may be necessary to add salt to a high enough concentration, i.e. 1–2 M, to bring about binding to the matrix.

2.3.4 Choice of eluent

As proteins are eluted by decreasing the salt concentration this amounts mainly to a choice of salt to apply the sample in initially. Many salts have been used for HIC, but 1–2 M NaCl is generally useful whilst for soluble proteins

1: Separation by charge, size, and hydrophobicity

1–2 M $(NH_4)_2SO_4$ is often employed. In addition to decreasing the salt concentration a simultaneous *increasing* concentration gradient of a gentle detergent, e.g. Triton X-100, CHAPS, octyl glucoside can also be used to assist elution of proteins from HIC media.

2.4 Affinity chromatography

Many chromatographic schemes utilize an affinity, or pseudo-affinity separation. Typically this occurs in the later stages of a purification scheme, owing to the generally more costly nature of affinity media, and is preceded by one or more 'non-specific' separation methods. However, in the most favourable instance, proteins can be purified to homogeneity in a single step. As Chapters 5–11 cover a range of affinity procedures they will not be considered further here.

3. Choice of hardware

3.1 Pre-packed or self-packed columns

Pre-packed columns offer great advantages in terms of time saved packing the column, testing the column, and so on; nowadays pre-packed columns are available in an extremely wide range of media. A further advantage is that the performance of pre-packed columns is in many cases substantially better than can be easily achieved with self-packed ones. A disadvantage is the cost, particularly for the larger scale columns, such as may be required during the initial stages of protein purification. In these cases it is definitely better to self-pack, especially since such columns may have a strictly limited life.

3.1.1 Types of column

Once the self-pack option has been chosen the question arises as to which type of blank column to use. (The actual mechanics of column packing are covered in the next section.) The simplest option is to use a simple glass tube, tapered at the bottom (*Figure 6*) so that it can be plugged with glass wool which serves to support the gel. For smaller columns, a disposable syringe barrel, minus the plunger, makes quite an effective blank column. Although equipment of this sort looks unsophisticated, when properly packed a good level of chromatographic resolution can be achieved.

The next and probably most popular option, is the commercial blank column (*Figure 7*). Pharmacia, Bio-Rad, and other companies make a wide range of blank column sizes. Routinely, blank columns are supplied with end-fittings, spare bed supports (a fine mesh at the base of the column which supports the gel beads), and tubing. Unless the blank is to be totally filled, it is also a very good idea to purchase the appropriate flow adapter. This is an adjustable plunger which allows the eluent to be applied directly to the top of the gel bed, with little or no dead volume above. Not only does a flow adapter allow

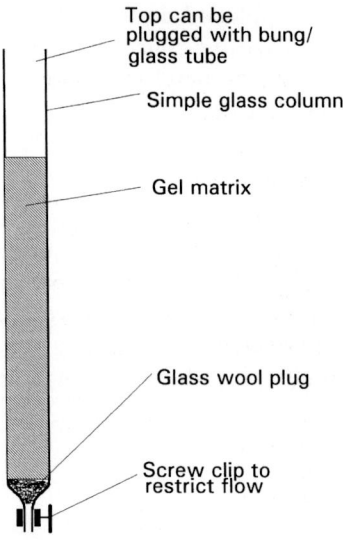

Figure 6. Simple glass chromatography column. The glass wool, used to retain the matrix can be replaced by a sintered plastic disc. A similar disc can also be placed on the the bed surface to protect it from disturbance when loading the sample. A syringe barrel will also function as a serviceable and inexpensive column.

different height columns to be poured, minimizing the dead volume can greatly improve the chromatographic resolution.

3.1.2 What size and shape of column?

The overall size of the column will of course depend on the amount of sample to be applied and in turn the volume chromatographic medium to be packed. The data sheet for a particular medium will give an indication of protein carrying capacity, i.e. mg protein bound/ml gel. Having roughly calculated the amount of medium needed, and hence the volume that the column must hold, the next choice is the shape of the column (long and thin, or short and fat?). Broadly speaking, for gel filtration, HIC, and chromatofocusing separations the column should be long and thin, whilst for ion exchange and affinity separations the column should be short and fat. For example, to pack a 100 ml gel filtration column, a 1.6 cm × 50 cm column would be used whilst 100 ml of ion exchange material would be more appropriately packed in a 2.6 cm × 20 cm column. (Note: chromatofocusing is a chromatographic procedure which is essentially isoelectric focusing on an ion exchange column (16). The method is capable of very high resolution.)

3.2 Fluidics

This term covers all of the equipment needed for eluent supply to and from the column, i.e. reservoir, tubing, and pump if used. At simplest, liquid flow

1: Separation by charge, size, and hydrophobicity

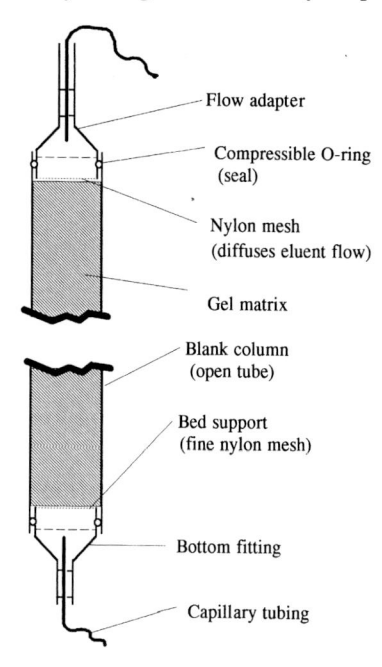

Figure 7. Typical commercially available column and fittings. In essence, this is identical to the column shown in *Figure 6*. An added facility is the flow adapter, which is an adjustable plunger that can be positioned just above the bed surface to permit sample application with little or no dilution.

through the column can be driven by gravity; the eluent reservoir is sited above the column. The flow rate through the column is then regulated by restricting the flow at the bottom of the column; a screw clip across a piece of silicone tubing is usually adequate. However, a peristaltic pump or high performance piston pump (e.g. as used for FPLC or HPLC) is desirable since it is then much easier to accurately control flow rate and starting/stopping the elution.

The tubing used will depend on the application, including the type and size of column. However, for most laboratory scale columns we have found 1/16" (~ 1.5 mm) Teflon or polypropylene tubing the best for low pressure (soft gel) chromatography. For FPLC and HPLC, where much higher pressures are encountered, PEEK tubing gives an excellent combination of flexibility, biocompatibility (proteins and peptides don't adsorb to, or denature on the tubing surface), and ease of cutting to size. The latter is not a trivial consideration. Tungsten tubing is biocompatible and very tough, but it requires special equipment and some skill to cut cleanly.

3.3 Detection

The simplest detection method is to collect fractions and measure their absorbance at 280 nm (for aromatic residues such as Phe, Tyr, and Trp) or some other appropriate wavelength if the target is coloured (e.g. 450 nm for cytochromes). This has the disadvantage that the measurements cannot be carried out in real time, nor can the peaks of absorbance due to the emerging proteins be used to trigger certain events, e.g. movement of a fraction collector.

More usually an optical detector is used which measures the absorbance of the eluate, usually at 280 nm for proteins. The absorbance is measured in a microcell, often with a cell volume of as little as 10–20 μl, allowing resolution of proteins that elute closely together. Typically with such optical detectors the wavelength monitored can be altered by changing an internal (removable) filter and the absorbance range (sensitivity) can be set to suit the amount of protein being separated. More sophisticated (and increasingly expensive) detectors are able to monitor two or more wavelengths simultaneously. Mostly, except for specialized applications, such sophistication is not required. Detection other than by absorbance is also an option and devices for measuring pH, conductivity (useful since it measures the *actual* salt gradient), and electrical potential are all available. These detectors are often used in conjunction with optical detection. A fuller discussion of detector devices is presented in Chapter 3.

3.4 Fraction collection

The simplest and by far the *most reliable* mode of fraction collection is manually, although it is sometimes tricky to collect exactly even-sized fraction. However, if the sample is really valuable, it is advisable to collect the key fraction by hand. Less tedious, and much more comfortable if chromatography is to be performed in the cold room, is the use of a fraction collector. Like detectors, these vary widely in sophistication and facilities on offer, varying from simple time-based detection to fully programmable capability. Be wary of purchasing anything too sophisticated, most of the facilities just will not be used and there is more of it to malfunction. Examples of typical 'workhorse' fraction collectors available at present are the FRAC100 (Pharmacia) and Redifrac (Bio-Rad). These can be used for simple time-based (min/fraction) or volume-based (ml/fraction), can do basic peak detection (change tubes when a protein peak is encountered), and control a peristaltic pump. The last feature is *extremely* useful since it permits unattended runs, e.g. overnight, to be carried out which will terminate at some pre-set volume, i.e. before the eluent runs out and the column runs dry (this is a chromatographer's nightmare, since it usually requires repacking the column (see Section 4).

4. Packing columns

Most chromatographic media for self-packing, e.g. Sephadex, Sepharose are intended for use at relatively low pressures. However, it is perfectly possible,

1: Separation by charge, size, and hydrophobicity

although a little trickier, to pack media which will withstand moderate pressures, i.e. up to a few hundred psi. (Note: psi = pounds per square inch. Other units which may be encountered are the Bar = 15 psi, and the Megapascal (MPa) = 150 psi.) Examples of such media are TSK media (Toso Haas) and various FPLC media such as Superose (Pharmacia). In this case a column extension (actually in effect a second column) must be coupled to the top of the column to be packed with a pressure-resistant connector. The extension serves as a reservoir for the gel slurry whilst packing is taking place. In either case, the basic procedure to be carried out is the same. *Protocol 3* covers the basic procedure for column packing.

Protocol 3. Packing a column

Equipment and reagents

- Blank column, flow adapter, and tubing
- Packing reservoir
- Peristaltic pump
- Matrix to be packed, e.g. Sephadex G25
- Side-arm flask
- Water powered vacuum pump
- 20 column volumes of buffer[a]
- 5 column volumes of buffer plus 5 mM NaN$_3$, or 25% (v/v) aqueous ethanol

Method

1. If necessary, remove the 'fines' from the chromatographic medium.[b] Removal of fines is extremely simple and is carried out by resuspending the gel in water equal to two to three times the settled gel volume, followed by leaving the larger gel beads to settle for 5–10[c] min before decanting off the fines. Usually the resuspension/settling/decanting steps are carried out two to three times.

2. To the settled gel (you should have sufficient volume to just about fill the column) in a side-arm flask, add enough buffer to give about a 50% slurry of the chromatographic gel.

3. At this stage it is good practice to remove as much dissolved air as possible by evacuating the side-arm flask for a few minutes using a water powered vacuum pump.

4. Resuspend the gel by stirring gently with a glass rod and *immediately* pour down a long glass rod placed in the blank column. This avoids any bubbles being produced. Once several centimetres of gel has been poured in, remove the long glass rod and pour in the rest of the slurry so that the packing reservoir or column extension is filled.

5. Attach the top fitting and start the pump.

6. Pump buffer through the column at the maximum recommended flow rate[d] (soft gels) or pressure/flow rate (medium gels). After a few column volumes of buffer have been pumped through the column, the top of the gel bed should have a flat level surface.[e]

Protocol 3. *Continued*

7. Fit the flow adapter, taking care not to disturb the surface of the gel bed. The surface of the adapter should be positioned down to within 1–2 mm of the gel surface.[f] If a flow adapter is not used it is good practice to float a circle of glass fibre filter disc (e.g. Whatman GF/A) or sintered plastic disc (Bio-Rad) onto the gel surface to minimize disturbance when loading (see below).

8. If the column is not to be used immediately, run through a few column volumes of packing buffer supplemented with 5 mM NaN_3 to inhibit bacterial growth. In some cases it may be recommended that the gel material be stored in 25% (v/v) ethanol for the same purpose. If appropriate store the column at 4°C.[g]

[a] Buffer composition will depend on the matrix packed, and the chromatography to be performed.
[b] 'Fines' are small fragments of the gel beads which, if not removed, will clog the column and severely restrict the flow.
[c] The actual settling time will depend on the grade of gel, with fine and medium grade gels taking longer than coarse grades. Recommended settling times are usually given on the accompanying data sheet, or may be in manufacturers' catalogues.
[d] It is usual to pack columns at slightly higher flow rates or pressures than they will eventually be used at. However, it is extremely important not to exceed recommended flow rate, particularly with soft gel matrices.
[e] If the top is not completely flat, resuspend the top 1–2 cm of matrix by repeatedly aspirating with a pipette.
[f] Some chromatographic media, in particular softer ion exchange gels, e.g. DEAE–Sepharose will shrink in the relatively high salt buffers used for elution of proteins. Don't adjust the flow adapter—the gel will expand again in the low salt concentration buffer used to re-equilibrate the gel.
[g] For softer gels, the packed column should not be brought back up to room temperature—the variation of gas solubility with temperature will mean that bubbles will form within the column and necessitate repacking.

5. Running the column

5.1 Equilibration

The first step in any chromatographic separation is to equilibrate the column with the eluent in which the sample is to be loaded. For gel filtration this will be the single eluent which will be used throughout the separation, whilst for ion exchange or affinity separations this is the low salt concentration buffer. For HIC it is the high salt buffer. Usually it is sufficient to run two to three column volumes through the column.

5.2 Loading the sample

It is good general practice to filter or centrifuge the sample prior to loading to remove fine particulate material which would block the column. Appropriate

1: Separation by charge, size, and hydrophobicity

centrifugation conditions are: microcentrifuge for 5 min at full speed, or 40 000 g for 10 min for larger samples. When filtration (preferable) is performed, sequential filtrations through 1 μm and 0.2 μm pore filters should subsequently be used.

Actual methods of sample application are various. For simple desalting, or affinity separations, and if a flow adapter is not used, the liquid within the column can be run down until the gel bed is just exposed. The sample is then applied and run into the surface of the gel. Finally the gel is *gently* overlayered with buffer and elution continued. A better method of application, if a flow adapter is in place, is to pump the sample onto the column using a peristaltic or piston pump. The sample can be placed in a purpose-built reservoir (10) and a three-way valve (Bio-Rad, Pharamacia) used to switch between this reservoir and the eluent. A simpler, but just as effective method is just to stop the pump and manually reposition the tubing to the eluent container, making sure the tubing is secured to the eluent container.

The most sophisticated, and best method of sample application is to use a sample loop. The sample is injected into a closed loop and at the appropriate point an injection valve is switched to feed the eluent flow through the loop and drive the sample onto the column. This is the method employed by high resolution, i.e. FPLC, HPLC systems. The loop should as near as possible be the same size as the sample to be injected and loops which are much larger than the sample should be avoided. For small sample (< 1 ml) loops can be constructed from an appropriate length of tubing. *Table 1* indicates the volume per length of some common tubing sizes.

For larger samples, application is usually by means of a 'Superloop' (10) which is a tube containing a floating piston and bypass channel (*Figure 8*). (Note: this is Pharmacia's product; equivalent devices are also available from other manufacturers, e.g. Waters.) The eluent drives the piston forward and pushes the sample onto the column. At the end of its travel, the eluent flows through a bypass channel to prevent pressure build-up. Samples of up to 50 ml can be applied to HPLC and FPLC columns using this device. (Note: once the

Table 1. Volume/length for typical tubing sizes

i.d. mm (inch)[a]	μl/inch	μl/cm
0.254 (0.01)	1.288	0.507
0.508 (0.02)	5.146	2.026
0.762 (0.03)	11.577	4.558
1.016 (0.04)	20.581	8.103
1.397 (0.055)	38.911	15.320
1.5[b] (0.059)	44.777	17.629

[a] In many cases, and particularly for PEEK HPLC tubing, sizes are given in inches.
[b] Polypropylene tubing of this size is often used for low pressure chromatography, e.g. with matrices like Sephadex.

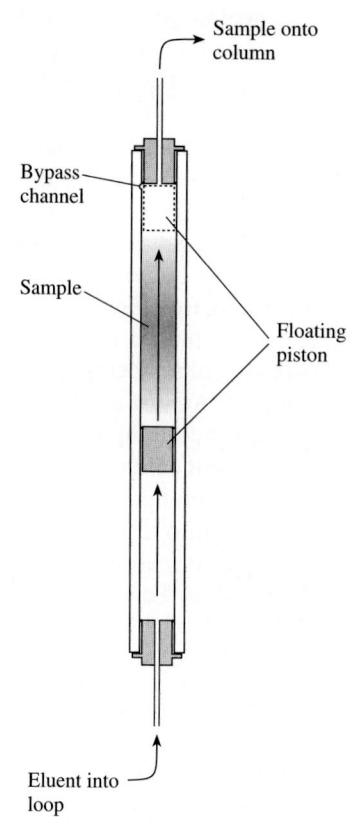

Figure 8. 'Superloop' for loading large samples onto columns. The eluent pushes the floating piston forward, displacing the sample onto the column. When all of the sample is loaded, a small channel in the wall of the superloop acts as a bypass to let the eluent through. The superloop is then switched out of the eluent flow before commencing any gradients.

sample is applied it is vital to switch the superloop out of circuit or else any salt gradients used for elution will be pumped into a large volume of eluent inside the superloop.)

5.3 Elution

The simplest elution conditions are with gel filtration where it is simply a matter of using a single eluent. However, with many chromatographic separations elution of the bound proteins is brought about by passing a gradient of salt or detergent through the column. The gradient can be formed in a number of ways. In more sophisticated systems the gradient is programmed via a control unit of some sort. Quite often, PC or Mac computers are used to control the chromatographic run in which the gradient is actually produced by the com-

1: Separation by charge, size, and hydrophobicity

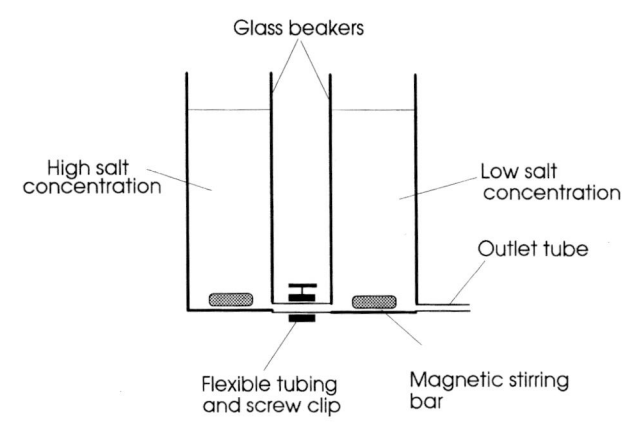

Figure 9. A simple gradient former prepared from two beakers or similar containers.

bined actions of a proportioning valve and mixing chamber. The major advantage of such systems is that complex gradients (see *Figure 4*) can easily be programmed to suit the chromatographic separation being performed.

For situations where programmable gradient mixing equipment is not available, perfectly acceptable linear gradients, sufficient for most purposes, can be generated by a home-made gradient maker. The simplest type consists of two containers, e.g. glass beakers, linked by a piece of flexible tubing (*Figure 9*) which can be clamped by a screw clip.

A more convenient device can be produced from a Perspex block into which two chambers are drilled, connected by a channel which can be closed by means of a stopcock (*Figure 10*). One of the chambers has a lead out tube

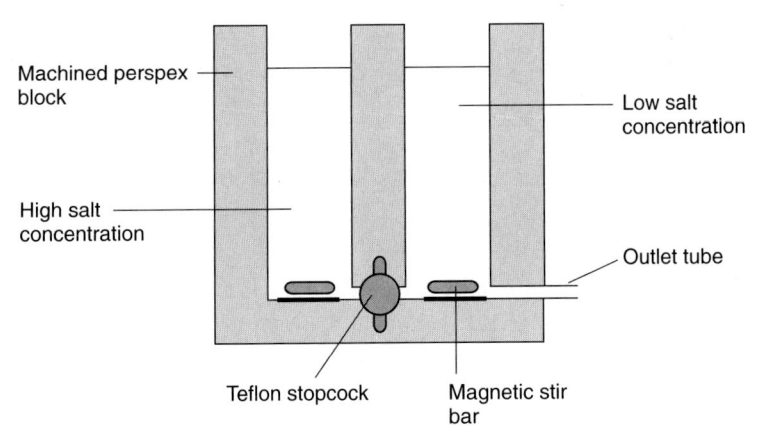

Figure 10. Commercially produced gradient maker. These are relatively simple for any competent machinist to produce. The chamber dimensions for a range of useful gradient volumes are given in *Table 2*.

19

Table 2. Dimensions of mixing chambers for gradient makers

Chamber dimensions w × h (mm)	Chamber volume (ml)	Max. gradient volume (ml)	Useful working range (ml)
20 × 80	25	50	10–40
25 × 100	49	98	20–80
33 × 140	120	240	50–200
40 × 200[a]	250	500	80–320
50 × 250	490	980	160–650
60 × 250	700	1400	280–1100

[a] Larger diameter chambers may be more easily produced from Perspex tubing.

from which the eluent is pumped. Purpose-made gradient makers of this type can be purchased in various sizes. However, a competent workshop should be able to produce one for substantially lower cost. *Table 2* provides chamber dimensions for a some useful gradient maker sizes.

To actually produce a gradient in either type of gradient maker the same procedure is followed. For example, for a 100 ml 0–2 M NaCl gradient:

(a) 50 ml 0 M NaCl buffer would be placed in the outer (lead out) chamber whilst 50 ml 2 M NaCl buffer would be placed in the inner chamber.

(b) The stopcock is then opened and the solution in the 0 M NaCl chamber stirred by means of a magnetic stirrer.

(c) Before starting the pump, it is necessary to clear air bubbles from the connecting channel. This is achieved by gentle pressure on one of the chambers with the palm of the hand.

(d) The pump is then started and as eluent is drawn from the low NaCl concentration chamber, hydrostatic pressure drives buffer through the connecting channel from the high NaCl concentration chamber into the low concentration chamber where it is immediately stirred in.

(e) The end-result, seen at the gradient mixer lead out tube, is a buffer solution which steadily increases from 0–2 M NaCl over 100 ml.

5.4 Column sanitation

Even with the appropriate precautions, most column matrices eventually become contaminated with non-specifically adsorbed material. In HPLC and FPLC columns this usually shows up as an anomalously high back pressure at the recommended flow rates. In order to clean the column, sequential washes with one to two column volumes of 2 M NaCl, water, 0.1 M NaOH, water, and 50% ethanol should be performed. However, data sheets for the chromatographic matrix or packed column should always be consulted to ensure any of the above are not deleterious. Finally, if the top of the column is obviously

1: Separation by charge, size, and hydrophobicity

discoloured, a few millimetres can be removed from the top of the gel bed followed by readjustment of the flow adapter to take up the newly created dead volume.

References

1. Sternweiss, P. C. and Pang, I.-H. (1990). In *Receptor-effector coupling: a practical approach* (ed. E.C. Hulme), pp. 1–29. Oxford University Press, Oxford.
2. *Gel filtration - principles and methods*. (1998). Pharmacia-Biotech.
3. Hagel, L. (1989). In *Protein purification: high resolution methods and applications* (ed. J.C. Jansen and L. Rydén), pp. 63–106. VCH Publishers Inc., NewYork.
4. Locascio, G. A., Tigier, H. A., and Battelle, A. M. del C. (1969). *J. Chromatogr.*, **40**, 453.
5. Andrews, P. (1970). *Methods Biochem. Anal.*, **18**, 1.
6. Potschka, M. (1987). *Anal. Biochem.*, **162**, 47.
7. Hagel, L. (1985). *J. Chromatogr.*, **324**, 422.
8. Himmelhoch, S. R. (1971). In *Methods in enzymology* (ed. W. B. Jakoby), Vol. 22, pp. 273–86. Academic Press, London.
9. Choudhary, G. and Horvath, C. (1996). In *Methods in enzymology* (ed. B. L. Karger and W. S. Hancock), Vol. 270, p. 47–82. Academic Press, London.
10. *Ion exchange chromatography - principles and methods*. (1998). Pharmacia-Biotech.
11. Levison, P. R., Mumford, C., Streater, M., Brandt-Nielsen, A., Pathirana, N.D., and Badger, S. E. (1997). *J. Chromatogr.*, **760**, 151.
12. Haff, L. A., Fägerstam, L. G., and Barry, A. R. (1983). *J. Chromatogr.*, **266**, 409.
13. Joseph, M. H. and Marsden, C. A. (1986). In *HPLC of small molecules: a practical approach* (ed. C.K. Lim), pp. 13–47. Oxford University Press, Oxford.
14. *Hydrophobic interaction chromatography*. (1998). Pharmacia-Biotech.
15. Hofstee, B. H. F. and Otillio, N. F. (1978). *J. Chromatogr.*, **159**, 57.
16. Sluyterman, L. A. Æ. and Elgersma, O. (1978). *J. Chromatogr.*, **150**, 17.

2

Practice and benefits of microcolumn HPLC

FRANK J. YANG

1. Introduction

The application of microcolumn HPLC have grown significantly in recent years, especially since the current trends in HPLC are to employ continuously smaller diameter columns in order to reduce solvent consumption, to enhance sample detection, to allow a direct interface with mass spectrometers, and to facilitate sample limited applications and hyphenated techniques. Microcolumn HPLC is the HPLC of choice for the most modern HPLC bench-top MS applications, being used for several ionization techniques including particle beam (1–3), continuous flow FAB (4–9), EI/CI (10–13), and electrospray and nanospray interfaces (14–20). The major driving force behind the rapid growth of microcolumn HPLC in the past few years has been the commercialization of bench-top low cost mass spectrometers with electrospray and nanospray interfaces. At reduced flow rates, a few µl/min and down to sub-µl/min, maximum sensitivity and baseline stability can be achieved. Thus, it can be expected that microcolumn HPLC-MS will become even more popular than capillary GC-MS in the near future as the ability to connect microcolumn HPLC directly to the mass spectrometer will be propitiated by the development of low cost, easy to use, high sensitivity, and high reliability mass spectrometer, and supported by the low flow rates used in microcolumn HPLC.

The role of microcolumn HPLC is also increasingly important in sample limited biological and bioengineering research where conventional analytical HPLC cannot provide detectable sample peak concentration. With a very limited sample quantity and sample concentration, capillary HPLC offers the best solution for the analysis of protein, peptide, and peptide tryptic digests in a few picomole to subpicomole concentration ranges. In neural science research, capillary column HPLC is the method of choice for the analysis of catecholamines and their metabolites.

Some major advantages of microcolumn HPLC include:

(a) Higher mass sensitivity than conventional HPLC for trace analysis.

(b) Same or higher sensitivity as conventional HPLC with smaller sample sizes for sample limited applications.

(c) Large reduction in solvent consumption and therefore solvent operation and toxic waste disposal cost.

(d) Very low flow rates used with microcolumns make it the method of choice for HPLC-MS, particularly the bench-top MS with electrospray and nanospray interfaces.

(e) Low flow rates used in microcolumns facilitate the unified approach to chromatography (21).

These major advantages result from the efficient packing of small particles such as 3 μm and 5 μm into a microcolumn leading to increased chromatographic efficiency. In general, the only difference between conventional and microcolumn HPLC is the inner diameter of the column used. A definition and a comparison of the different sizes of columns are given in *Table 1*.

As a result of the smaller column diameters in microcolumn HPLC, the flow rate is reduced and the eluted peak volume (flow rate multiplied by peak width at half-height) is subsequently reduced by a factor equal to the ratio of the cross-sectional areas of the different sized columns. For example, a 20 times reduction in peak volume would be measured in a 1 mm i.d. column at 50 μl/min flow rate with respect to a 4.6 mm i.d. analytical column at 1000 μl/min. A 100 μm i.d. packed fused-silica column operated at 0.5 μl/min would result in a more than 2100 times reduction in peak volume and therefore a 2100 times enhancement in sample detection. This increase in peak concentration allows microbore columns to perform trace analysis never before possible with conventional columns. In addition, the coupling of packed capillary columns to concentration-dependent detectors such as absorbance, refractive index, and electrochemical detectors enhances the application range of HPLC to trace levels of sample components and impurities. The applicable detection limit for a refractive index detector when combined with a packed capillary column is capable of reaching nanograms instead of the microgram level measurable with

Table 1. A comparison of column types[a]

Type	Column i.d.	Flow rate	Loading	Detection enhanced
Conventional	4.6 mm	0.5–2 ml/min	2–10 mg	1
Narrow bore	2.1 mm	200–400 μl/min	0.5–2 mg	5
Microbore	0.8–1 mm	25–60 μl/min	50–500 μg	20–25
Packed capillary	100–500 μm	1–15 μl/min	1–50μg	80–2000
Nanobore	< 100 μm	< 1 μl/min	< 1 μg	2000–10 000
Open tubular	< 3–50 μm	0.01–1 μl/min	< 0.01 μg	2000–20 000

[a] Typical column diameters, flow rates, loading capacity, and enhancement of sample detection (compared with a conventional column) are shown.

a 4.6 mm i.d. column. Such a capillary column HPLC-RI detector system may provide a universal detection and analysis system for samples lacking chromophores.

There are few drawbacks in microcolumn HPLC with respect to the conventional column HPLC. In principle, microcolumn HPLC can replace all applications currently used with 4.6 mm i.d. columns. The slow adaptation of microcolumn HPLC technique in the routine quality control laboratories may be attributed to three major reasons:

(a) Performance of a conventional pump designed for conventional columns but adapted with a flow splitter (so as to produce low μl/min flow rates) cannot meet the stringent requirement for high long-term reliability and reproducibility.
(b) The lack of a dedicated and properly designed microcolumn HPLC system that ensures reproducible and reliable operation for routine microbore column HPLC applications.
(c) The lack of recognition of the different requirements in the practice and the instrumental design for microcolumn HPLC.

2. Practice of microcolumn HPLC

Microcolumn HPLC has the same column chemistry and relative retention of sample components as standard columns. However, the practice of microcolumn HPLC requires more attention to pump performance for the low flow rates required of microcolumns, the instrumental band broadening contribution, and the other considerations that are needed to ensure optimum performance for microcolumns.

2.1 Preparation and routine care of microcolumns

Many investigators (22–35) have discussed microcolumn packing procedures. The procedure in general follows the stepwise pressurization technique patented by Yang in 1982 (23, 25). A general procedure for the packing of the microbore column is given in *Protocol 1*.

Protocol 1. Microcolumn preparation and testing

Reagents
- 10% to 30% (w/v) of packing material in appropriate solvent

Method
1. Connect the column outlet end to a 2 μm filter and the inlet end to the slurry packing reservoir.

Protocol 1. Continued

2. Mix 10% to 30% (w/v) of packing material into a solvent slurry such as acetone, methanol, isopropyl alcohol, THF, or a mixture of slurry solvents.
3. Degas the packing slurry and rapidly fill the packing slurry reservoir with the packing solvent.
4. Rapidly pressurize the slurry reservoir to around 100–300 atms depending on the length and diameter of the column.
5. Gradually increase the pressure to the packing slurry to 500–600 atms after the packing material has been filled to about 70% of the column length.
6. Keep the final packing pressure for at least 30–50 min under ultrasonic agitation. After ultrasonic agitation has stopped, allow the column pressure to slowly decline as to prevent back-flow disruptions in the column bed.
7. Install column end filters and unions.
8. Condition the column with test solvent for at least 60 min.

A well prepared microcolumn needs to be tested for column packed bed stability in addition to column separation efficiency and peak symmetry. A stable packed bed is critical for high column efficiency and long column lifetime. A simple visual indication is the lack of air gaps after letting the bed dry. Stability can also be determined by a characteristic pressure drop across the column at a given flow rate and test solvent composition. For example, a 15 cm × 1 mm i.d. column packed with 5 µm silica particles gives a typical pressure drop of around 800 psi at 50 µl/min flow rate, while a 25 cm × 320 µm packed fused-silica column with 5 µm silica particles yields about 1500 psi at 5 µl/min. A stable packed bed should not be affected by the repeated rapid pressurizations and depressurizations that are normally practiced to reduce gradient regeneration time between runs. The uniformity and stability of the packed bed is advantageous over a 4.6 mm i.d. column and allows reverse flow operation of the column without loss of separation efficiency or change of retention time for the sample components.

A typical test report for column performance is shown in *Figure 1* that also includes information about column length and diameter, detector parameters, solvent composition and flow rate, column pressure drop, retention time, peak area, peak height, peak name, skew factor at 10%, peak resolution, and plate number. *Figure 1* indicates that the column has a 1120 psi pressure drop at 50 µl/min flow rate for a 60:40% $ACN:H_2O$ that met the criteria for a stable packed bed. The 25 cm length column has an excellent theoretical plate number approaching 23 000 with a skew factor of nearly 1.0. It should be noted that the testing of the packed fused-silica column requires great care in re-

2: Practice and benefits of microcolumn HPLC

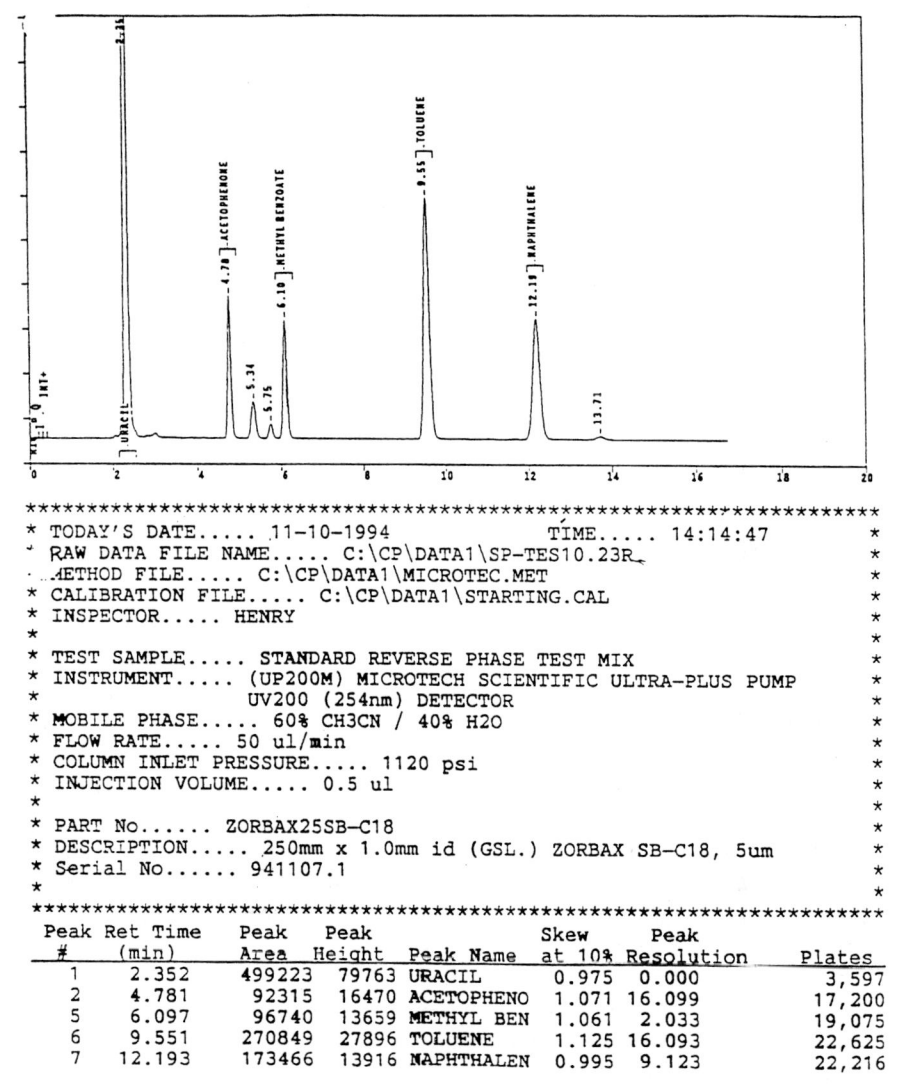

0.0 to 20.0 min. Low Y=8.0 High Y=60.0 mv Span=52.0

```
***************************************************************
* TODAY'S DATE..... 11-10-1994             TIME..... 14:14:47  *
* RAW DATA FILE NAME..... C:\CP\DATA1\SP-TES10.23R             *
* METHOD FILE..... C:\CP\DATA1\MICROTEC.MET                    *
* CALIBRATION FILE..... C:\CP\DATA1\STARTING.CAL               *
* INSPECTOR..... HENRY                                         *
*                                                              *
* TEST SAMPLE..... STANDARD REVERSE PHASE TEST MIX             *
* INSTRUMENT..... (UP200M) MICROTECH SCIENTIFIC ULTRA-PLUS PUMP*
*                 UV200 (254nm) DETECTOR                       *
* MOBILE PHASE..... 60% CH3CN / 40% H2O                        *
* FLOW RATE..... 50 ul/min                                     *
* COLUMN INLET PRESSURE..... 1120 psi                          *
* INJECTION VOLUME..... 0.5 ul                                 *
*                                                              *
* PART No...... ZORBAX25SB-C18                                 *
* DESCRIPTION..... 250mm x 1.0mm id (GSL.) ZORBAX SB-C18, 5um  *
* Serial No...... 941107.1                                     *
***************************************************************
```

Peak #	Ret Time (min)	Peak Area	Peak Height	Peak Name	Skew at 10%	Peak Resolution	Plates
1	2.352	499223	79763	URACIL	0.975	0.000	3,597
2	4.781	92315	16470	ACETOPHENO	1.071	16.099	17,200
5	6.097	96740	13659	METHYL BEN	1.061	2.033	19,075
6	9.551	270849	27896	TOLUENE	1.125	16.093	22,625
7	12.193	173466	13916	NAPHTHALEN	0.995	9.123	22,216

Figure 1. Packed microbore column test report.

ducing the system dead volume. The use of a 50 µm i.d. fused-silica tubing for connecting the column to the injector and to the detector is highly recommended. Additionally, the use of on-column detection (24, 26) and split flow injection (36) can achieve the optimum plate number measured for the packed capillary columns.

Microcolumns are inherently more easily plugged due to the reduction in column cross-sectional area and thus the total number of pores in the column end filters. A 1 mm i.d. column has 21 times less pores than a 4.6 mm i.d. column. A 250 μm packed fused-silica column has 341 times less pores in comparison to a 4.6 mm i.d. column. a 50 μm i.d. packed capillary column may have less than ten 2 μm pores at the inlet/outlet filters. It is important to observe any change in column pressure drops after a few sample injections. A gradual increase in column head pressure indicates a gradual build-up of contaminants at the inlet of the column. Also, it is strongly recommended to reverse the flow in the column, allowing partially plugged pores to be cleaned. A rule of thumb is to reverse the flow direction of the column whenever a 25% increase in column pressure drop is observed. This procedure is very useful and important for packed capillary HPLC, particularly in the chromatography of biological samples. The above column routine maintenance relies on monitoring the column pressure as an indicator of inlet frit blockage. It is therefore difficult to observe this effect in a split flow operation if a constant pressure is maintained at the inlet of the column. It is also important to filter out contaminants such as sample impurities, particulates, subcellular particles, and salts prior to sample injection.

The effect of room temperature variation on retention and the handling of packed fused-silica columns should be acknowledged and controlled. The effect of room temperature variation can be eliminated by a thermostat control of the column temperature or by increasing the thermal mass of the column by assembling it within a higher thermal mass tubing. *Figure 2* shows the Col-Guard for Micro-Tech Scientific column, serving to provide thermal stability as well as excellent mechanical strength for ease of handling. It is also advisable not to connect the end of the packed fused-silica column directly inside the rotor of the injector. This approach minimizes band broadening contribution from the injector although it increases the risk of either plugging the internal channel or scoring the rotor due to internal leakage from the back-flow of column packing material into the injector rotor. A 5–10 cm by 50 μm i.d. fused-silica tubing is recommended for interfacing microcolumn to injector and detector.

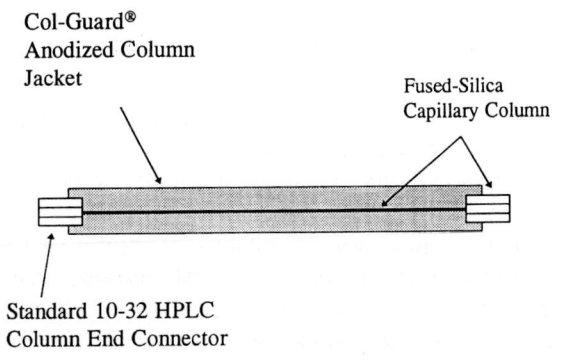

Figure 2. Col-Guard anodized column jacket diagram for fused-silica capillary columns.

2.2 Injection techniques

To demonstrate the benefit of detection enhancement for microcolumns, a conventional 4.6 mm i.d. column and a microcolumn using the same amount of sample will be compared. As predicted by theory, the initial sample zone dispersion should be less than 10% of the peak volume to maintain column performance. Earlier, it was noted that microcolumns yield a reduced peak volume and thus an increase in peak concentration as opposed to 4.6 mm i.d. columns. On the other hand, to analyse the same sample volume requires the application of one of the following sample zone focusing techniques to reduce the large sample zone dispersion resulting from a large sample injection in microcolumns.

2.2.1 On-column focusing sample solvent effect

The choice of sample solvent for sample clean-up or dilution needs to be more carefully considered in microcolumn HPLC than in conventional column HPLC. A proper choice of sample solvent may allow solute molecules to be focused at the inlet end of the column even with a large volume injection. The sample solvent on-column focusing technique as introduced by Yang (23) in 1982 involves injecting a sample prepared in a solvent with a weaker elution strength than that of the initial elution solvent in either isocratic or gradient elution HPLC. The sample solvent on-column focusing effect is demonstrated in *Figures 3b* (23) where a distinctive improvement in both peak resolution and peak concentration were obtained for the fast eluting thio-containing pesticides in 13% H_2O:87% CH_3OH instead of 100% CH_3OH as shown in *Figure 3a*. The sample solvent on-column focusing effect was also utilized in a protein/peptide tryptic digest application where small amounts of a more concentrated tryptic sample was diluted in a weak solvent A containing 0.1% TFA in H_2O to a concentration below a few picomoles for injection into a microbore column. This technique allowed sample volumes as high as 20 μl to be injected into a 1 mm i.d. packed column and a 5 μl sample to be injected into a packed capillary column.

2.2.2 On-column focusing: effect of a secondary solvent in the sample loop

Frequently, the sample solvent is unknown, cannot be changed, or cannot be diluted. For such situations, an alternative approach is to fill the sample loop with a weaker solvent (relative to the initial elution solvent) followed by the sample for analysis. Secondary solvent on-column solute focusing effect is obtained after injection due to a quick replacement of the sample solvent with a large amount of a weaker secondary solvent. An example is shown in *Figure 4* for which a 5 μl sample loop containing 4 μl of 0.1% TFA in H_2O and a 1 μl peptide tryptic digest sample is injected into a 15 cm × 320 μm fused-silica column packed with 5 μm Zorbax C18 300 Å silica particles. Excellent peak

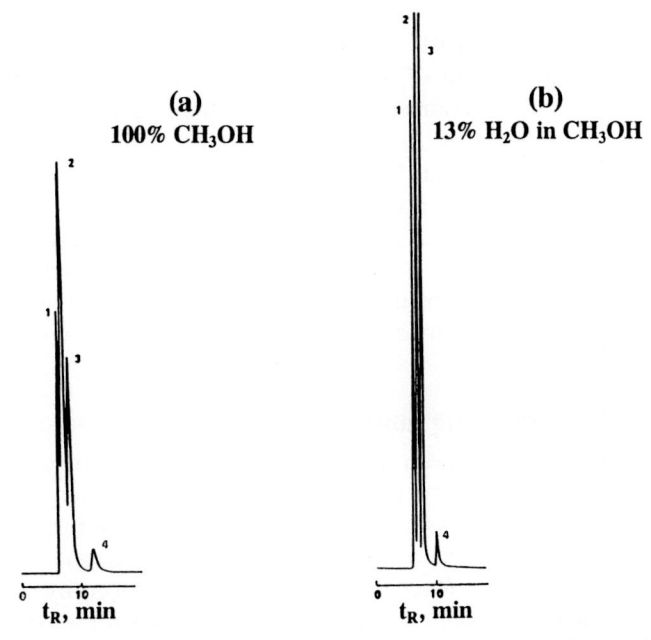

Figure 3. On-column focusing by utilizing a weaker sample solvent. Column is a 50 cm × 250 μm C18 5 μm reverse-phase particles; mobile phase is 100% CH_3OH. Sample solvent is (a) 100% (v/v) methanol, and (b) 87% CH_3OH, 13% H_2O (v/v). Samples are: 1, benzenethiol; 2, hexanethiol; 3, octanethiol; 4, dedecanethiol. Sample volume was 0.6 μl and the detector was a flame photometric detector in sulfur detection mode.

shape and baseline resolution for three pairs of peptide tryptic digests were measured. This approach is particularly useful for handling an unknown sample or unknown sample solvent matrix.

2.2.3 On-injector pre-column focusing

On-injector pre-column focusing is particularly important when microbore column HPLC is used for ultratrace analysis and where a large sample volume is needed to ensure sufficient detection of sample peaks. As shown in *Figure 5*, a large volume of sample can be loaded into a pre-column installed on the sample loop of an injector. Solute molecules are then focused within the pre-column. After accumulating an adequate amount of solute molecules for adequate detection or for micro-purification, the injector is then switched from sample load to inject position for chromatographic elution. This technique allows sample volumes as large as 1000 μl to be injected provided that there is a proper selection of packing material, sample solvent, and wash solvent to ensure that there is no sample breakthrough from the pre-column. This technique can also be adapted for routine sample desalting and can be automated to handle multiple samples.

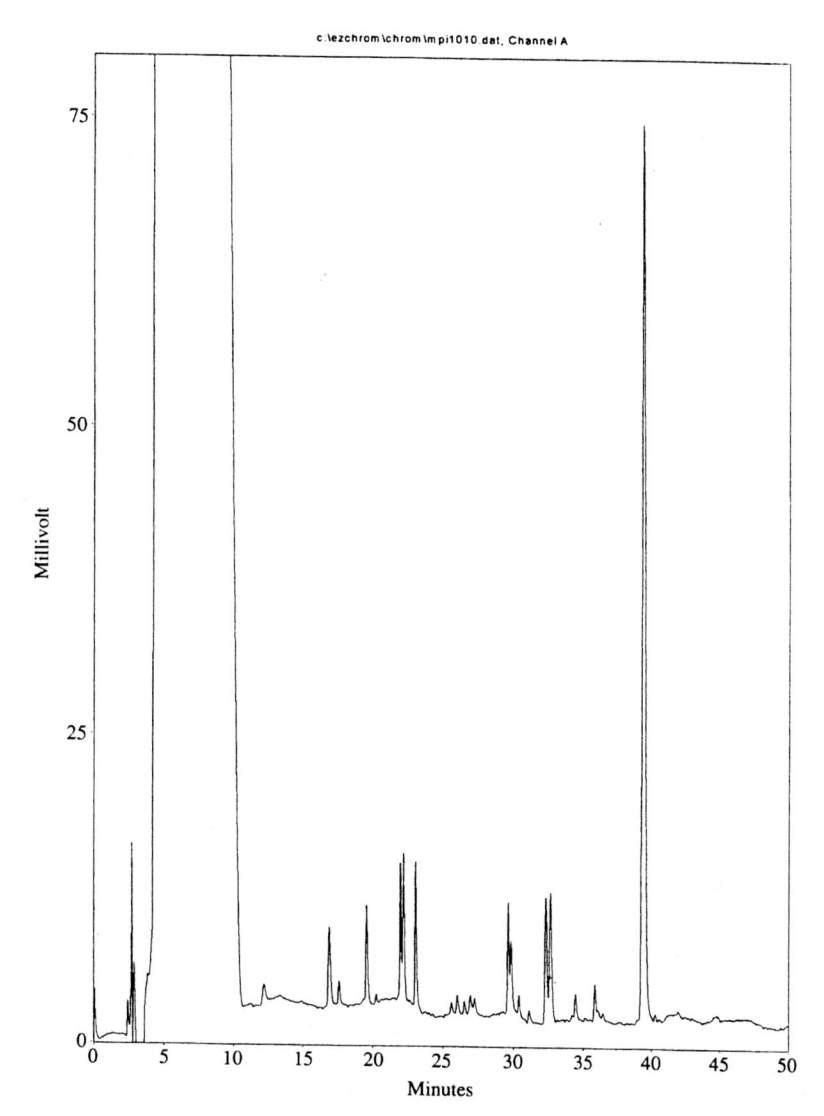

Figure 4. The effect of a weak solvent in a sample loop on the focusing of solute molecules at the front of a microcolumn.

2.2.4 On-column focusing in a gradient microbore column HPLC

A large sample size can be handled in a gradient elution HPLC method. Initial sample zone spreading as a result of large sample size injection can be refocused at the inlet of the column by applying a weak initial solvent elution strength. Sample sizes as large as 50 μl can be injected into a 15 cm × 1 mm

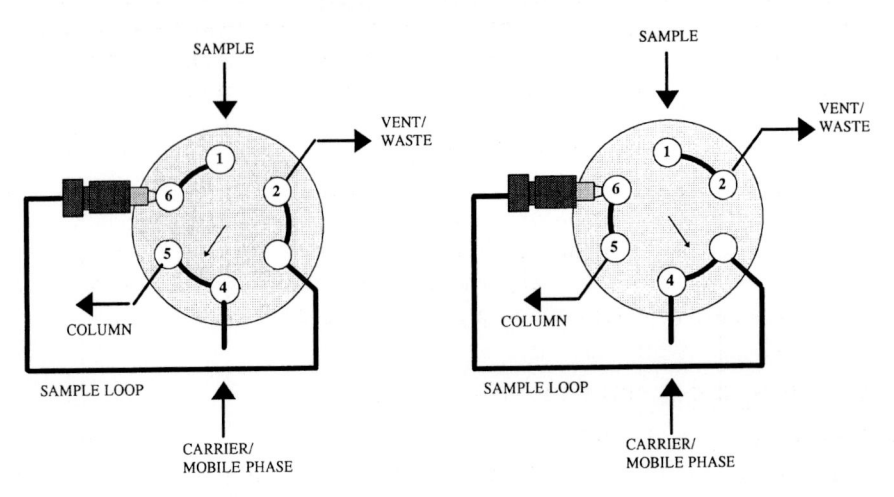

Figure 5. Schematic flow diagram for an on-injector pre-column concentration set-up. The pre-column or guard column is installed to port 4 of a 6 port high pressure valve. The sample is loaded from port 1 and concentrated on the pre-column. At the appropriate time, the valve is switched to the 'inject' postion and the sample is then eluted from the pre-column into the microcolumn.

i.d. microbore packed column. Sample size between 20 nl to 10 µl can normally be injected into a packed capillary column. For larger sample size or sample containing salts, the technique of on-injector pre-column focusing should be applied.

2.3 Detector considerations

As the peak volume in microbore column is reduced, it is important to optimize the detector flow cell volume to minimize any band broadening due to a large flow cell volume. A rule of thumb is that the cell volume should be less than 10% of the peak volume. This suggests that a cell volume should be less than 0.83 µl for a peak volume of 8.3 µl eluted from a 15 cm × 1 mm i.d. column given a 50 µl/min flow rate times a peak width of 10 sec. Similarly the cell volume should be less than 0.167 µl for a peak volume of 1.67 µl eluted from a packed capillary column assuming 10 µl/min flow rate times a peak width of 10 sec. The on-column detection technique (26) provides flow cell volumes down to the nanolitre range and its widely utilized for capillary electrochromatography, capillary zone electrophoresis, and capillary column HPLC. Many other microflow cells for UV/visible, fluorescence, electrochemical, diode array, and laser-induced fluorescence detectors have been recently commercialized for both microcolumn HPLC and CZE applications.

The modern bench-top mass spectrometer is an ideal detector for microcolumn HPLC, especially since the enhancement in peak concentration can be fully realized in conjunction with mass spectrometer detection. Direct coupling of a packed capillary column with an inner diameter of less than 320 μm at a flow rate below 5 μl/min to an electrospray or nanospray interface is the method of choice for maximum sensitivity and versatility.

Packed capillary column HPLC may also be directly coupled to flame-based detectors such as helium plasma discharge detector, flame photometric detector, thermoionic selective detector (thermionic nitrogen/phosphorus detector, see *Figures 3a* and *3b*), etc. At 5 μl/min and lower flow rates, the coupling of the above highly sensitive and selective detectors to the packed capillary columns will greatly expand the application of HPLC into the analysis of many classes of compounds otherwise impossible today.

2.4 Pumping system considerations

2.4.1 Conventional pumping systems

As discussed earlier, one of the major obstacles towards the acceptance of microcolumn HPLC in routine applications has been the lack of a reliable and reproducible micro-pumping system. Conventional reciprocating pumping systems designed for standard analytical columns cannot deliver the low flow rates required for microbore column applications. Small piston volume, micro-syringe pumps are commonly adapted for low flow rate applications in microcolumn HPLC today. They offer pulseless solvent delivery while possessing excellent control at low flow rates. However, the microsyringe pump may require frequent refills when operating at higher flow rates for 1 mm i.d. microbore columns. Thus it is more difficult to set-up for multiple autosampler runs. In addition, during the syringe refill cycle, pump flow to the microcolumn is temporarily halted. A long refill time could result in a great loss of column head pressure and system flow equilibration. At very low flow rates and at high pressures for small inner diameter packed capillary columns, a long re-equilibration period may be required after the syringe refill step. The use of multiple microsyringe pumps with a synchronized pump switching control may allow flow compensation for syringe refills.

For gradient microcolumn HPLC, the majority of applications today are still utilizing the split flow technique introduced by Van der Wal and Yang (36) in 1983. The split flow technique, based on either constant pressure or constant split ratio at the splitter, allows a conventional reciprocating HPLC pumping system to be used for microcolumn HPLC. Some difficulties encountered using the conventional pumping system in microcolumn HPLC are discussed below.

In a low pressure proportioning conventional HPLC reciprocating pumping system as shown in *Figure 6*, the gradient step is defined by the piston volume. For example, a piston volume of 100 μl has a gradient step change of 100 μl since a new composition change at the inlet port of the pump head can occur

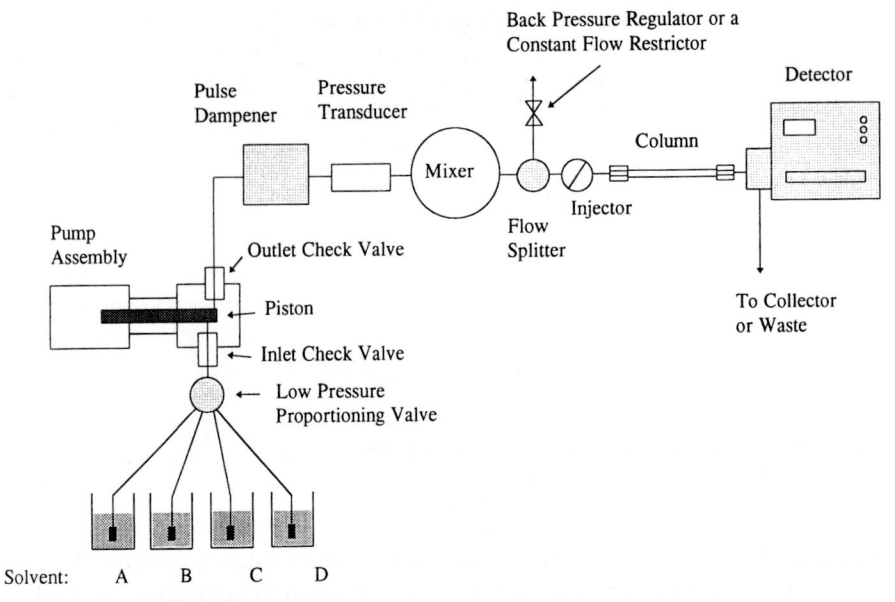

Figure 6. A flow diagram of a conventional low pressure proportioning HPLC system.

only after the piston has emptied its 100 µl solvent in the piston. A typical stairway gradient profile as shown in *Figure 7* is obtained. This shows that with increasing piston volume and decreasing flow rate, gradient step change over time increases. For a piston volume of 100 µl at a flow rate of 5 µl/min, a 20 min gradient step change time is required. This slow change over in gradient step cannot produce a linear gradient and cannot perform fast microcolumn HPLC. Therefore, it is necessary to reduce the piston volume or to raise the flow rate to 500 or 1000 µl/min for example in order to reduce the proportioning step time for a better gradient resolution. This high flow rate cannot be reliably split down to a few µl/min for packed capillary column application due to the instability of the splitter at a high flow rate and at a high split ratio as a result of 'Coanda effect' (37, 38). In addition, the benefit in solvent saving for routine applications utilizing microcolumn HPLC cannot be achieved.

The conventional HPLC reciprocating pumping systems designed for standard (e.g. 4.6 mm i.d.) columns have a dead volume on the order of 800 µl to 5 ml. The large dead volume causes a long delay for the solvent gradient to reach the inlet of the column, particularly since a low flow rate is required for microcolumn HPLC. For example, a 160 min gradient delay time would be expected for a 800 µl dead volume at a 5 µl/min flow rate. In addition, a long gradient regeneration time may be required since near ten volumes of the purge solvent are needed to purge the system dead volume and generate the initial solvent condition. Again, this suggests that a conventional pump

2: Practice and benefits of microcolumn HPLC

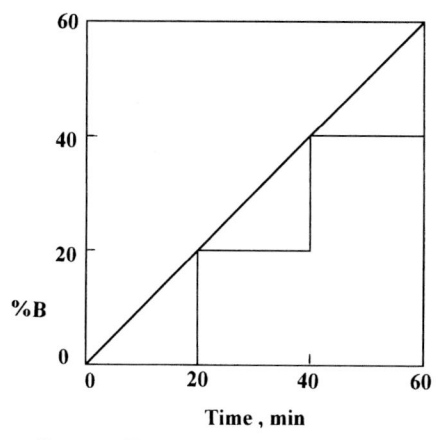

Figure 7. Stairway gradient profiles for 5 μl/min flow rate using a 100 μl piston volume and 0–50% linear gradient in 50 min.

utilizing low pressure gradient proportioning may be used for microcolumn HPLC if a flow splitter is used and if the pump is operating at high flow rates to reduce both the gradient delay time and the gradient regeneration time.

The conventional HPLC reciprocating pumping system designed for 4.6 mm i.d. columns is not capable of delivering flow rates at 0.1 μl/min and below, as required for packed capillary column HPLC. Most of the reciprocating low pressure proportioning pumping systems may have a minimum flow rate set point at either 1 or 10 μl/min and thus it is difficult to obtain a reproducible gradient without splitting in gradient elutions at microbore column flow rates.

2.5 Flow splitting

The flow splitting technique as introduced by Van der Wal and Yang (36) has been adapted by many researchers who have modified conventional HPLC pumping systems for microcolumn HPLC. However, the technique suffers from poor long-term flow stability and reliability in addition to variation in column flow rates. In addition, when a splitter is used, there is not much benefit in saving solvent consumption and disposal cost. However, the flow splitting technique is an easy and inexpensive (< US $50) approach to evaluate the value of microcolumn HPLC for applications such as detection enhancement of trace impurities, sample limited applications, and LC–MS. Two simple and inexpensive approaches for setting-up a flow splitter are presented below.

2.5.1 Constant column back pressure regulator

As shown in *Figure 6*, a constant back pressure regulator (Tascom Company) was installed into the tee union (ZTIC; Valco, Houston, TX, USA) for the pump and the microbore column before the injector. A constant column head pressure can be precisely and accurately regulated to maintain a desirable flow rate through the column. The advantage of the set-up include ease in:

- adjusting column flow rates
- purging and cleaning the flow path from clogging
- routine maintenance and diagnosis for column plugging

However, in gradient elution, the flow rate in the column changes according to the change in viscosity gradient composition. The viscosity of an $H_2O:CH_3OH$ mixed solvent gradually increases to almost 40% higher than that of pure water until the solvent composition surpasses 50% methanol. At a composition higher than 70% methanol, the viscosity of the solvent system gradually decreases to about 30% less than pure water. The variation in viscosity increases the peak retention time in comparison to that obtained for a constant flow gradient system. Also, it is difficult to know the degree of clogging on the inlet of the column and is therefore difficult to assess when a reverse flow clean-up should be performed in order to extend the life of a column. This approach also suffers from a gradual increase in retention time after several injections of real samples as the inlet of the column becomes clogged by the particulates from the injector rotor such as subcellular particles from the biological samples, lipids, and salts.

2.5.2 Constant split ratio/flow rate

Figure 6 also shows an inexpensive and easy method for setting-up a constant split ratio flow splitting system. A fixed length, e.g. 50 cm of a 50 μm i.d. fused-silica tubing was selected to give a desirable flow rate for a microbore column operation pressure and connected to a Valco zero dead volume tee joint. A 5 cm × 50 μm i.d. fused-silica tubing was used to connect between the tee and the injector where the microbore column was installed. The system is then operated under a constant split ratio regardless of the solvent viscosity change in the course of elution gradient. The advantages of this set-up are its low cost (less than $50 US), and ease in set-up and use.

However, a constant split ratio or flow rate in the column can only be maintained if the flow restriction through both the restrictor tubing and the microcolumn is invariable. Such an assumption and expectation are difficult to fulfil under real world conditions since there is less room in smaller diameter columns for the large number of pores required for the free passage of liquid—8000 times less for a 50 μm i.d. packed fused-silica column than for a 4.6 mm i.d. column. In practice, for packed fused-silica columns, it is often found that constant flow restriction cannot be maintained after several injections of real samples. Another constraint is that the split ratio during and after sample injection could change if a large sample volume containing a radically different viscosity solvent from the HPLC mobile solvent is injected. If a large split ratio is used, the flow to the microcolumn with high restriction may also be turned off as a result of the Coanda effect (37, 38). This effect is caused by the outgoing flow stream preferring one channel over the other and then switching due to minor disturbances such as changes in flow rate, temperature, or viscosity of the sample and mobile solvent. In general, a split ratio below 10:1 may provide a more reliable flow split operation. For a packed capillary column at a flow rate of 5 μl/min, a conventional pump should be able to generate a gradient at a maximum of 50 μl/min in order to obtain a long-term reproducible result for split flow applications.

2.6 New generation microbore column pumping systems

With an increasing interest in sub-μl/min and μl/min flow rates for nanospray and electrospray (HPLC-MS applications) the new generation micro-pumping systems (39, 40) have been developed to meet the following requirements:

(a) μl/min to sub-μl/min flow rates.
(b) Generation of a solvent composition gradient at a few μl/min and at below μl/min for isocratic applications without flow splitting.
(c) Mixing noise-free baseline at the detection limits of the detectors.
(d) Small gradient delay time and gradient regeneration time.
(e) Performance requirement in retention time and peak area reproducibility in both isocratic and gradient applications.

These requirements are difficult to meet utilizing conventional pump technology based on either conventional gear drive or cam drive designs. Fortunately the advances in computer disk drive technology in the 1990s has revolutionized the technology for controlled precision, accuracy, and robustness in the disk driver motor. The adaptation of the disk driver technology with a closed loop digital motion control algorithm for the microcolumn HPLC pumping system produces an inherently simple, accurate, and robust design with nanolitre resolution for packed capillary column HPLC applications. A microcolumn HPLC pumping system based on the integrated closed loop digital motion control principle was designed and patented by Yang and Fan (39, 40).

In this system a high resolution digital encoder is mounted on the shaft of a motor that is directly driving the piston of the pump. A ceramic piston with a piston volume of 25 μl is chosen for microbore column flow rates from 0.1 μl/min to 1000 μl/min. A 100 μl piston volume can also be mounted for conventional flow rates from 1 μl/min to 5 ml/min. The use of a digitally controlled motor with 3072 steps per turn and a capacity for pumping 25 μl for every 6.5 turns allows low flow rate capability and precise control. This high resolution digital control is equivalent to a digital control resolution of 25 μl/(6.5 × 3072 steps) or 1.25 nl resolution. Such a high control capability at ultra-low flow rates allows an accurate and precise delivery of solvent at 0.1 μl/min and below.

The micro-reciprocating pump designed above is a single piston pumping system that offers simplicity in operation and pulse-free continuous solvent delivery. Multiple pump modules can be combined with a high pressure mixer to facilitate precise composition proportioning and mixing for gradient microcolumn HPLC. In microcolumn HPLC gradient elution, some critical considerations are unique to microbore column HPLC:

- gradient step resolution
- mixing efficiency
- solvent compressibility and solvent flow isolation
- gradient delay time
- gradient regeneration
- gradient reproducibility

2.6.1 Gradient step resolution

Gradient step resolution measured by a minimum solvent composition per cent change per minute, is the key factor in determining the degree of peak resolution in a gradient run for a complex sample. A pumping system that can produce a lower composition change per minute provides a higher resolution for closely eluted components. To obtain a 0.1% gradient step resolution at 5 µl/min flow rate, each pumping system should be capable of delivering 0.005 µl/min accurately and reproducibly. Such a stringent requirement may only be achievable by using the advanced digital control principle in conjunction with a high precision motor and direct driver assembly.

2.6.2 Mixing efficiency

For microcolumn HPLC, conventional low pressure proportioning cannot be adapted as discussed earlier due to the large gradient step required by the large piston volume and a long gradient delay time from the use of a large dead volume. In order to reduce gradient delay time and to produce a fast micro-gradient step, a high pressure mixing method is required. In general, a dynamic mixer is required for flow rates above 5 µl/min and a small volume static mixer is effective for flow rates below 5 µl/min. A comprehensive study on mixing efficiency of both static and dynamic mixers was reported by Schultz et al. in 1982 (41). It was clearly indicated that a dynamic mixer has substantially higher efficiency for solvent mixing than a static mixer of equal volume. In general, a static mixer requires a mixing volume that is five to ten times the flow rate to produce effective mixing. A dynamic mixer may require about two to four times the flow rate. *Figures 8A* and *8B* compare the effect of a static and dynamic mixing using the same mixing volume. *Figure 8A* clearly shows that at 150 µl static mixer is too small to mix H_2O with methanol at 50 µl/min effectively. *Figure 8B* shows reproducible (five consecutive gradient runs) and smooth gradient profiles where the stirring bar inside the mixer was turned on to produce dynamic mixing.

2: Practice and benefits of microcolumn HPLC

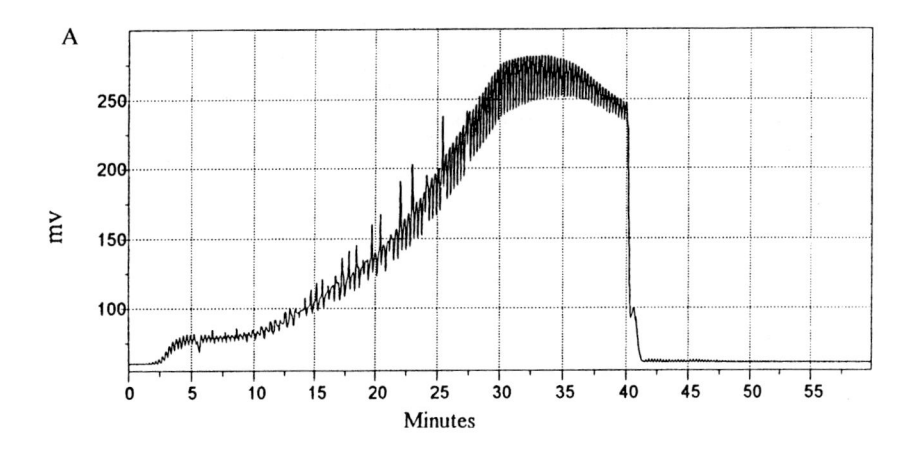

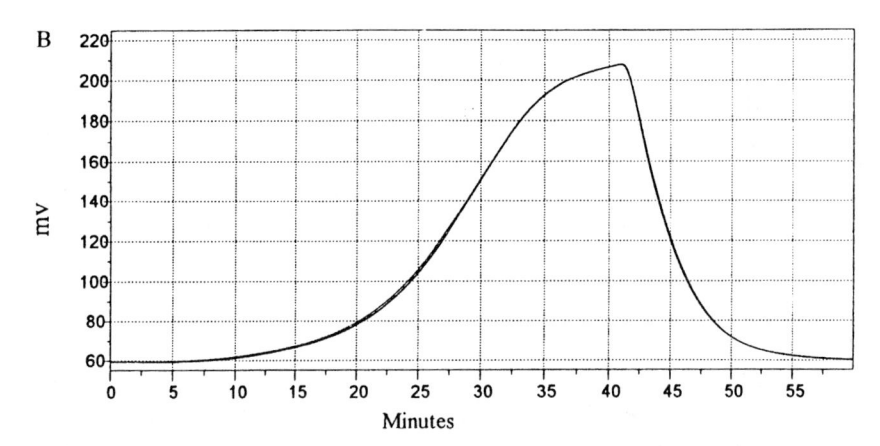

Figure 8. Mixing noise: a micro-gradient with and without a dynamic mixer. Solvent A is H_2O; solvent B is CH_3OH. The gradient was from 0% B to 100% B (v/v) in 30 min and held for 20 min at a flow rate of 50 μl/min. (A) A 150 μl mixing volume static mixer was used. (B) A dynamic mixer with a stirring bar inside the mixer was turned on for effective mixing.

To improve mixing efficiency for high flow rates, a conventional approach is to increase the mixer volume. Such an approach is not ideal due to the need for continuously larger mixer volumes and longer gradient delay time. The mixing efficiency can be greatly improved by adapting multiple mixers in series. This approach improves mixing exponentially with each additional mixer in series without increasing the mixing volume by much. For example, when a 20 μl primary mixer is coupled to a 100 μl secondary mixer, the two mixers in series have a total volume of only 120 μl, however the combination can effectively mix solvents at flow rates as high as 150 μl/min.

2.6.3 Solvent compressibility and flow isolation

With a conventional HPLC reciprocating pumping system utilizing low pressure proportioning and mixing, solvent channels may be isolated by the low pressure proportioning valves. No solvent cross-talk or back-flow into each other's flow path can occur except when a high pressure mixer is used to form a solvent composition gradient. Fortunately for conventional HPLC operated at a high flow rate, a back-flow of solvent into each other's flow path can be quickly purged out and thus has no noticeable effect on gradient initial composition, gradient delay time, or gradient regeneration time. For microcolumn HPLC where a high pressure mixer is used, it is important to prevent any solvent back-flow due to the fact that it requires a long time to purge out the contaminant region at a low flow rate. The degree of solvent back-flow depends upon relative solvent compressibility, flow path internal volume, and operation pressure. In *Table 2*, compressibility factors and viscosity for several common HPLC solvents including water are given. Methanol is shown to be two times more compressible than water. When water and methanol are used in a reverse-phase micro-HPLC with a high pressure mixer, water will be pumped beyond the mixer tee into the methanol flow path. The compression of methanol will depend on the pressure at the mixer tee. A typical gradient elution chromatogram as shown in *Figure 9* reveals retained solute peaks being eluted much later than the solvent peak since the modifier solvent required more than 40 minutes to purge out water that had reversed back into the methanol flow path.

This delay in solute peak elution is even worse if one of the solvent flow paths that communicates with the other solvent flow paths at the high pressure mixer leaks. A leakage in a cross-communicating solvent flow is very difficult to detect without a flow isolation check valve. With individual solvent flow path isolation check valves, diagnosis of pump failures such as pump seal

Table 2. Physical properties of some common HPLC solvents

Solvent	Compressibility $\times 10^4$ (atm^{-1})	Viscosity cp	Pressure dependence of viscosity $\times 10^3$ (atm^{-1})
Water	0.46	0.8	0.053
Methanol	1.23	0.52	0.47
Ethanol	1.1	1.003	0.585
Acetonitrile	0.96	0.345	–
Chloroform	0.97	0.519	0.625
Carbon tetrachloride	1.07	0.845	1.25
n-Hexane	1.61	0.296	1.15
n-Heptane	1.42	0.355	1.09
n-Octane	1.2	0.483	1.12
Diethyl ether	1.87	0.212	1.11
Dichloromethane	0.99	0.39	–

2: Practice and benefits of microcolumn HPLC

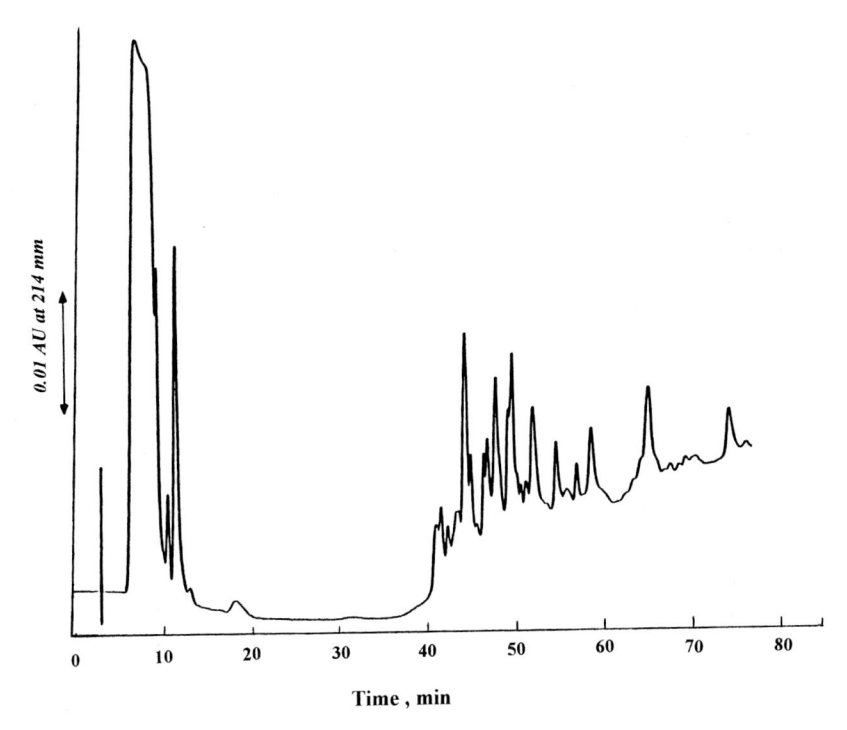

Figure 9. Gradient delay due to solvent back-flow. Solvent A is 0.1% (v/v) TFA in H$_2$O; solvent B is 0.1% (v/v) TFA in acetonitrile. The gradient from 0% B to 60% B (v/v) in 30 min and held at 60% B (v/v) for 60 min. The flow rate was 4 µl/min and UV detection was at 214 nm. The column was a 15 cm × 320 µm i.d. packed fused-silica column and the sample was a myoglobin tryptic digest.

leakage, pump head check valve failure, or connector leakage are easily made. Solvent isolation for a high pressure mixer is therefore very important for preventing solvent back-flow and for rapid gradient regeneration. A simple approach for achieving flow isolation is to build individual check valves into each solvent flow path in the mixer. The check valves on the mixer ensure zero dead volume interface between check valve and the mixer chamber and allows total solvent flow isolation until mixing within the mixer chamber.

2.6.4 Gradient delay time

Conventional pumping systems utilizing low pressure proportioning for gradient forming causes a long gradient delay time and cannot be adapted for microcolumn HPLC applications. Gradient delay time in microcolumn HPLC at low flow rates can be reduced by using a small volume, high pressure mixer with flow isolation check valves as discussed in the previous section. Gradient delay times for such high pressure mixers and check valve systems are equivalent to the ratio of the volume of the mixer to the flow rate. By reducing the

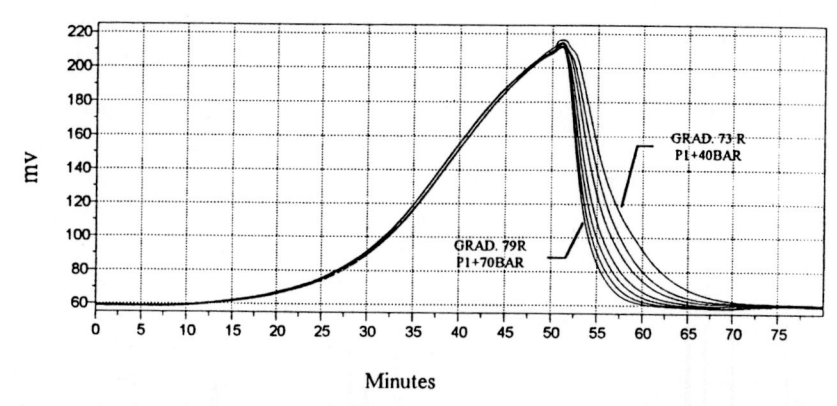

Figure 10. Comparison of gradient regeneration time at 40, 45, 50, 55, 60, 65, and 70 bar for pump 1 (H_2O) and 32 bar for pump 2 (CH_3OH). The gradient was from 0% to 100% B (v/v) in 30 min and then held for 20 min, and the flow rate was 10 µl/min. A single stage dynamic mixer of 20 µl volume was used.

volume of the mixer, the gradient delay time for micro-HPLC can be reduced to under two minutes. The volume of the mixer should be as small as possible but large enough to ensure complete mixing. It should be noted that the volume of the mixer also affects gradient linearity. In general, an exponential gradient profile is normally obtained for a larger mixer volume to flow rate ratio. The larger the mixer volume, the slower the rise of the initial gradient slope. *Figure 10* shows a typical exponential gradient profile for a high pressure mixing system. A slow increment in the first part of the gradient provides a grater resolution than a linear gradient for those early eluting components. A linear gradient in a high pressure mixing system, however, can be obtained by using a linear gradient forming software program.

2.6.5 Gradient re-equilibration time

Re-equilibration time may be excessively long for microcolumn HPLC using a high pressure dynamic mixer. The time required for regeneration of the initial gradient depends on the ratio of the total system volume (the sum of mixer volume, injector volume, and column volume) to the flow rate. The smaller the ratio of the system volume to the regeneration flow rate, the faster the gradient regeneration time. A gradient regeneration time of more than 80 minutes was measured for a system volume of 100 µl. A 26 µl system volume (including a 20 µl mixer volume, a 1 µl injector volume, and a 5 µl column volume) may require 25 minutes re-equilibration period (for a fivefold purging of the system volume) when it is operated at 5 µl/min flow rate. Further reduction in gradient regeneration time may be achieved by using a solvent pre-pressurization technique as shown in *Figure 10*. Purge flow rates may be increased during gradient re-equilibration time to reduce the time needed for regenerating the initial solvent composition.

2: Practice and benefits of microcolumn HPLC

Protocol 2. Solvent pre-pressurization technique

1. Determine the system pressure (e.g. 2000 psi) for the column at a given flow rate (e.g. 50 µl/min) and solvent composition (e.g. 100% H_2O:0% acetonitrile).

2. Set the primary solvent pump (e.g. pump 1 for pumping H_2O) to a gradient re-equilibration pressure (e.g. 2500 psi) above the system pressure and the modifier pump, pump B, to equal the system pressure (e.g. 2000 psi). This step allows residual solvent in the system at the end of the run to be purged out at a higher flow rate (e.g. 50 µl/min × 2500/2000). Also, it allows the modifier solvent flow path to reach the system pressure (e.g. 2000 psi) from the final system pressure (1600 psi) for zero gradient delay time.

3. At the end of the gradient elution run, the pump will increase pressure for each pump to that set in step 2 (2500 psi and 2000 psi respectively). The system is then purged at 62.5 µl/min with 100% solvent 1 (i.e. H_2O) until the pressure decays to 2000 psi (flow rate for 100% solvent 1 being 50 µl/min). The system is then purged at the flow rate and solvent composition set for the initial condition of the gradient elution method until the end of the pre-set equilibration time.

4. Sample is then injected and the gradient elution initiated.

5. Pressurization process begins again in step 3 after the end of each run.

Figure 10 (A. Witzel, personal communication) shows gradient traces for seven consecutive gradient runs at 10 µl/min from 100% H_2O to 100% methanol in 30 min with a 20 min hold time at 100% methanol. The system pressure and the modifier pressure for a fused-silica test-tubing was 32 bar. The effect of the initial H_2O pressure at 40, 45, 50, 55, 60, and 70 bar on the gradient regeneration time was evaluated. A reduction of the gradient regeneration time from 25 to 12.5 min was obtained for a regeneration purge pressure from 40 to 70 bar, respectively. Gradient regeneration can be most effectively reduced by using a programmable automated purge valve located between the mixer and the injector. The purge valve was turned on at the end of the gradient run for purging the mixer. The purge flow rate can be set to a high value to reduce the re-equilibration time.

2.6.6 Gradient reproducibility

The optimum goal for microcolumn HPLC is to offer the above discussed benefits such as solvent saving, sample limited applications, trace component detection, micro-LC-MS, and more. However, a good long-term gradient reproducibility is essential for the acceptance of the microcolumn HPLC as a robust technique for routine applications. The new generation high precision

Table 3. Reproducibility date for the Ultra-Plus micro-pumping system at 2.9 µl/min flow rate[a]

min Injection no.	Isocratic CH$_3$CN/H$_2$O (58:42) Retention time	Gradient CH$_3$CN/H$_2$O (30:70–90:10) 90 Retention time
1	63.72	77.64
2	63.8	77.65
3	63.83	77.61
4	63.75	77.57
5	63.73	77.63
6	63.84	77.66
7	63.81	
8	63.74	
9	63.74	
10	63.69	
mean	63.76	77.63
std. dev.	0.051	0.033
% RSD	0.08%	0.042%

[a] The test sample is toluene, the column is 100 cm × 250 µm fused-silica column with HS C18, 5 µm

micro-pumping system in conjunction with hardware innovations such as micro-dynamic multistage mixer, flow isolation check valves, and fast gradient re-equilibration techniques, offer an important tool for making new advances into the new frontier in microseparation techniques.

Microbore column HPLC has demonstrated, as shown in *Table 3*, an excellent reproducibility in both isocratic and gradient elutions (J. N. Alexander IV, personal communication). A 0.08% RSD for retention time reproducibility at 2.9 µl/min flow rate was obtained for an isocratic elution of toluene test probe with a 63.76 min retention time. For gradient elution at 2.9 µl/min, a 0.042% RSD was obtained for the toluene test probe eluted at 73.63 min. Results as shown here demonstrated that microbore column HPLC using direct drive digital closed loop control can achieve an equal or more reproducible result than a conventional HPLC for a 4.6 mm i.d. column at a 1 ml/min flow rate. *Figure 11* shows an r-globulin tryptic digest map at 50 µl/min flow rate for four consecutive runs. Notice the high resolution and retention time and area reproducibility for the tryptic digest map at 0.5%/min gradient increment. These results clearly demonstrate that the micro-HPLC system based on computer disk drive technology presented here is superior to the conventional cam or belt drive pumping system for low flow rate control. The new generation microbore column HPLC system can work reliably and provide reproducible retention and peak area result.

3. Conclusion

Microbore column HPLC is emerging as a necessary tool for sample limited applications in biological research and for modern, bench-top mass spectro-

2: Practice and benefits of microcolumn HPLC

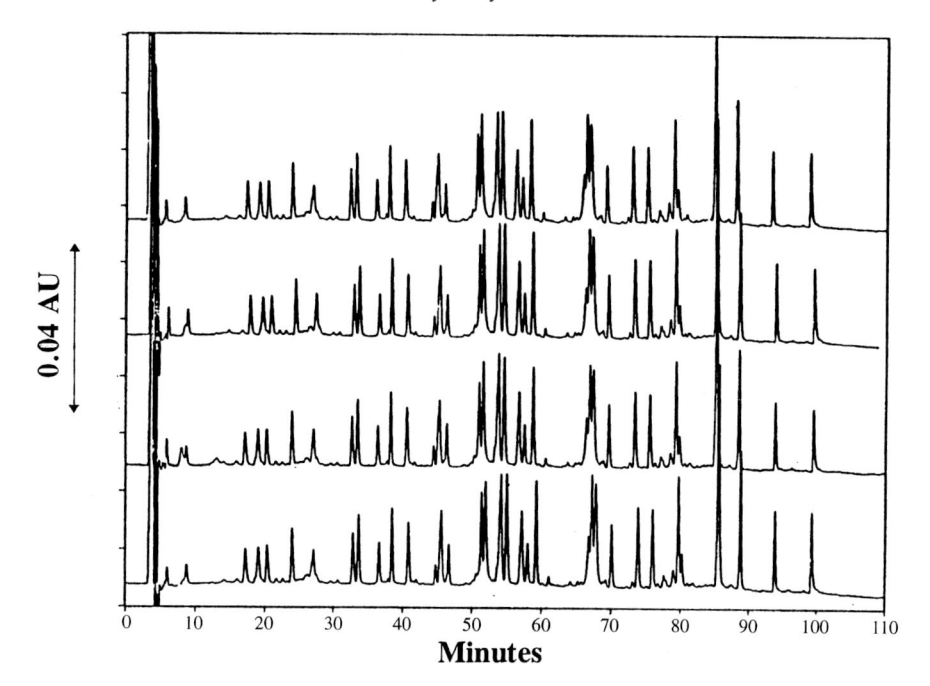

Figure 11. Peptide tryptic digest map at 50 µl/min without flow splitting using a 15 cm × 1 mm i.d. microbore column packed with 5 µm Zorbax C18 stable bond 300 Å pore size particles. The sample was 10 µl of 60 pmol tPA digest. The UV detector was set at 214 nm and column temperature was 50°C. Solvent A was H_2O:ACN:TFA, 98:2:0.1; solvent B was ACN:H_2O:TFA, 90:10:0.09. The gradient was from 96% B (v/v) to 60% B (v/v) in gradient steps of 0.5% B/min.

meters. As the small bench-top mass spectrometer becomes more affordable, many applications that traditionally utilize 4.6 mm i.d. conventional columns with UV/vis, fluorescence, or diode array detector may be replaced by microcolumns. Routine quality control applications using the same packing material as that for the 4.6 mm i.d. column can provide an equal or better separation efficiency at a much lower operation cost. For sample limited requirements, packed capillary columns offer the most sensitive mass detection. Sample amounts of a few nanolitres are sufficient for packed capillary HPLC analysis. Also, sample concentrations as low as a few femtomoles can be measured using a packed capillary column HPLC-MS system or with a UV detector equipped with a small flow cell. In the practice of micro-HPLC, the new generation digitally controlled micro-pumping system demonstrates excellent performance in terms of retention time and peak area reproducibility. Gradient elution at 3 µl/min without flow splitting can also be performed. Microcolumn HPLC has been widely adapted for protein/peptide tryptic digest mapping. The combination of recent technological advances in pumping system, in microcolumns, and in low cost bench-top mass spectrometers have established a broad way for the acceptance of microcolumn HPLC for routine sample analysis. It is however, necessary to recognize the difference in both system hardware requirements and application techniques between microcolumn and conventional column HPLC. Special attention to the system dead volume, flow path leakage, injection techniques, column temperature stability, column care, detector interface volume, flow cell volume, solvent flow path cross-talk, gradient delay time, gradient regeneration time, solvent prepressurization, etc. is required. For example, it is particularly useful to reverse column flow to purge out column contaminants whenever a 20–30% increase in column inlet pressure is observed. New generation microcolumn HPLC systems need to incorporate all of the above considerations into the system design. A trouble-free and reproducible microcolumn HPLC application can be expected as new advances in microcolumn HPLC frontier research is achieved.

References

1. Cappiello, A. and Bruner, F. (1993). *Anal. Chem.*, **65**, 1281.
2. Cappiello, A., Famiglini, G., and Bruner, F. (1994). *Anal. Chem.*, **66**, 1461.
3. Cappiello, A. and Famiglini, G. (1994). *Anal. Chem.*, **66**, 3970.
4. Mosely, M. A., Deterding, L. J., Tomer, K. B., and Jorgenson, J. W. (1990). *Anal. Chem.*, **63**, 1467.
5. Coutant, J. E., Chen, T. M., and Ackerman, B. L. (1990). *J. Chromatogr.*, **529**, 265.
6. Pleasance, S., Thibault, P., Mosely, M. A., Deterding, L. J., Tomer, K. B., and Jorgenson, J. W. (1990). *J. Am. Soc. Mass Spectrom.*, **1**, 312.
7. Gagne, J. P., Carrier, A., Varfalvy, L., and Bertrand, M. J. (1993). *J. Chromatogr.*, **647**, 13.
8. Gagne, J. P., Carrier, A., Varfalvy, L., and Bertrand, M. J. (1993). *J. Chromatogr.*, **647**, 21.
9. Tomer, K. B., Perkins, J. R., Parker, C. E., and Deterding, L. (1991). *J. Biol. Mass Spectrom.*, **20**, 783.
10. Alborn, H. and Stenhagen, G. (1987). *J. Chromatogr.*, **394**, 35.
11. Stenhagen, G. and Alborn, H. (1988). *J. Chromatogr.*, **474**, 285.
12. Barefoot, A. C. and Reiser, R. W. (1987). *J. Chromatogr.*, **398**, 217.
13. Barefoot, A. C. and Reiser, R. W. (1989). Biomed. Environ. Mass Spectrom., **18**, 77.
14. Hail, M., Lewis, S., Jardine, I., Liu, J., and Novotoy, M. (1990). *J. Microcol. Sep.*, **2**, 285.
15. Deterding, L. J., Parker, C. E., Perkins, J. R., Mosely, M. A., Jorgenson, J. W., and Tomer, K. B. (1991). *J. Chromatogr.*, **554**, 329.
16. Huang, E. C. and Henion, J. D. (1991). *Anal. Chem.*, **63**, 732.
17. Hunt, D. F., Alexander, J. E., McCormack, A. L., Martino, P. A., Michel, H., Shabanowitz, J., *et al.* (1991). In *Techniques in protein chemistry II* (ed. J. J. Villagranca), p. 441. Academic Press, San Diego.
18. Griffin, P. R., Coffman, J. A., Hood, L. E., and Yates, J. R. III. (1991). *Int. J. Mass Spectrom. Ion Proc.*, **111**, 131.

19. Schutz, G. A., Alexander, IV J. N., and Poli, J. B. (1996). *The 44th ASMS Conference Proceedings*, March 12–16, p. 909.
20. Tomer, K. B., Mosely, M. A., Deterding, L. J., and Parker, C. E. (1994). *Mass Spec. Rev.*, **13**, 431.
21. Yang, F. J. (ed.) (1989). *Microbore column chromatography: a unified approach to chromatography*. Marcel Dekker, New York.
22. Yang, F. J. (1982). *J. Chromatogr. Sci.*, **20**, 241.
23. Yang, F. J. (1982). *J. Chromatogr.*, **236**, 284.
24. Yang, F. J. (1983). US Patent 4375163.
25. Yang, F. J. (1984). US Patent 4483773.
26. Yang, F. J. (1981). *JRC & CC*, **4**, 83.
27. Ishii, D., Takeudi, T., and Wada, H. (1988). In *Introduction to microscale high-performance* liquid chromatography (ed. D. Ishii), pp. 33–67. VCH, New York, NY.
28. Kennedy, R. T. and Jorgenson, J. W. (1989). *Anal. Chem.*, **61**, 1128.
29. Karlson, K., and Novotny, M. (1988). *Anal. Chem.*, **6**, 1662.
30. Malik, A., Li, W., and Lee, M. L. (1993). *J. Microcol. Sep.*, **5**, 361.
31. Tong, D., Bartle, K. D., Clifford, A. A., and Edge, A. M. (1995). *J. Microcol. Sep.*, **7**, 265.
32. Hsieh, S. and Jorgenson, J. W. (1996). *Anal. Chem.*, **68**, 121.
33. Vissers, J. P. C., Vanden Hoef, E. C. J., Classens, H. A., Laven, J., and Cramers, C. A. (1995). *J. Microcol. Sep.*, **7**, 239.
34. Boughtflower, R. J., Underwood, T., and Maddin, J. (1995). *Chromatographia*, **41**, 398.
35. Hoffmann, S. and Blomberg, L. (1987). *Chromatographia*, **24**, 46.
36. Van der Wal, S. J. and Yang, F. J. (1983). *HRC & CC*, **6**, 216.
37. Miller, C. J. (1967). *Electron. World*, **77**, 23.
38. Chavez, S. P. and Richards, C. G. (1969). *Fluid. Q.*, **2**, 40.
39. Yang, F. J. and Fan, C. (1993). US patent 5253981.
40. Yang, F. J. (1997). US Patent 5630706.
41. Schwartz, H. and Brownbee, R. G. (1984). *Am. Lab.*, **16 (10)**, 43.

3

Detection devices

RAYMOND P. W. SCOTT

1. Introduction

In chromatography, the detection device is as essential to a successful analysis as the distribution system that produces the separation, and over the years there has been a continuous synergistic development between these two essential components of the chromatograph. As the columns and thin-layer plates have become more efficient, the solute bands have become reduced in size, and the solute concentrations smaller. These changes have educed the development of detectors, with higher sensitivities, and smaller sensor volumes. The improved detectors, in turn, have evoked improvements in column technology which have produced higher separating efficiencies. It is clear that the design of the detecting device is critical in high resolution chromatography.

1.1 Detector specifications

Accurate performance criteria, or specifications, must be available for any detecting device, to enable the user to determine its suitability for a specific application. This is necessary, not only to compare its performance with alternatives supplied by other instrument manufacturers, but also to determine the optimum chromatography system with which it must be used to achieve the maximum efficiency (1). The specifications should be presented in a standard form and in standard units, so that detectors that function on widely different principles can be compared. The major detector characteristics that fulfil these requirements together with the units in which they are measured are summarized in *Table 1*.

There are other specifications such as: total detector–system dispersion, sensor dimensions, sensor volume, time constant of overall response, together with flow, pressure, and temperature sensitivity. However, only the major specifications given in *Table 1* will be discussed here and for further details the reader is directed to ref. 2.

1.1.1 The dynamic range of the detector

There are two important ranges that are specified for a detector, and these are the dynamic range, and the linear dynamic range. The dynamic range extends

Table 1. Detector specifications

Dynamic range	(D_R) g/ml (e.g. 3×10^{-9} to 6×10^{-5})
Response index	(r) dimensionless
Linear dynamic range	(D_{LR}) g/ml (e.g. 1×10^{-8} to 2×10^{-5})
Detector response	(R_C) volts/g or (specific units of measurements/g)
Detector response	(N_D) usually in millivolts but may be in specific units (e.g. refractive index units)
Sensitivity or minimum detectable concentration	(X_D) g/ml (e.g. 3×10^{-8}) but may be in specific units (e.g. absorption units)

from the minimum detectable concentration (i.e. the sensitivity) to that concentration at which the detector no longer responds to any concentration increase. The dynamic range of a detector (D_R), for which 720 000 would be a typical value, is usually given in the form:

$$D_R = 7.2 \times 10^5$$

The manufacturer should give the dynamic range in dimensionless units as it is a ratio of concentrations and independent of the units used. Employing the minimum detectable concentration in conjunction with the dynamic range, the maximum concentration detectable can then be calculated.

1.1.2 Detector linearity

Detector linearity is important when a detector is used for quantitative analysis. It is defined as the concentration range over which the detector response is linearly related to the concentration of solute passing through it. That is:

$$V = AC_m$$

where V is the output of the detector, C_m is the concentration of solute being measured, and A is a constant.

Because of the imperfections in mechanical and electrical devices, true linearity is a hypothetical concept, and in practice, detectors only approach this ideal response. It is therefore important to have some measure of linearity that is specified in numerical terms, so that the proximity of the detector response to true linearity is known. Fowlis and Scott (3) proposed a method of measuring detector linearity. They assumed that for a closely linear detector the response could be described by the following power function:

$$V = AC_m^r \qquad [1]$$

where (r) is defined as the response index and the other symbols have the meanings previously ascribed to them.

It follows that for a truly linear detector, r = 1, and the proximity of (r) to unity will indicate the extent to which the response of the detector deviates from true linearity. The basic advantage of defining linearity in this way, is that if the detector is not perfectly linear, and a value for (r) is known, then a correction can be applied to accommodate the non-linearity. There are alternative methods for defining linearity which, in the author's opinion, are less precise and less useful. The recommendations of the ASTM, E19 committee (4) on linearity measurement were as follows:

'the linear range of a PD (photometric detector) is that concentration range of the test substance over which the response of the detector is constant to within 5% as determined from a linearity plot, – the linear range should be expressed as the ratio of the highest concentration on the linearity curve to the minimum detectable concentration.'

This method for defining detector linearity is perfectly satisfactory, and ensures a minimum linearity from the detector, and an acceptable quantitative accuracy. However, the specification is significantly 'looser' than that given above, and there is no means of correcting for any non-linearity that may exist. If the response index is not available, it is strongly advised that the response index of all detectors that are to be used for quantitative analysis is determined (5).

1.1.3 The linear dynamic range

The linearity of most detectors deteriorates at high concentrations, and thus the linear dynamic range of a detector will always be less than the dynamic range. The symbol for the linear dynamic range is usually taken as (D_{LR}), and if the dynamic range is 5600, it would be be specified as follows:

$$D_{LR} = 5.6 \times 10^{-3} \text{ for } 0.98 < r < 1.02.$$

The lowest concentration in the linear dynamic range is usually equal to the minimum detectable concentration, or the sensitivity of the detector. The largest concentration in the linear dynamic range would be that where the response factor (r), falls outside the range specified.

1.1.4 Detector response

Detector response can be defined as the voltage output for unit change in solute concentration. Alternatively, it can be taken as the voltage output that would result from unit change in the physical property that the detector measures, e.g. refractive index or dielectric constant. The detector response (R_c) can be determined by injecting a known mass of solute (m) onto the column and measuring the response. Assuming the concentration of the solute at the peak maximum is twice the average peak concentration, then:

$$R_c = \frac{hWQ}{sm}$$

where h is the peak height; w is the peak width in cm at 0.607 h, which is the height of the points of inflection of the Gaussian curve; Q is the flow rate through the sensor in ml/min; s is the chart speed in cm/min.

The response of a detector will differ between different solutes as well as between different detectors. Consequently, the response of two detectors of the same type and geometry can only be compared using the same solute and the same mobile phases.

1.1.5 Detector noise

Detector noise is the term given to any perturbation on the detector output that is not related to an eluted solute. It is a basic property of the detecting device, and determines the ultimate sensitivity, or minimum detectable concentration, that can be achieved. Detector noise is arbitrarily divided into three types: short-term noise, long-term noise, and drift, all three of which are depicted in *Figure 1*.

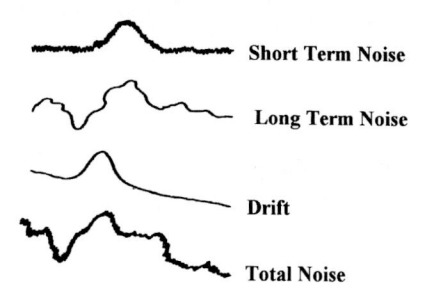

Figure 1. Different types of detector noise.

Short-term noise consists of baseline perturbations that have a frequency that is significantly higher than the eluted peak. Short-term noise can be easily removed by appropriate noise filters. Its source is usually electronic, arising from either the detector sensor system or the amplifier. Long-term noise consists of baseline perturbations that have a frequency that is similar to that of the eluted peak. This type of detector noise is the most significant and damaging, as it is indiscernible from very small peaks in the chromatogram. Long-term noise cannot be removed by electronic filtering, without affecting the profiles of the eluted peaks. In *Figure 1*, the peak profile can be easily discerned above the high frequency noise, but is lost in the long-term noise. Long-term noise usually arises from temperature, pressure, or flow rate changes in the sensing cell. Long-term noise ultimately limits the detector sensitivity or the minimum detectable concentration. Drift results from baseline perturbations that have frequencies that are significantly larger than that of the eluted peak. Drift results from changes in ambient temperature, mobile phase flow rate, and pressure or variations in solvent composition. In general, the sensitivity of the detector should never be set above the level where the combined noise exceeds 2% of the full scale deflection (FSD) of the recorder, or appears as more than 2% FSD of the computer simulation of the chromatogram.

1.1.6 Measurement of detector noise

The detector noise is defined as the maximum amplitude of the combined short- and long-term noise, measured over a period of ten minutes. The detector must be connected to a column and mobile phase passed through it during measurement. The detector noise is obtained by constructing parallel lines, embracing the maximum excursions of the recorder trace over the defined time period, as shown in *Figure 2*.

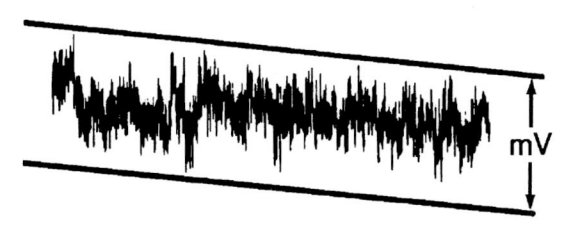

Figure 2. The measurement of detector noise.

The distance between the parallel lines measured in millivolts is taken as the measured noise (v_n), and the noise level (N_D) is calculated in the following manner:

$$N_D = v_n A = \frac{v_n}{B}$$

where v_n is the noise measured in volts from the recorder trace; A is the attenuation factor; B is the alternative amplification factor (A = 1/B).

Detector sensitivity, or minimum detectable concentration (MDC), is defined as the minimum concentration of solute passing through the detector, that can be unambiguously discriminated from noise. The size of the signal that will make it distinct from the noise (the signal-to-noise ratio) is generally accepted as two. Thus for a concentration-sensitive detector, the detector sensitivity (X_D) is given by:

$$X_D = \frac{2N_D}{R_c} \text{ (g/ml)}$$

R_c and N_D being determined in the manner previously described.

It should be noted that the sensitivity of a detector is not the minimum mass that can be detected. The minimum detectable mass is the system mass sensitivity, which would also depend on the characteristics of the apparatus as well as the detector and, in particular, the type of column employed.

2. The classification of detectors

Detectors have been classified in a number of ways, and at present there is no consensus of opinion as to which is most appropriate. The different basis of classification are as follows, but throughout this chapter the bulk and solute property classification will be used.

2.1 Bulk property and solute property detectors

One method of detector classification, and one that is probably the most frequently used, is based on the function of their response. Bulk property detectors measure some bulk physical property of the mobile phase, e.g. refractive index or dielectric constant. Bulk property detectors are very susceptible to changes in mobile phase composition or temperature and can not be used for gradient elution in liquid chromatography. They are also very sensitive to changes in pressure and flow rate, and consequently the operating conditions of the chromatograph needs to be very carefully controlled. Solute property detectors function by measuring some property of the solute that the mobile phase does not possess, or has to a very reduced extent. An example of a solute property detector would be the UV detector in liquid chromatography, which detects only those substances that absorb in the UV. This classification is rarely precise, as it is often difficult to employ a phase system that includes a mobile phase to which the detector has no response whatsoever, relative to that of the solute. For example, the UV detector may be employed with a mobile phase that contains small quantities of a polar solvent such as ethyl acetate. At higher wavelengths, ethyl acetate will exhibit little or no absorption and the detector may well behave as a true solute property detector. However, if the wavelength is reduced, e.g. to detect peptides 214 nm, then the ethyl acetate will absorb quite strongly and the system will begin to resemble that of a bulk property detector. It is evident that this type of classification is not clear cut and can be confusing, as a given detector can exhibit both characteristics depending on the conditions under which it is operated.

2.2 Mass-sensitive and concentration-sensitive detectors

Some detectors respond to changes in solute concentration, while others respond to the change in mass passing through the sensor per unit time. Concentration-sensitive detectors provide an output that is directly related to the concentration of solute in the mobile phase passing through it. The UV absorption detector would be typical of a concentration-sensitive detector. Mass-sensitive detectors respond to the mass of solute passing through it per unit time and is thus independent of the volume flow of mobile phase. However, there is only one liquid chromatography detector that is mass-sensitive (6) (the transport detector), which at the time of writing this book is not commercially available.

2.3 Specific and non-specific detectors

Specific detectors respond to a particular type of compound or a particular chemical group. The fluorescence detector would be a typical specific detector that responds only to those substances that fluoresce. Non-specific detectors respond to all solutes present in the mobile phase and their catholic performance makes them a very useful and popular type of detector. Unfortunately, non-specific detectors tend to be relatively insensitive. The refractive index detector is probably the most non-specific detector available in liquid chromatography, but as already mentioned it also has the least sensitivity of the commonly used detectors.

3. Detecting devices for high efficiency columns

The essential characteristics of a detecting device that is to be used with high efficiency columns can be summarized as follows:

- high sensitivity
- low dispersion
- low sensor volume

In the chromatographic column, mixtures are resolved by moving the individual solute bands away from each other, and at the same time constraining the band dispersion, so that the solutes are eluted discretely. The movement of the peaks apart (the column selectivity), is controlled solely by the choice of the phase system. The peaks are kept narrow almost exclusively by column design. Both factors make demands on the detector format and in particular the dimensions of the sensor.

In order to reduce peak dispersion, low stationary phase loading must be used which, in turn, reduces the size of the sample and, consequently demands higher detector sensitivities. Due to the parabolic velocity profile that occurs in tubular conduits, serious peak dispersion can occur if such conduits are used to connect the column to the sensor. The variance dispersion increases as the fourth power of the tube radius and linearly with the tube length, as shown by the Golay equation (7). It follows that all tubing should be as short as possible, and have as minimum a diameter that is commensurate with the tubing not becoming blocked.

The effect of sensor volume on peak size and shape is a little complex. The detector responds to an average value of the total amount of solute in the sensor cell. In the extreme, the sensor volume or cell could be large enough to hold two closely eluted peaks and thus give a response that would appear as though only a single solute had been eluted, albeit distorted in shape. This extreme condition rarely happens, but serious peak distortion and loss of resolution can still result. This is particularly so if the sensor volume is of the same order of magnitude as the peak volume. The problem can be particularly severe when microbore columns are being used. The elution profile of a peak eluted from a column 3 cm long, 3 mm internal diameter, packed with particles 3 μm in diameter is depicted in *Figure 3*.

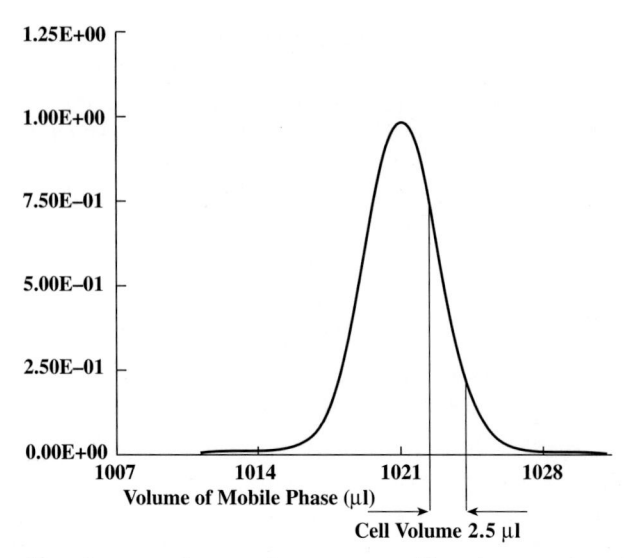

Figure 3. The effect of sensor volume on detector output. The column used was 3 cm × 3 mm, packed with 3 μm diameter packing material. The column efficiency was 5000 theoretical plates and the solute k' value was 2.

The peak is considered to be eluted at a capacity ratio (k') of 2, and it is seen that the peak width at the base is about 14 μl wide. The sensor cell volume is 2.5 μl and the portion of the peak in the cell is included in the figure. It is clear the detector will respond to the mean concentration of the slice contained in the 2.5 μl sensor volume. It is also clear that, as the sensor volume is increased (or alternatively the column is made more efficient, and the peak volume is reduced), an even greater part of the peak will be contained in the cell.

Thus the output will be an average value of an even larger portion of the peak, which will cause serious peak distortion. The effect of a finite sensor volume can be easily simulated with a relatively simple computer program, as shown in *Figure 4*. The example given, although not the worst case scenario, is a condition where the sensing volume of the detector can have a very serious effect on the peak profile and, consequently, the resolution. The column is a small bore column, and thus the eluted peaks have a relatively small peak volume, which is commensurate with that of the sensing cell. It is seen that even a sensor volume of 1 μl has a significant effect on the peak width, and it is clear that if the maximum resolution is to be obtained from the column, then the sensor cell volume should be no greater than 2 μl. It should also be noted that the results from the use of a sensor cell having a volume of 5 μl are virtually useless, and that many commercially available detectors do, indeed, have sensor volumes as great as, if not greater than, this. It follows that if high

3: Detection devices

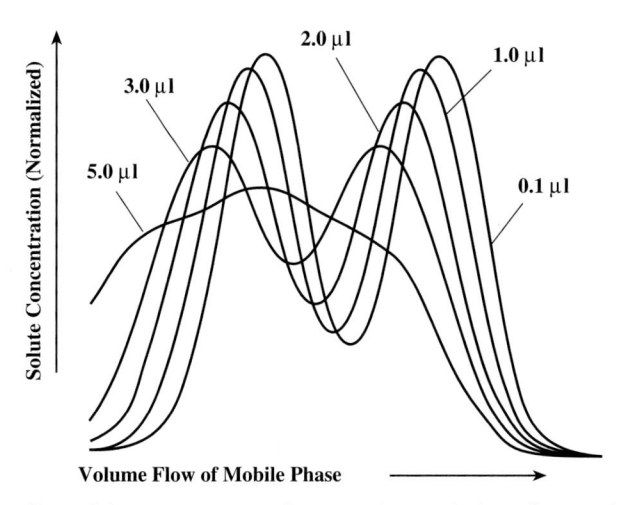

Figure 4. The effect of detector sensor volume on the resolution of two solutes. The two solutes have elution times that differ by 4 sec on a 15 cm × 1 mm microbore column packed with a 5 μm diameter matrix. k' for the first peak is 1.

efficiency columns are to be employed, sensors with such volumes must be carefully avoided.

4. Light absorption detectors

The two most commonly used liquid chromatography detectors in high resolution liquid chromatography, are the UV absorption detector and the fluorescence detector. The former detects all substances that absorb in the UV range of wavelengths, and the latter, although specific, is one of the more sensitive types of detectors. Utilizing techniques such as derivatization, and wavelength selection for the excitation and emission light, the fluorescence detector can also be made applicable to a significant range of different chemical types. There are two forms of UV detector, the multi-wavelength UV detector, and the fixed wavelength UV detector, the latter being slightly more sensitive and considerably less expensive than the former. The relationship between the intensity of UV light transmitted through a cell (I_T) and the concentration of solute in it (c) is given by Beer's Law:

$$I_T = I_o e^{-kcl}$$

or

$$\ln(I_T) = \ln(I_o) - kcl$$

where I_o is the intensity of the light entering the cell; l is the path length of the cell; k is the molar extinction coefficient of the solute for the specific wavelength of the UV light.

Differentiating:

$$\frac{d\left(\log \frac{I_T}{I_o}\right)}{dc} = -kl$$

It is seen that the sensitivity of the detector, as measured by the transmitted light, will be directly proportional to the value of the extinction coefficient (k), which is characteristic of the substance being detected, and the path length of the cell (l). It follows that there will be restraints on the reduction in the size of the sensor cell for use with high efficiency columns. In order to achieve a minimum sensor volume, without decreasing the cell length, its radius must be reduced. Such a procedure, will however, reduce the amount of light falling on the photocell, which will, in turn, reduce the signal-to-noise. Ultimately, this will lead to a loss of sensitivity. It follows that here is a limit to the reduction of the cell dimensions, and consequently the sensor volume. Limiting the volume of the sensor to a practical minimum will also limit the minimum dimensions that can be used for the high efficiency column (8).

4.1 The fixed wavelength UV detector

A diagram of the fixed wavelength detector is shown in *Figure 5*. It consists of a simple cylindrical sensor cell with angular inlets and outlets. These are set at an angle to the axis of the cell, to ensure a circular motion in the mobile phase within the cell. This swirling motion breaks up the parabolic velocity profiles within the liquid, which would cause band dispersion and loss of chromatographic efficiency. The sensor cell is terminated by quartz windows and at one end, a low pressure mercury arc lamp is situated to provide UV light at

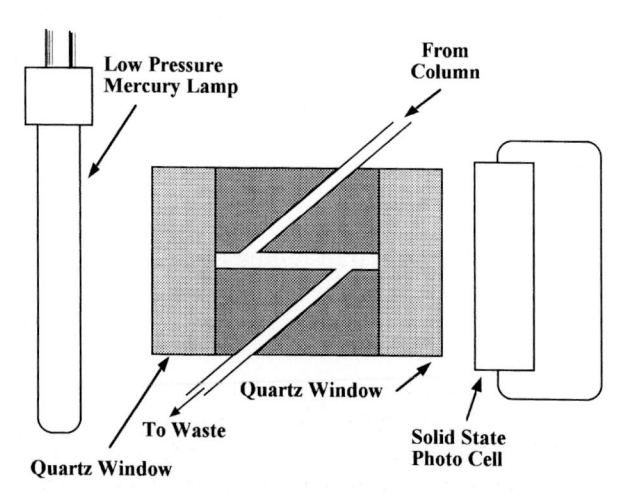

Figure 5. Fixed wavelength UV detector.

3: Detection devices

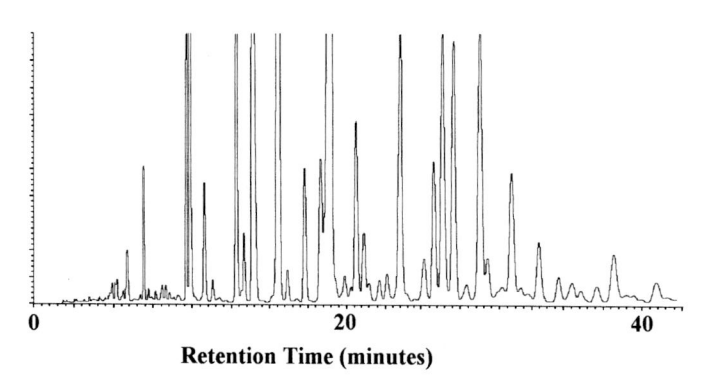

Figure 6. Chromatogram of an aromatic hydrocarbon extract from coal obtained from a high efficiency small bore column.

254 nm. At the other end a solid state photosensor is installed, that monitors the light passing through the cell. The output from the photocell is amplified, and fed to either a recorder, or the A/D converter of a data acquisition system.

Other light sources are available that emit light at lower wavelengths are the low pressure cadmium lamp (229 nm), and the low pressure zinc lamp (213 nm and 307 nm). The fixed wavelength detector lends itself to miniaturization, and was employed to demonstrate the separation obtained from the first high efficiency liquid chromatography column, having in excess of 250 000 theoretical plates (9).

The chromatogram obtained is shown in *Figure 6*. The column was 2 m long, 1 mm internal diameter, and packed with a reverse-phase matrix having a particle diameter of 10 μm. Inlet pressure was 6000 psi, the mobile phase an acetonitrile/water mixture, and the chromatogram was developed isocratically. The fixed wavelength UV detector used had a volume of about 2.7 μl. The column efficiency was about 250 000 theoretical plates, and it should be noted that the separation time is over 40 hours. Modern high efficiency columns are considerably faster, but unfortunately, assuming an optimum column design, if the inlet pressure has a practical maximum limit, then the efficiency can only be increased further by accepting longer retention times. The small volume fixed wavelength UV detector can also be used to monitor very fast separations, providing the detector electronics have a sufficiently fast response. An example of a high speed separation monitored by such a detector (10) is shown in *Figure 7*.

The separation was monitored by a fixed wavelength UV detector, similar in design to that employed to display the separation from the high efficiency column shown in *Figure 6*. As a result of its simple design, and the ease in which it can be miniaturized, the fixed wavelength UV detector is often chosen to monitor the fastest separations and those exhibiting the highest

Raymond P. W. Scott

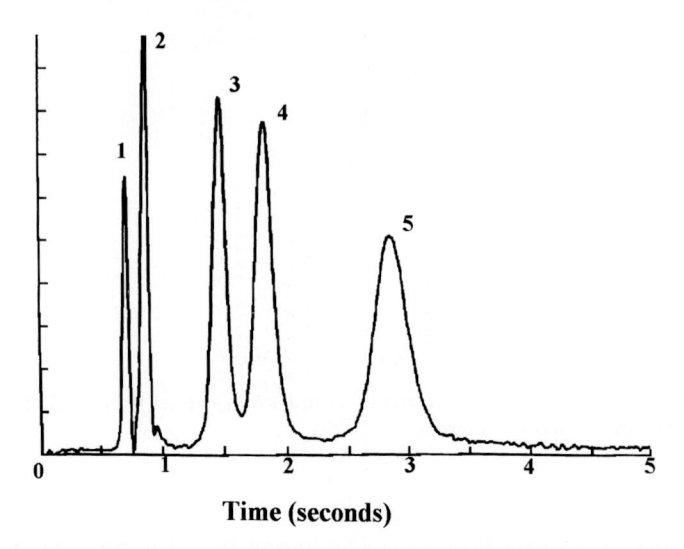

Figure 7. High speed separation of a five component mixture. Packing: Hypersil 3-m. Column internal diameter was 2.6 mm and length 2.5 cm. Mobile phase: 2.2% (v/v) methyl acetate in *n*-pentane. Linear velocity: 3.3 cm/sec. 1, *p*-xylene; 2, anisole; 3, nitrobenzene; 4, acetophenone; 5, dipropyl phthalate.

resolution. This type of detector has also been used in electro-chromatography and capillary electrophoresis. Although, strictly not chromatography, electrophoresis is a high resolution technique. Capillary electrophoresis and electrochromatography are mostly carried out in fused-quartz capillary tubes, coated with polyimide. The on-column UV detector is formed by removing a small portion of the coating (about 2 mm) from the end of the tube, which is then threaded through a metal block, in the manner shown in *Figure 8*.

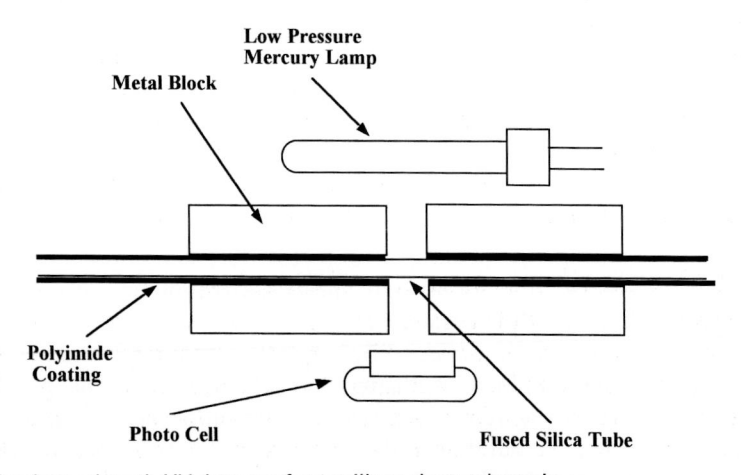

Figure 8. Fixed wavelength UV detector for capillary electrophoresis.

3: Detection devices

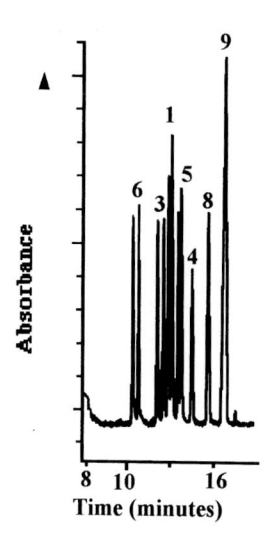

Figure 9. Electropherograms of phenoxy acid herbicides monitored by on-column UV absorption detection. 1, dichloroprop; 3, mecoprop; 4, 2-(3-chlorophenoxy)propionic acid; 5, 2-(4–chlorophenoxy)propionic acid; 6, silvex; 7, 2-(2-chlorophenoxy)propionic acid; 8, 2-phenoxypropionic acid. Capillary 57 cm (50 cm to detection window). Running electrolyte: 25 mM sodium phosphate, 600 mM borate pH 5, containing 10 mM 2,3,6-tri-*O*-methyl-β-carbodiimide.

The clear portion of the electrophoresis column is aligned with another hole, drilled normal to the tube. Light from a low pressure mercury lamp passes diametrically across the tube, and falls onto a photoelectric cell. As the separated solutes pass the window in the tube, light is adsorbed, and the output of the photocell falls. The path length of the effective sensor cell is exceedingly short (the diameter of the tube, *c.* 0.25 mm) and so the system is not very sensitive. However, as the concentrations of solutes separated by electrophoresis are relatively high, detection is effective, and only the dynamic range of response is restricted. An example of a separation monitored by an on-column fixed wavelength UV detector (11), is shown in *Figure 9*.

It is seen that despite the short path length of the sensor cell, a good separation is achieved. This is due to the high solute concentrations being obtained from the sharp component bands that result from high resolution systems.

4.2 The multi-wavelength UV detector

The multi-wavelength UV detector functions on exactly the same principle as the fixed wavelength detector, but incorporates a monochromator that allows absorption at a specific wavelength to be monitored. This type of detector also incorporates a semiconductor diode array, that allows absorption to be simultaneously monitored over a wide range of different wavelengths.

The diode array detector utilizes a broad spectrum light source such as a deuterium or xenon lamp. Light from the lamp is focused by means of an achromatic lens through the sample cell and onto a holographic grating. The dispersed light from the grating is arranged to fall on a linear diode array. The resolution of the detector ($\Delta\lambda$) will depend on the number of diodes (n) in the array, and also on the range of wavelengths covered ($\lambda_2 - \lambda_1$). Thus:

$$\Delta\lambda = \frac{\lambda_2 - \lambda_1}{n}$$

It is seen that the ultimate resolving power of the diode array detector will depend on how narrow the individual photocells can be commercially fabricated. A diagram of a diode array detector is shown in *Figure 10*.

The array may contain many hundreds of diodes, and the output from each diode is regularly sampled by a computer and stored on a hard disc. At the end of the analysis, the output from any diode can be selected from memory and a chromatogram produced from the UV adsorption at the wavelength monitored by that particular diode. The output of a single diode can also be monitored in real time, so that the chromatogram can be followed as the separation develops. A spectrum of any solute can be obtained by recalling from memory the output of all the diodes taken at its retention time. In addition the normalized spectra can be ratioed across the peak, and if the peak is pure, a constant value will be obtained confirming peak purity. Finally, the spectrum from the front of the peak, and one from the back of a peak, can confirm the presence (and possibly the identity) of two different solutes, even though they have not been resolved by the chromatographic system. In practice, due

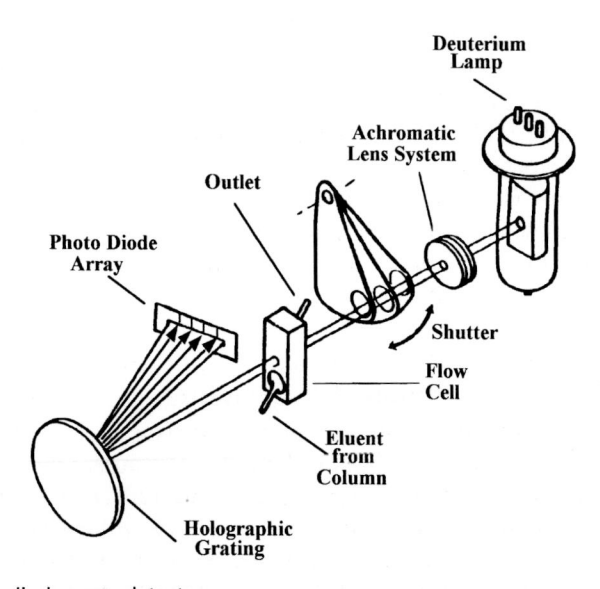

Figure 10. The diode array detector.

to the limited information provided by the UV spectrum of a compound, and the great similarity between the majority of UV spectra, the diode array detector is not often used for solute identification. The device is more often employed to provide selective wavelength monitoring, so that a wide range of compounds, that absorb at vastly different wavelengths, can all be monitored during a single separation process. The cell volume of the diode array detector can be reduced to 2 μl or less, and so it can be used with high efficiency columns. To date, it does not appear to have been used in electro-liquid-chromatography.

5. Fluorescence detectors

Fluorescence detectors are some of the most sensitive types of liquid chromatography detectors available, and can be used to monitor substances that naturally fluoresce and also those that form fluorescent derivatives. The fluorescence detector can be extremely versatile as, with the aid of appropriate monochromators, both the excitation light and the emission light can be selected. The single wavelength fluorescence detector is the more widely used, particularly in electro-liquid-chromatography, where a laser is often employed as the excitation source. The multi-wavelength alternative, which in reality is a fluorescence spectrometer, is more versatile, generally less sensitive, and far more expensive.

Most fluorescent detectors are configured in such a manner that the fluorescent light is viewed at an angle (usually at right angles) to the direction of the exciting incident light beam. This arrangement minimizes the amount of incident light that may provide a background signal to the fluorescent sensor. The fluorescence signal (I_f) is given by:

$$I_f = \phi I_o (1 - e^{-kcl})$$

where ϕ is the quantum yield (the ratio of the number of photons emitted and the number of photons absorbed); I_o is the intensity of the incident light; c is the concentration of the solute; k is the molar absorbance; l is the path length of the cell.

The fluorescence signal is not linearly related to solute concentration, and thus the signal must be modified for quantitative analysis. This is usually carried out by an appropriate non-linear amplifier or by the computer software after data acquisition.

5.1 The single wavelength fluorescent detector

The simple fluorescence detector, a diagram of which is shown in *Figure 11*, utilizes a single excitation wavelength, but monitors fluorescent light of all wavelengths (i.e. within the wavelength sensitivity of the photocell). The UV lamp is usually a low pressure mercury lamp, which provides relatively high intensity UV light at 253.7 nm. Many substances that fluoresce will, to a lesser or greater extent, be excited by light of this wavelength. The excitation light is focused by a quartz lens through the cell, and another lens, situated normal

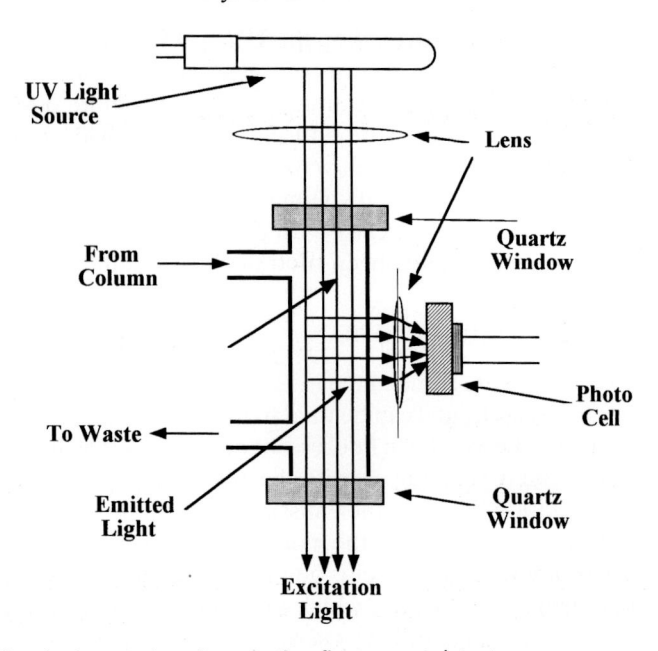

Figure 11. The single wavelength excitation fluorescent detector.

to the incident light, focuses the fluorescent light onto a photocell. Typically, a fixed wavelength fluorescence detector will have a sensitivity (minimum detectable concentration at an excitation wavelength of 254 nm), of about 1×10^{-9} g/ml, a linear dynamic range of about 10^3, where the response index (r) will lie between about 0.96 and 1.04.

A recent exciting use of the single wavelength fluorescence detector, as a monitor for electro-liquid-chromatography, is given by the work of Hereren *et al.* (12), who used laser-induced fluorescence, in conjunction with electro-liquid-chromatography to provide high sensitivity, and high column efficiencies.

Electro-liquid-chromatography differs from normal liquid chromatography, in that the mobile phase flow is electro-osmotically driven, as opposed to the use of pressure. Electro-osmotic flow is not accompanied by the parabolic velocity profile normally associated with Newtonian flow. This significantly reduces the resistance to mass transfer contribution to band dispersion, and thus provides higher efficiency. The layout of the apparatus used by Hereren is shown in *Figure 12*. The column that was used was 75 cm long, 75 μm internal diameter, and the detection window was 50 cm from the inlet. Light from the laser was focused onto a optical fibre bundle, the output from which was passed through a lens and onto a plane mirror, to be focused onto the column detection window.

Fluorescent light emitted from the window area was focused by means of a lens, through a filter to remove any incident light, through a pin-hole, and onto

3: Detection devices

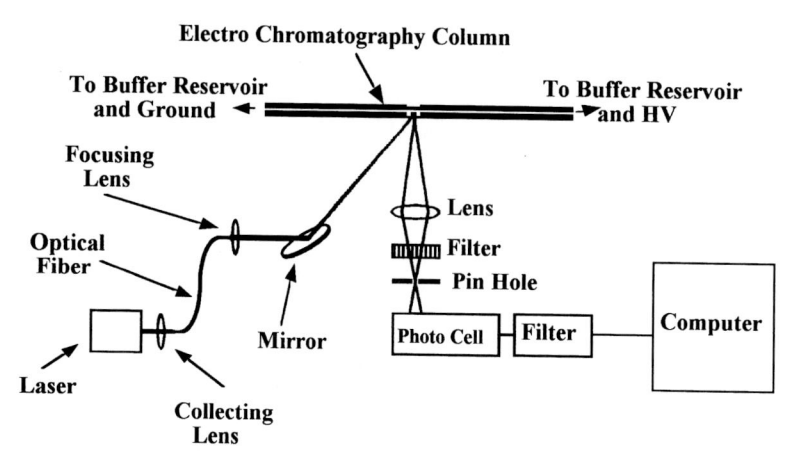

Figure 12. Apparatus for electro-liquid-chromatography using laser excited fluorescence detection.

the photocell. The output from the photocell was amplified, passed through a low-pass filter, and acquired by a computer. The results obtained from the system when used for the separation of a mixture of derivatized amino acids is shown in *Figure 13*. Two chromatograms are shown, one monitored by absorption, and the other by fluorescence measurements.

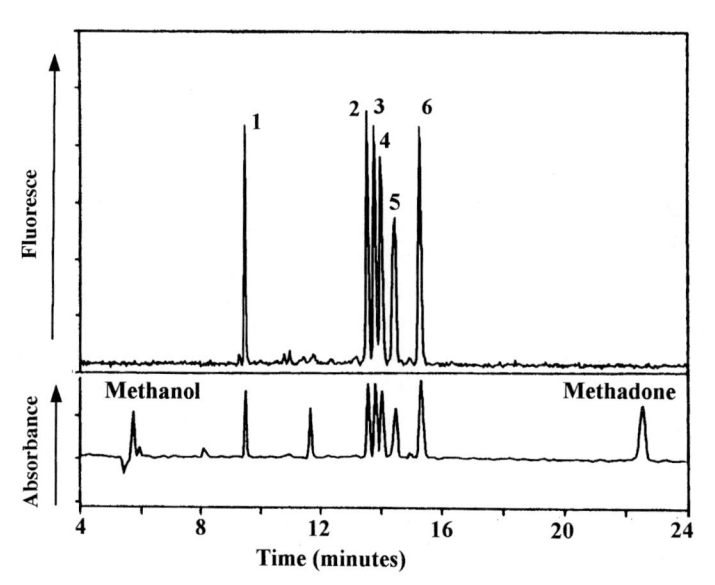

Figure 13. The separation of some fluorescein isothiocyanate derivatized amino acids by electro-liquid-chromatography using laser excited fluorescence detection. The fluorescent derivatives were as follows: 1, arginine; 2, glutamine; 3, phenylalanine; 4, asparagine; 5, serine; 6, glycine.

The derivatizing reagent was fluorescein isothiocyanate, and each amino acid was present at a level of 10 µM, which is a relatively large sample for this type of system. Nevertheless, it is seen that the fluorescence monitor appears to give a far greater signal-to-noise than the absorbance monitor, and the resolution provided by the microbore column is more than adequate. The feasibility of the use of the laser to enhance fluorescence detection in high resolution chromatography has been clearly established, and the technique offers potentially very high sensitivities with further development. The great advantage of the system is the extremely small sensor volume, that allows the very low dispersion columns to be successfully operated.

5.2 The multi-wavelength fluorescence detector

The multi-wavelength fluorescence detector is far more expensive and complicated, and usually incorporates two monochromators. One monochromator provides a selected excitation wavelength and the other, a selected emission wavelength. Some instruments have a diode array sensor that analyses the emission light and permits on-the-fly emission spectra to be taken. Unfortunately, the commercially available devices have relatively large sensor cells and consequently must be used with some caution with high efficiency columns. The multi-wavelength fluorescence detector does provide more detailed spectral data for confirming solute identity, and the sensitivity can be enhanced by selecting optimum excitation and emission wavelengths, for solutes of particular interest. However, unless custom modified to suit high efficiency columns, the multi-wavelength fluorescence detector is not recommended for high resolution liquid chromatography.

6. Electronic detectors

Electronic detectors, which includes the electrical conductivity detector, and the electrochemical detector, can be constructed with very small sensor volumes and still exhibit very high sensitivity. Consequently, they are ideal for use in high resolution liquid chromatography. The disadvantages are that the electrical conductivity detector can only detect those substances that are ionized, and the electrochemical detector can only detect those substances that can be chemically oxidized or reduced. Nevertheless, these two types of detectors have found extensive use with high efficiency columns.

6.1 The electrical conductivity detector

The electrical conductivity sensor is simple in the extreme, and consists merely of two electrodes, situated in the eluent from the chromatography column. A diagram of the sensor is shown in *Figure 14*.

The electrodes (usually made of stainless steel) are placed in the arms of a Wheatstone bridge and an AC potential (1–10 kHz) applied across the bridge.

3: Detection devices

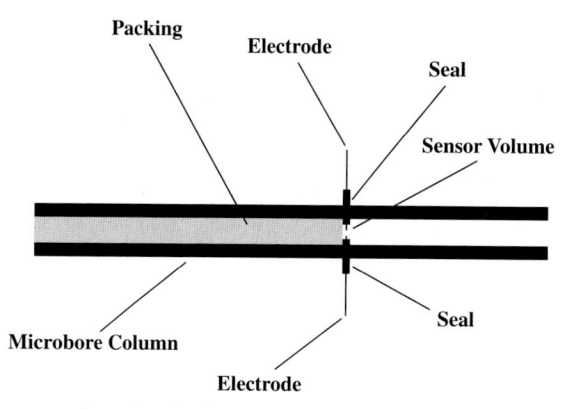

Figure 14. The sensor of an electrical conductivity detector.

The bridge is initially balanced, and as an ionic solute is eluted from the column, the impedance between the plates changes. The resulting out-of-balance signal is processed and passed either to a recorder, or to the data acquisition input of a data processor. It is essential to employ an AC potential in the bridge circuit, to avoid the electrodes being polarized. Care must be taken to eliminate any buffer ions that may be present in the mobile phase prior to detection. This can be achieved by the use of a suppresser column between the chromatography column and the detector. It is usually packed with an appropriate ion exchange material, or, if an organic buffer is being used, and inorganic ions are being separated, then a reversed-phase suppresser column can be used. An example of the use of the electrical conductivity detector is given by the separation of a mixture of alkali and alkaline earth cations, at levels of a few parts per million, and is shown in *Figure 15*. The cations lithium, sodium, ammonium, potassium, magnesium, and calcium were present in the original mixture at concentrations of 1, 4, 10, 10, 5, and 10 p.p.m. respectively.

A proprietary ion exchange column, IonPacCS12, was used and the mobile phase consisted of a 20 nM methanesulfonic acid solution in water. The separation is an interesting example of the use of the ion suppression technique. The methanesulfonic acid solution, if passed through the detector, would have had a high electrical conductivity and thus, give a large signal on the detector which would swamp the signal from the ions being determined. Thus, after the mobile phase leaves the column and consequently, after the methanesulfonic acid has achieved its purpose, and helped produce the desired separation, the agent must be removed. This is necessary so that the mobile phase actually entering the detector contains little or no ions other than those being analysed. The methanesulfonic acid can be removed by passing the mobile phase through a short reverse-phase column before entering the detector. The reverse-phase column will remove any organic material present in the mobile

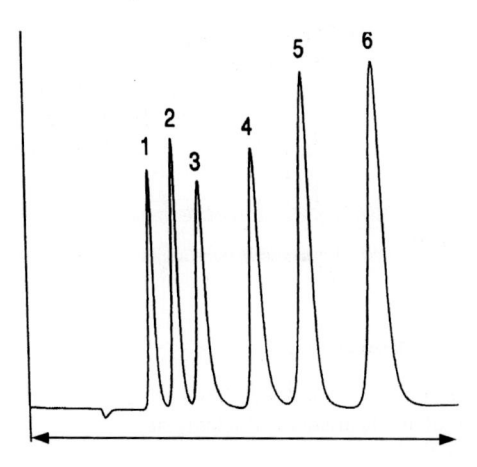

Figure 15. The separation of alkali and alkaline earth cations monitored by the electrical conductivity detector. The peaks represent: 1, Li^+; 2, Na^+; 3, NH_4^+; 4, K^+; 5, Mg^{2+}; 6, Ca^{2+}.

phase by dispersive adsorption. This technique of ion suppression is frequently used in ion exchange chromatography when using the electrical conductivity detector. It should be pointed out, however, that any suppresser system introduced between the column and the detector, that has a finite volume, is a possible source of band spreading. Consequently, in high resolution liquid chromatography, the connecting tubes and suppression column must be very carefully designed, to eliminate or reduce this dispersion to an absolute minimum.

6.2 The electrochemical detector

This detector responds to substances that are either oxidizable or reducible, and the electrical output results from an electron flow caused by the reaction that takes place at the surface of the electrodes. In the electrochemical detector, three electrodes are employed, the working electrode, where the oxidation or reduction takes place, the auxiliary electrode, and the reference electrode which compensates for any changes in the background conductivity of the mobile phase. At the surface of the working electrode, the reaction of the solutes is extremely rapid and proceeds almost to completion. This results in the layer close to the electrode being virtually depleted of reactant. Consequently, a concentration gradient is established between the electrode surface and the bulk of the solution. This concentration gradient results in the solute diffusing into the depleted zone, at a rate that is proportional to the concentration of the solute in the bulk of the mobile phase. As a result, the current generated at the surface of the electrode will be determined by the rate at which the solute reaches the electrode, and consequently, will exhibit a linear response with respect to solute concentration. The electrodes can take a num-

3: Detection devices

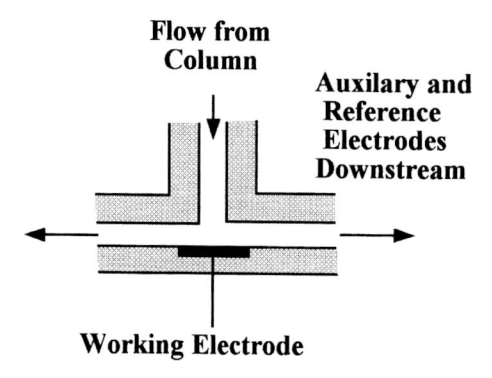

Figure 16. Typical electrode configuration of an electrochemical detector.

ber of different geometric forms and an example of a common arrangement is shown in *Figure 16*.

The example given is a wall jet electrode, where the column eluent is allowed to impinge directly onto the working electrode which is situated opposite the jet. This arrangement not only increases the velocity of the liquid passing over the electrode and thus the transfer coefficient, but also provides scrubbing action on the surface of the working electrode, which can reduce the need for frequent cleaning. The volume of the sensor can be kept exceedingly small, and is thus very suitable for use with high efficiency columns. The material used for constructing the working electrode needs to be mechanically rugged, and have long-term stability. The first substance used for electrode construction was carbon paste, made from a mixture of graphite and some suitable dielectric substance. Vitreous or 'glassy' carbon is an excellent material for electrode construction, particularly if it is to be used with organic solvents, and is probably the most popular contemporary electrode material available. It can be readily cleaned mechanically, and performs particularly well relative to other alternative materials, when operated at a negative potential.

Electrochemical detection can impose certain restrictions on both the type of chromatography that can be employed, and the mobile phase that can be used. Reversed-phase chromatography, however, is ideally suited to electrochemical detection. Nevertheless, certain stringent precautions must be taken for the effective use of the detector. The mobile phase must be completely free of oxygen, which can be removed by sparging the solvent reservoir with helium. The solvents must also be free of metal ions, or a very unstable baseline will result. The electrochemical detector is extremely sensitive but suffers from a number of drawbacks three of which are as follows. First, the mobile phase must be extremely pure and, as already stated, must be free of oxygen and metal ions. A more serious problem arises, however, from the adsorption and accumulation of the oxidation or reduction products on the surface of the working electrode. The detector must be regularly dissembled and cleaned,

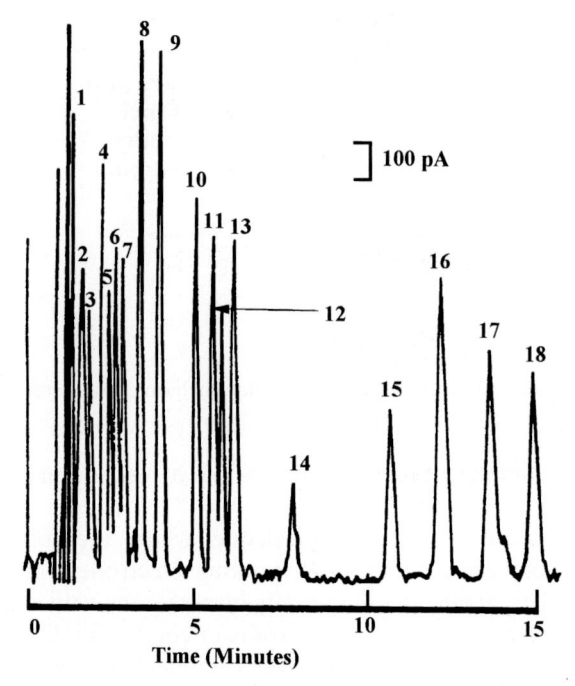

Figure 17. The separation of some catecholamines monitored by an electrochemical detector. The column used was a HC-3 C18 (100 mm × 4.6 mm). The mobile phase comprised an aqueous solution of 100 nM formic acid, 0.35 nM octane sulfonic acid, 1 nM citric acid, 0.1 nM EDTA, 5% (v/v) acetonitrile, 0.25% (v/v) diethylamine, adjusted to pH 3.1 with KOH. Flow rate was 1 ml/min and oxidative amperometric detection was performed using with glassy carbon electrode at 100 mV potential versus a Ag/AgCl electrode. Peaks represent: 1, 3,4 dihydroxymendelic acid 200 pg; 2, L-dopa 600 pg; 3, vanillymendelic acid 400 pg; 4, norepinephrine 200 pg; 5, α-methyl dopa 600 pg; 6, 3-methoxy-4-hydroxyphenylglycol 400 pg; 7, epinephrine 200 pg; 8, 3,4-dihydroxybenzylamine 200 pg; 9, normetanephrine 400 pg; 10, dopamine 200 pg; 11, metanephrine 400 pg; 12, 3,4-dihydroxyphenylacetic acid 200 pg; 13, N-methyl dopamine 400 pg; 14, tyramine 1 ng; 15, 5-hydroxyindole-3-acetic acid 200 pg; 16, 3-methoxytyramine 400 pg; 17, 5-hydroxytryptamine 200 pg; 18, homovanillic acid 400 pg. (Courtesy of the Perkin-Elmer Corporation.)

usually by a mechanical abrasion procedure. Much effort has been put into reducing this contamination problem but, although diminished, the problem has not been completely eliminated.

An example of the use of the electrochemical detector to monitor the separation of a series of catecholamines is shown in *Figure 17*. It is seen that the detector operates at an extremely high sensitivity, some of the substances being present at a level of 100 pg.

6.3 The multi-electrode array detector

The advent of the porous carbon electrode permitted the development of the electrode array detector. This electrode is made of porous graphitic carbon,

3: Detection devices

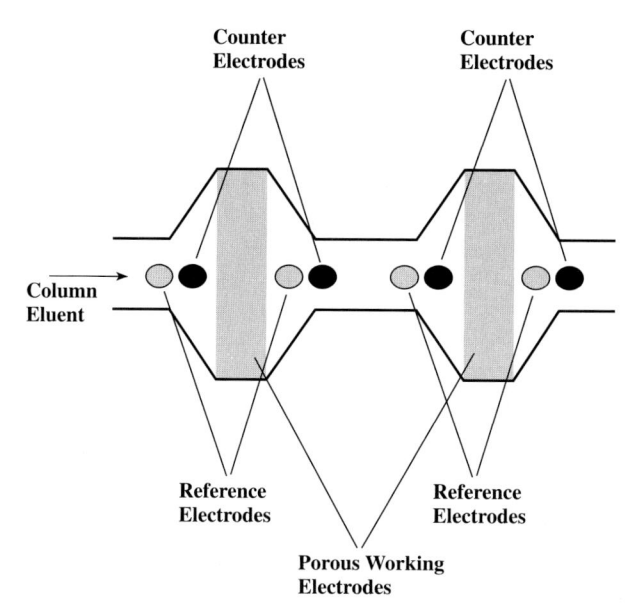

Figure 18. The coulometric electrode system employing porous graphitic carbon electrodes.

which has a very high surface area, is mechanically robust and, more important, is permeable to the mobile phase. As a consequence, flow through electrodes can be constructed. As the surface area is greatly in excess of that required for efficient electrochemical reaction, it can suffer excessive contamination before it fails to function. In fact up to 95% of the surface can be contaminated before it requires cleaning. In use, the porous electrode offers such a large surface area to the solute, that 100% of the material is reacted. Consequently the electrochemical reaction is no longer amperometric, but now coulometric; this is an important difference and makes the array system practical. The electrode system is shown diagramatically in *Figure 18*.

Each electrode unit consists of a central porous carbon electrode, on either side of which is situated a reference electrode and an auxiliary electrode. As the pressure drop across the porous electrode is relatively small, these electrode units can be connected in series, forming an array. Normally up to 16 units can be placed in series. The array operates with a progressively greater potential being applied sequentially to the electrodes of each consecutive unit. This results in all the solutes migrating through the array, until each reaches the unit that has the required potential to permit its oxidation or reduction. From the point of view of high resolution liquid chromatography, the electrode array detector gives improved apparent chromatographic resolution, in a similar way to that of the diode array detector, or any other spectroscopic detection system. Two peaks that have not been chromatographically resolved, and are eluted together, can still be shown as two peaks

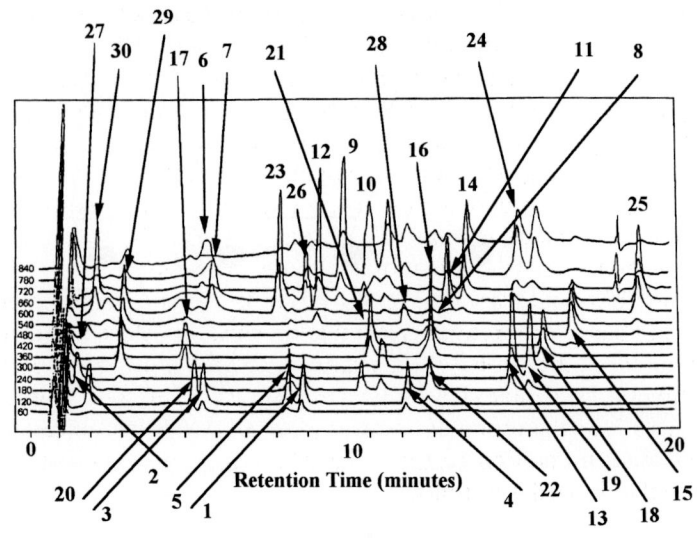

Figure 19. The separation of 30 neuroactive substances monitored by an electrochemical array. The mobile phase was 1% (v/v) to 40% (v/v) methanol in 0.1 M KH_2PO_4 pH 3.4 buffer with ion pairing. Peaks represent: 1, dihydroxyphenylacetic; 2, dihydroxyphenylethylene glycol; 3, L-dopa; 4, dopamine; 5, epinephrine; 6, guanine; 7, guanosine; 8, homovanallic acid; 9, hydroxybenzoic acid; 10, hydroxyindoleacetic acid; 11, hydroxyphenylacetic acid; 12, hydroxyphenyllactic acid; 13, hydroxytryptophan; 14, kynurenine; 15, melatonin; 16, metenephrine acid; 17, methoxyhydroxyphenyl glycol; 18, methoxytyramine; 19, n-methylserotonin; 20, norepinephrine; 21, normetenephrine; 22, salsolinol; 23, octopamine; 24, serotonin; 25, tryptophan; 26, tyrosine; 27, uric acid; 28, vanillic acid; 29, vanylmandelic acid; 30, xanthine. (Courtesy of the *Analyst*.)

that are resolved electrochemically, and can be quantitatively estimated. An advantage of the array system is that high oxidation potentials can be used, without the high background currents and noise that usually accompany such operating conditions. The electrodes that are operating at high voltages are 'buffered' by the previous electrodes, operating at lower voltages, which results in reduced background currents and noise. An example of the application of the detector to monitoring the separation of a number of neurochemical substances is shown in *Figure 19* (13).

It is seen that the electrode array system adds another dimension to high resolution liquid chromatography. However, it must also be remembered that in order to use the detector, the solutes must be amenable to electrochemical reaction, and be capable of separation using a mobile phase that will conduct an ion current.

7. Light scattering detectors

Light scattering detectors are fairly recent on the commercial market, and have become popular because they offer molecular weight measurements as

3: Detection devices

well as simple detection. The overall sensitivity of these detectors appears to be very similar to that of the refractive index detector, with about the same linearity. However, the most important characteristic of this detector is not its propensity for accurate quantitative analysis, but its proficiency in providing molecular weight data for extremely large molecules. There are two different types of light scattering detectors, low angle light scattering detectors, and multi-angle light scattering detectors. Both have fairly large sensor volumes, but the latter can be made as small as 3 µl, and thus could be used as detectors with many high efficiency liquid chromatography columns.

In the multiple angle laser light scattering detector, measurements are made at a number of different angles, none of which are close to the incident light. This reduces the problem associated with scattering from particulate contaminants in the sample. Data taken at a series of different angles to the incident light allows the root mean square (rms) of the molecular radius $\langle r^2 \rangle^{1/2}$ to be calculated in addition to the molecular weight of the substance. The relationship that is used is as follows:

$$\frac{cK}{R\phi} = a\langle r^2 \rangle^{1/2} \sin(\theta)^2 + bM_w$$

where c is the concentration of solute; K is the substance optical constant, $R\phi$ is the Rayleigh constant; r is the root mean square radius; θ is the scattering angle; Mw is the molecular weight of the solute; a, b are constants.

In fact, theory can provide explicit functions for (a) and (b) but values for these constants are usually obtained by calibration (*Figure 21*). The total number of different angles at which the scattered light is measured differs from one instrument to another, and commercial equipment that measures the intensity of the scattered light at 16 different angles is available. It is clear that the greater the number of data points taken at different angles, the more precise the results will be. A diagram of a multiple angle laser light scattering detector, which measures the light scattered at three different angles, is shown in *Figure 20*.

This device contains no mirrors, prisms, or moving parts and is designed such that the light paths are direct, and not 'folded'. Light passes from the laser (wavelength 690 nm) directly through a sensor cell. Light scattered from the centre of the cell, passes through three narrow channels, to three different photocells, set at 45°, 90°, and 135° to the incident light. Thus, scattered light is continuously sampled at three different angles during the passage of the solute through the cell. A continuous analogue output is provided from the 90° sensor, and all the sensors are sampled every 2 sec. The molecular weight range extends from 10^3 to 10^6 Daltons, and the rms radii from 10–50 nm. The total cell volume is about 3 µl, and the scattering volume is 0.02 µl. The detector has a sensitivity, defined in terms of the minimum detectable excess Rayleigh ratio, of 5×10^{-8} cm^{-1} which is difficult to translate into normal concentration units, but appears to be equivalent to a minimum detectable concentration of about 10^{-6} g/ml.

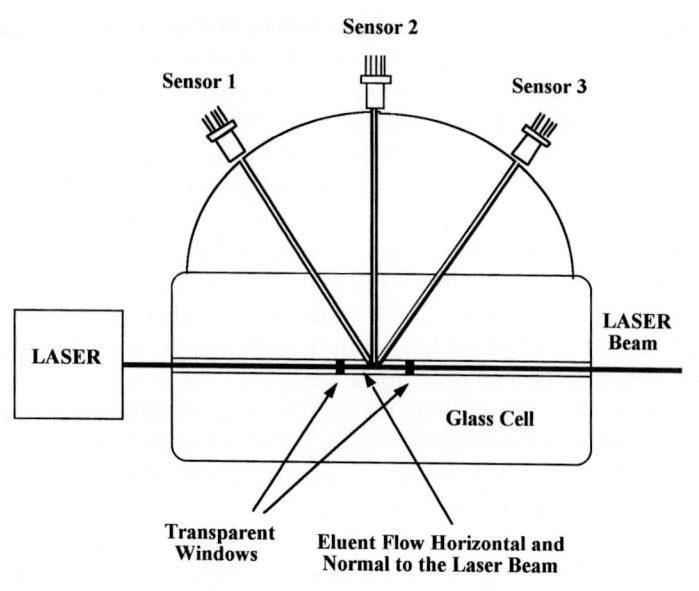

Figure 20. The multiple angle laser light scattering detector (miniDawn®). (Courtesy of Wyatt Technology Corporation.)

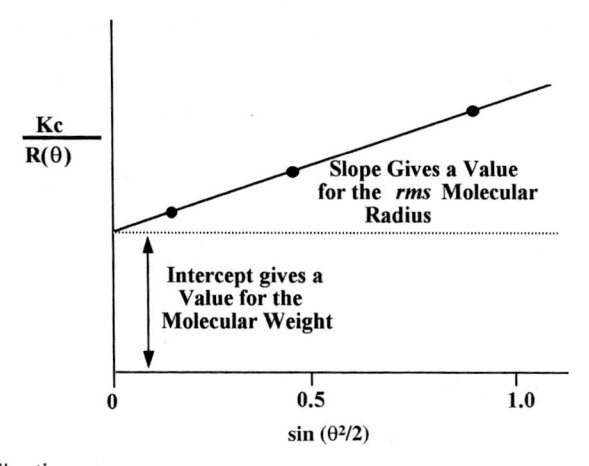

Figure 21. Calibration curves.

The relationship between the intensity of the scattered light, the scattering angle, and the molecular properties are as follows:

$$\frac{cH}{R\phi} = 2cA_2 + \frac{1}{M_w P(\phi)}$$

where $P(\phi)$ describes the dependence of the scattered light on the angle of scatter, and the other symbols have the meanings previously attributed to them. Employing appropriate reference standards, graphs of the form shown in *Figure 21* can be constructed to evaluate constants (a) and (b) and thus permit the measurement of the molecular weight and molecular radius of unknown substances.

There are many other detection devices available that, with appropriate modification could be employed with high efficiency columns, but the ones described here are recommended as simple, reliable, and available at reasonable cost. In general, high resolution chromatography requires detecting devices that have small sensor volumes, minimum lengths of column to detector tubing, and fast sensor and electronic response. At this time, progress towards even higher column resolution than is presently available is not inhibited by the lack of suitable detecting devices.

References

1. Scott, R. P. W. (1992). *Liquid chromatography column theory*, p. 175. John Wiley & Sons, Chichester.
2. Scott, R. P. W. (1996). *Chromatography detectors*, p. 17. Marcel Dekker Inc., New York.
3. Fowlis, I. A. and Scott, R. P. W. (1963). *J. Chromatogr.*, **11**, 1.
4. Scott, C. G. *ASTM*, E19 No. E689-79.
5. Scott, R. P. W. (1996). *Chromatography detectors*, p. 27, Marcel Dekker Inc., New York.
6. Scott, R. P. W. and Lawrence, J. G. (1967). *Anal. Chem.*, **39**, 830.
7. Golay, M. J. E. (1958). *Gas chromatography 1958* (ed. D. H. Desty), p. 36. Butterworths, London.
8. Scott, R. P. W. (1992). *Liquid chromatography column theory*, p. 185. John Wiley & Sons, Chichester.
9. Scott, R. P. W. and Kucera, P. (1979). *J. Chromatogr.*, **169**, 951.
10. Katz, E. and Scott, R. P. W. (1982). *J. Chromatogr.*, **253**, 159.
11. Mechref, Y. and El Rassi, Z. (1996). *Anal. Chem.*, **68(10)**, 1771.
12. von Hereren, F., Verpoorte, E., Manz, A., and Thormann, W. (1996). *Anal. Chem.*, **68(13)**, 2044.
13. Svendsen, C. N. (1993). *Analyst*, **118 (Feb.)**, 123.

4

Capillary electrophoresis of peptides and proteins

BI-HUANG HU and LENORE M. MARTIN

1. Introduction

The goal of this chapter is to describe a variety of approaches that may be adopted by the biomedical researcher or bioanalytical chemist to achieve successful separations of peptides and proteins. The list of modes commonly used for the analysis and separation of peptides and proteins by capillary electrophoresis (CE) includes: capillary zone electrophoresis (CZE), micellar electrokinetic capillary chromatography (MEKC), capillary isoelectric focusing (cIEF), and capillary gel electrophoresis (CGE). This list is constantly growing and appears unlimited. The application of CE to answer specific questions which commonly occur in peptide and protein research will also be discussed in this chapter.

Electrophoresis is the migration of a charged species under the influence of an electric field. Traditionally, electrophoresis has been the primary tool for the separation of peptides and proteins (1), based upon the fact that peptides and proteins are electrically charged molecules. Traditional electrophoretic techniques, such as slab gel electrophoresis, or isoelectric focusing in tube gels, are widely used for the separation of peptides and proteins, even though traditional techniques suffer from significant problems, such as long preparation and analysis times, poor reproducibility, low efficiency, low resolution, and a difficulty in automation.

CE is a rapidly growing separation technique with the advantages of short analysis times, high efficiency, high resolution, and full automation (2). Due to the traditional importance of electrophoretic techniques for the separation of peptides and proteins, it was natural that CE was exploited to produce high resolution separations of peptides and proteins. According to the theory of CE, macromolecules should separate with high efficiencies. Thus it was anticipated that a high efficiency should be obtained for peptides and proteins based on their low diffusion coefficients (3).

One difficulty which hindered the application of CE to peptides and large proteins resulted from the silanol groups on the inner surface of fused-silica capillaries. Large peptides and proteins possess both numerous charges, and hydrophobic patches, which cause them to adsorb to the fused-silica capillary walls. Diverse interactions are responsible for this adsorption. These interactions include Coulombic (charge–charge) interactions, hydrogen bonding, the hydrophobic effect, and Van der Waals attractions between the analytes and the capillary inner walls. Adsorption results in peak tailing, poor efficiency, poor reproducibility, and errors in quantitation. The universality of this phenomenon dictates that the most important task for the bioanalyst is to choose separation conditions that minimize adsorption to the capillary wall. A stable and reproducible deactivation of the capillary wall is frequently necessary to achieve separations of large peptides and proteins by CE. Several techniques to accomplish this goal have been explored and much progress has been made.

2. The capillary

Fused-silica has been widely used as the capillary material in CE. Fused-silica capillaries are transparent to UV–visible light, heat transferable, and when coated with a protective layer of polyimide, are extremely flexible and easy to handle. These capillaries allow easy on-line detection by removal of a small portion of the polyimide coating, which creates an optical window. Inexpensive fused-silica capillaries of defined inner and outer diameters are commercially available in ten metre lengths. Optical windows may be created using a disposable lighter to burn off the coating and a tissue to wipe it clean. Cutting the capillary into the desired lengths also requires removal of the polyimide coating and one must pay attention to ensure that the ends are flat and no debris has clogged the capillary opening.

2.1 Choosing a capillary diameter and length

At present, approaches to the separation of peptides and proteins have concentrated on the use of open-tubular, very narrow bore fused-silica capillaries of 20–100 μm i.d. (and varied outer diameters). Theoretically, the capillary i.d. should be as small as possible, because reducing the internal diameter will increase the capillary inner surface area-to-volume ratio. The larger area-to-volume ratio will allow the capillary to dissipate heat more efficiently. Dissipation of the heat generated by the electric field counteracts the Joule heating effect and permits a higher electrical field to be applied, leading to short analysis times and higher resolution of peaks. Although a very small capillary of 5 μm i.d. has been used for the separation of the proteins from a single cell, the capillary lifetime was pretty short, about one hour, on average (4). Therefore, the routine use of small capillaries (i.d. < 10 μm) is limited by detector sensitivity, sample loading, and capillary clogging. Generally, capillaries of 50 μm i.d. have been widely used for the separation of peptides and proteins.

4: Capillary electrophoresis of peptides and proteins

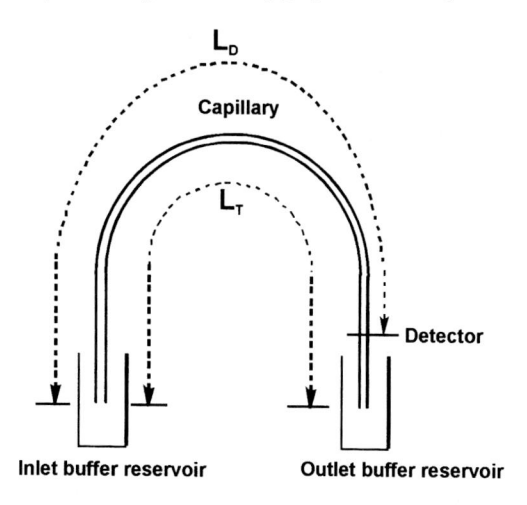

Figure 1. Schematic of a capillary electrophoresis instrument showing the effective length (L_D) and total length (L_T) of capillary relative to the detector.

In capillary gel electrophoresis, or for the purpose of micropreparative work, capillaries of 100 μm i.d. may be used.

Capillary length (see *Figure 1*) may refer to effective capillary length (L_D) or total capillary length (L_T). Decreasing the capillary length at a constant applied voltage (usually the highest your power supply can provide) will cause an increase in the strength of the applied electric field (increased efficiency) and will decrease the residence time of the analytes in the capillary (shorter analysis times). In order to use the capillary more efficiently, the effective lengths should be as close as possible to the total lengths ($L_D \sim L_T$) and the shortest possible capillary should be used; usually, the effective lengths are 5–30 cm shorter than the total lengths (5, 6). Effective capillary lengths (L_D) used in the separation of peptides and proteins range from 7–100 cm, and total capillary lengths (L_T) from 12–120 cm (7). These two parameters (L_D and L_T) are somewhat dependent on the instrument dimensions. For example, the Dionex capillary electrophoresis system requires that the minimum capillary total length be 45 cm, and the distance from the detector window to the capillary outlet be exactly 5 cm.

2.2 Modifying the capillary wall

Fused-silica capillary inner walls contain free silanol groups that will ionize under aqueous conditions above pH 2. One phenomenon associated with a negatively charged capillary surface, known as the electro-osmotic flow (EOF), is a bulk flow of liquid in the capillary under the influence of an applied electric field which causes uniform movement of all species (negatively charged, positively charged, and neutral) in the same direction. EOF does not broaden peaks because the electric field is relatively uniform over the entire cross-section of the capillary (unlike the parabolic flow profile found in HPLC). The EOF can be manipulated to enhance a given separation by modifying either the capillary wall or the separation buffer. The apparent observed electrophoretic mobility ($\bar{\mu}_{app}$) of a solute is related to the intrinsic electrophoretic mobility ($\bar{\mu}$, in the absence of EOF) by *Equations 1, 2,* and *3* (8).

$$\bar{\mu}_{app} = \bar{\mu} + \bar{\mu}_{eof} \qquad [1]$$

$$\bar{\mu} = |\bar{\mu}_{app}| - |\bar{\mu}_{eof}| = \text{difference in velocity/E} \qquad [2]$$

Solving for the electrophoretic mobility of our analyte we get:

$$\bar{\mu} = \left(\frac{L_D L_T}{V}\right)\left[\left(\frac{1}{t_{app}}\right) - \left(\frac{1}{t_{eof}}\right)\right] \qquad [3]$$

where t_{app} is the observed migration time of an analyte in minutes, t_{eof} is the migration time of an uncharged molecule in minutes, L_D and L_T are the capillary lengths from the inlet to detector, and inlet to outlet, respectively. V is the applied voltage in volts.

In addition to EOF, there is another important phenomena arising from the use of charged capillary walls. In CE separations of peptides and proteins, the negatively charged capillary wall strongly interacts with any positive charges in these analytes, resulting in poor efficiency and reproducibility. To achieve a successful separation of positively charged (basic) peptides and proteins, the wall interactions must be minimized.

2.2.1 Coating the capillary wall

Coating the inner surface of the capillary is a fundamental and effective approach to minimize solute adsorption onto the capillary wall. The capillary wall is either modified through covalent bonding, or by simple adsorption of a hydrophilic or hydrophobic substance. The effect of coating the inner capillary wall is to mask any negative charges resulting from ionization of the silanol groups. Depending on the coating reagents used, such coated capillaries may have little or no EOF. Elimination of the EOF is required for CGE and is beneficial for cIEF techniques. Reversed EOF is also useful in some cases.

i. Covalently bonded coatings

Covalently bonded coatings are usually obtained by the use of a bifunctional organosilane reagent. One functional group of the reagent reacts specifically with the silanol group on the capillary inner surface to form a Si-O-Si-C linkage, and the other functional group either:

(a) Permits the covalent attachment of a polymeric coating.
(b) Reacts with a specific compound.

The main function of polyacrylamide coatings is to eliminate the EOF. The most widely used coating method was first proposed by Hjertén (see *Figure 2*), and his method is described in *Protocol 1* (9).

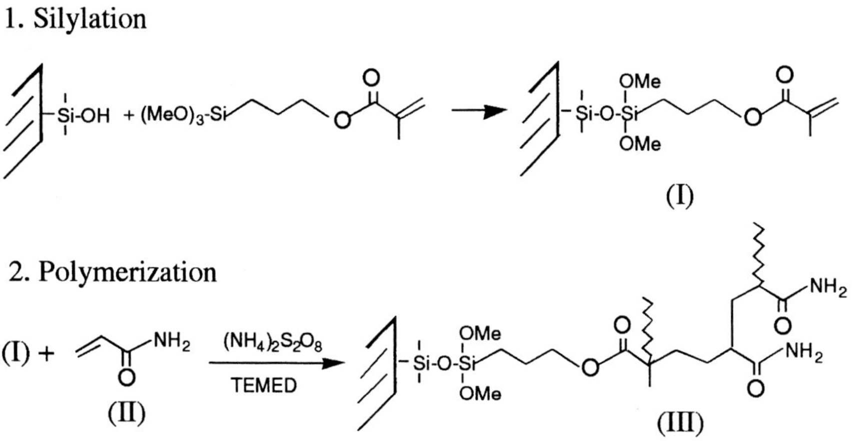

4: Capillary electrophoresis of peptides and proteins

1. Silylation

2. Polymerization

Figure 2. The chemistry of attaching linear polyacrylamide coatings to the inner capillary wall using Si-O-Si-C linkages. First, the bifunctional organosilane reagent MAPS reacts with free silanol groups on the wall to form the anchoring Si-O-Si-C bonds. Secondly, these attached monomers (I) participate in the polymerization of acrylamide (II) to covalently attach the polymers (III) to the capillary wall.

Protocol 1. Coating capillaries with linear polyacrylamide via a Si-O-Si-C linkage[a]

Equipment and reagents
- 50 μm i.d. fused-silica capillary (Polymicro Technologies)
- 18.2 MΩ H$_2$O (Milli-Q water purification system, Millipore)
- Ferrules and adapters for attaching a syringe to the capillary (Alltech)
- Toluene (Burdick and Jackson, HPLC grade, low water content)
- Nitrogen gas, pre-purified
- Oven that reaches 110°C (or gas chromatograph)
- 3-methacryloxypropyltrimethoxysilane (MAPS) (Fluka)
- N,N,N',N'-tetramethylethylenediamine (TEMED) (Bio-Rad)
- Ammonium persulfate (Bio-Rad)
- Acrylamide (Bio-Rad)
- Sodium hydroxide
- Hydrochloric acid
- Methanol (HPLC grade)
- 0.45 μm filter (Rainin)

Method

1. Fill the capillary (approx. 1 m long) with a 1 M NaOH solution using a syringe, and let it stand for 1 h at room temperature. Rinse the capillary with water for 10 min, then rinse with an 0.1 M HCl solution for 30 min, and wash with water for 10 min.[b] Finally, dry the capillary in an oven[c] at 110°C for 4 h while gently flushing with nitrogen.

2. Flush the dried capillary with a 10% (v/v) solution of MAPS in toluene using a syringe, and allow this solution to flow through the capillary[d] for 1 h. Then let the solution stand in the capillary for 1 h at room temperature. Purge the capillary with nitrogen, wash it with toluene for

Protocol 1. *Continued*

10 min, wash with methanol for 10 min, and then wash with water for 15 min.

3. Prepare a 3% (w/v) aqueous acrylamide solution and degas it for 15 min in a side-armed Erlenmeyer flask, by sonicating the flask in an ultrasonic bath under reduced pressure. Immediately before using, add 1 µl TEMED and 1 mg ammonium persulfate per ml of the degassed solution, and filter the solution with a 0.45 µm filter.

4. Immediately fill the capillary with the solution made in step 3, using a syringe, and allow it to stand for 1 h. Remove the solution from the capillary using suction and rinse the capillary with water for 15 min. Finally, purge the capillary with nitrogen.

[a] Modified from the methods in refs 9 and 10.
[b] May be done using the automatic mode of the CE instrument.
[c] A gas chromatographic oven works well.
[d] By pressure or siphoning.

With prolonged use, the unsatisfactory hydrolytic stability of the siloxane bond (Si-O-Si-C) under alkaline conditions was observed to limit the utility of such a coating. Another capillary treatment procedure, in which the polymeric coating is attached to the capillary wall via a direct Si-C linkage, was proposed (11) (see *Figure 3*). This improved coating procedure is detailed in

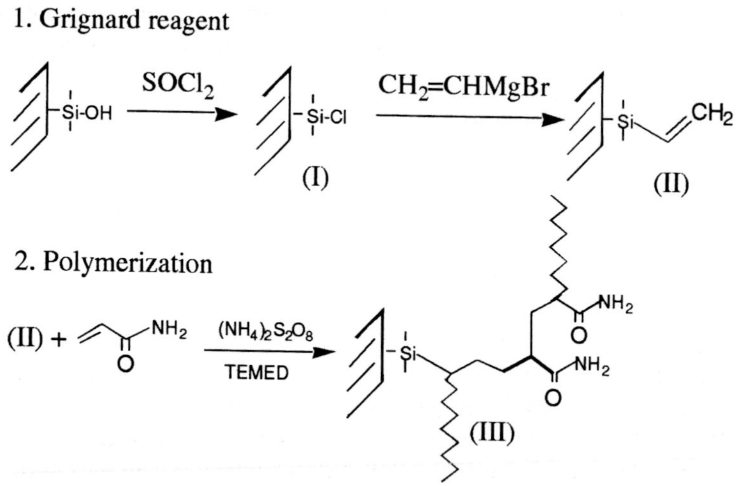

Figure 3. The chemistry of attaching linear polyacrylamide coatings to the inner capillary wall through Si-C linkages. In dry conditions, the silyl chloride (I) reacts with vinyl magnesium bromide to firmly anchor the polymer via a Si-C bond (II). The polymerization yields covalently attached polyacrylamide (III).

4: Capillary electrophoresis of peptides and proteins

Protocol 2. Linear polyacrylamide coatings attached to the capillary through a Si-C linkage remain stable over the pH range 2–10.5. Polyacrylamide coatings are well suited for capillary electrophoretic separations of a wide variety of proteins (11–13).

Protocol 2. Coating capillaries with linear polyacrylamide via a direct Si-C linkage[a]

Equipment and reagents
- Thionyl chloride (Fluka)
- Vinyl magnesium bromide (0.25 M solution in THF) (1 M from Aldrich)
- Tetrahydrofuran (THF) (Aldrich)
- Equipment and reagents from *Protocol 1*

Method

1. Using a syringe, fill a capillary (approx. 1 m long) with a 1 M NaOH solution and let it stand for 1 h at room temperature. Wash the capillary with water for 30 min and dry in an oven[b] at 110°C for 6 h while flushing it with nitrogen.

2. Pass SOCl$_2$ through the dried capillary for 5 min using a suction pump, then seal the capillary at both ends using a small propane torch, and keep it at 70°C for 6 h. Break open the seals and pass nitrogen through the capillary for 5 min.

3. Introduce a diluted Grignard reagent, 0.25 M vinyl magnesium bromide in THF, into the capillary by suction for 5 min, seal the capillary again, and keep it at 70°C for 6 h. Open the seals and rinse the capillary with THF for 10 min, then wash with methanol for 10 min, and then wash with water for 15 min.

4. Finally, follow *Protocol 1*, steps 3 and 4.

[a] Modified from the methods given in refs 11 and 13.
[b] A gas chromatographic oven works well.

Simple silylation of the capillary wall with an organosilane reagent is useful as a method for coating the inner surface of the capillary. For example: 3-aminopropyltrimethoxysilane-modified capillaries are widely used in CE-MS (4). To obtain this type of cationic coating, use the procedures outlined in *Protocol 1*, steps 1 and 2, except substitute 3-aminopropyltrimethoxysilane for MAPS. Capillary walls modified with 3-aminopropyltrimethoxysilane will have a net positive charge at pH 7 or below, so that the EOF is reversed (14). The presence of positive charges on the capillary walls helps to reduce peptide and protein adsorption through Coulombic repulsion. This cationic coating approach is especially useful for the separation of basic proteins.

In summary, the instability of the Si-O-Si-C bond, and the relatively limited

Table 1. Covalently bonded coatings used in the CE separations of peptides and proteins

Coating type	Synthesis	Remarks	Reference
Si-O-Si-C linkage			
Linear polyacrylamide	1. MAPS[a] 2. Acrylamide polymerization	Stable at pH 4–7 Widely used Limited lifetime	9, 15
Poly(vinylpyrrolidinone)	1. MAPS 2. 1-Vinyl-2-pyrrolidinone polymerization	Stable at low pH Residual adsorption	16
Poly(ethylene glycol)	1. GOPS[b] 2. PEG-600	Suitable pH 3–5	17
Poly(AAEE)[c]	1. Triethoxysilane 2. Allyl methacrylate 3. AAEE polymerization	pH range 3–8.5 Stable at pH 8.5 Lifetime 70 h	10
Cross-linked polyacrylamide	1. MAPS 2. Acrylamide and bisacrylamide polymerization	pH range 2–10 Residual adsorption	18
Cross-linked polyacrylamide	1. Poly (methylvinylsiloxanediol) and a silane cross-linker 2. Acrylamide polymerization 3. Formaldehyde	pH range 2–9 High efficiency Useful for both acidic and basic proteins	19
Epoxy polymer	1. GOPS 2. EGDE[d] polymerization 3. Glycidol polymerization	pH 5–10 Lifetime 120 h	20
Glycero-glycidyloxypropyl	1. GOPDS[e] 2. Glycerol		16
Maltose	1. Triethoxyaminopropylsilane 2. Maltose	EOF depends on pH Low efficiency	21
Arylpentafluoro group	1. 3-APS[f] 2. Pentafluorobenzoyl chloride	Good efficiency at neutral pH	22
α-Lactalbumin bonded	1. 3-APS 2. Glutaraldehyde 3. α-Lactalbumin	Good efficiency at neutral pH	23
3-Aminopropylsilane	1. 3-APS	Reversed EOF Used in CE-MS	4, 14, 24
Epoxy diol	1. GOPS 2. HCl solution	pH 3–5 Low efficiency	21
Si-C linkage			
Linear polyacrylamide	1. Thionyl chloride 2. Vinyl magnesium bromide 3. Acrylamide polymerization	Stable at pH 2–10.5 Elimination of EOF	11, 13

[a] 3-Methacryloxypropyltrimethoxysilane.
[b] 3-Glycidyloxypropyltrimethoxysilane.
[c] Acryloylaminoethoxyethanol.
[d] Ethylene glycol diglycidyl ether.
[e] (3-Glycidyloxypropyl)diisopropylethoxysilane.
[f] 3-Aminopropyltrimethoxysilane.

4: Capillary electrophoresis of peptides and proteins

organic coverage achieved with this coating make it less than satisfactory. Coatings attached via direct Si-C linkage promise to yield increased column lifetimes. Many examples of different covalently-anchored coatings used for the separation of peptides and proteins are listed in *Table 1*.

ii. Adsorbed coatings

Adsorbed coatings are promising alternatives to covalently bonded coatings. The capillary surface is shielded by the adsorption of a polymer or surfactant through ionic interactions (25), hydrogen bonding (26), or hydrophobic interactions (27, 28), if the capillary wall has been previously silylated. The advantage of adsorbed coatings is their ease of preparation. To coat the capillary via adsorption, the polymer or surfactant solution is passed through the capillary: organic synthesis is not required in this process. Coatings can be easily regenerated by passing additional polymer reagent through the capillary. The procedure in *Protocol 3*, for the preparation of a poly(ethylene imine) (PEI) coating, is modified from ref. 25.

Protocol 3. Coating a capillary through adsorption of poly(ethylene imine)[a]

Equipment and reagents

- Capillary electrophoresis system
- 50% (w/v) poly(ethylene imine) in water (PEI, M_r 600 000–1 000 000, Fluka)
- 50 μm i.d. fused-silica capillary, 0.45 μm filter, and 18.2 MΩ water (see *Protocol 1*)

Method

1. Take a capillary (L_T = 60 cm), make a detection window. Install the capillary according to the CE instrument manual.
2. Flush the capillary with a 1 M NaOH solution for 20 min, at 1 bar of pressure, then flush with a 0.1 M NaOH solution for 10 min, and finally wash with water for 10 min, using the CE instrument in automatic mode.
3. Prepare a 10% (w/v) aqueous solution of PEI, filter the solution with a 0.45 μm filter, and degas it in an Erlenmeyer flask with side-arm for 15 min by sonicating in an ultrasonic bath under reduced pressure.
4. Flush the prepared capillary with the degassed PEI solution for 10 min at 1.5 bar of pressure, then leave the solution in the capillary for 1 h. Rinse the capillary with water for 10 min, and finally rinse with running buffer for 15 min.
5. Reverse the polarity of the CE instrument before sample analysis!

[a] Modified from ref. 25.

$[-NHCH_2CH_2-]_x[-N(CH_2CH_2NH_2)CH_2CH_2-]_y$

PEI

$[-N^+(CH_3)_2-(CH_2)_6-N^+(CH_3)_2-(CH_2)_3-]_n \cdot 2Br^-$

Polybrene

Figure 4. The chemical structures of the cationic polymeric capillary coatings: poly(ethylene imine) (PEI) and hexadimethrine bromide (Polybrene).

The cationic PEI coating is stable in the pH range 3–10. This type of coated capillary has a reversed EOF at a wide range of buffer pHs, because PEI establishes a stable, net positively charged, amine layer on the capillary inner wall. Alternatively, capillaries with a reversed EOF have been prepared either using a polymeric coating agent obtained from Applied Biosystems Inc. (ABI) called Micro-Coat (29), or a 1–5% (w/v) aqueous solution of Polybrene (30). Polybrene® (hexadimethrine bromide, Aldrich) is a cationic polymer similar to PEI, but without titrateable protons (see *Figure 4*).

Neutral capillary coatings such as poly(ethylene oxide) (PEO), hydroxyethylcellulose (HEC), and hydroxypropylmethylcellulose (HPMC) have also proved useful for peptide and protein CE. Adsorption of the coating to the capillary was accomplished by first filling the capillaries with a 1 M HCl solution and then flushing with the desired polymer solutions containing 0.1 M HCl (26). The PEO coating is stable in the pH range 3–7, and may be used for the separation of basic proteins. HEC and HPMC coatings are more unstable than PEO, due to their low affinity for adsorption onto silica surfaces.

Several methods produce coatings with the characteristics of both an adsorbed coating and a covalently bonded coating. One method for producing a more stable PEI coating involves the adsorption of PEI to the capillary inner wall followed by cross-linking of the adsorbed PEI with ethylene glycol diglycidyl ether (31). The resulting coating is stable in the pH range 2–12. In some reported coating methods, the capillary wall was first silylated using octadecyltrichlorosilane or dichlorodimethylsilane. The second step involved flushing the silylated capillary with surfactants such as Brij 35, Tween (27, 28), or Pluronics (28, 32). Alternatively, the second step can utilize polymer coating solutions such as methylcellulose or poly(vinyl alcohol) (28). Good run-to-run reproducibility can be obtained by first repeatedly flushing a new capillary with these coating solutions, secondly, a wash for quickly replenishing the coating is added to each analysis.

2.2.2 Using buffer additives

The buffer additives discussed here act as dynamic coatings: shielding the analyte from the capillary inner wall during a run. The addition of additives into the running buffer is a simple and easy way to minimize peptide and protein adsorption to the capillary wall. This dynamic approach may actually be considered a fortified, adsorbed coating since additional additive added to the running buffer will dynamically coat the capillary wall and will preserve that coating. The advantages of this type of coating are the easy preparation and optimization of capillaries achieved by dissolving the additive in the running buffer. The main disadvantage of this approach is that the buffer additive may interact with the analytes in an unforeseen manner.

i. Surfactants and polymers

Surfactant additives present in the running buffer at concentrations above their critical micelle concentrations (CMC) can form micelles. This is the fundamental feature of micellar electrokinetic chromatography (MEKC), one of the CE modes (see Section 5.3). Some surfactants cannot only form micelles but can also adsorb to the capillary wall, modifying the capillary wall surface. Cationic and non-ionic surfactants are frequently used to dynamically coat the capillary wall through either ionic or hydrophobic interactions (33). MEKC has permitted the separation of neutral molecules by CE, based on their partition equilibrium into and out of the micelles. Another term for this partition technique is 'pseudo-stationary phase CE'.

The addition of cationic surfactants to the running buffer causes a reversal of EOF, similar to the effect of the cationic coatings described above. Reversal of the EOF results from the formation of a bilayer or hemimicelle of the cationic surfactant on the capillary wall (34). For example, using Fluorad® (*Figure 5*), at concentrations above the CMC, a constant reversed EOF was obtained (33). When using buffer additives which reverse the EOF, the polarity of the electrodes should be switched by simply reversing the polarity of the CE high voltage power supply (often called 'negative polarity'). This ensures that the direction of the EOF still goes from the injection end toward the detection end.

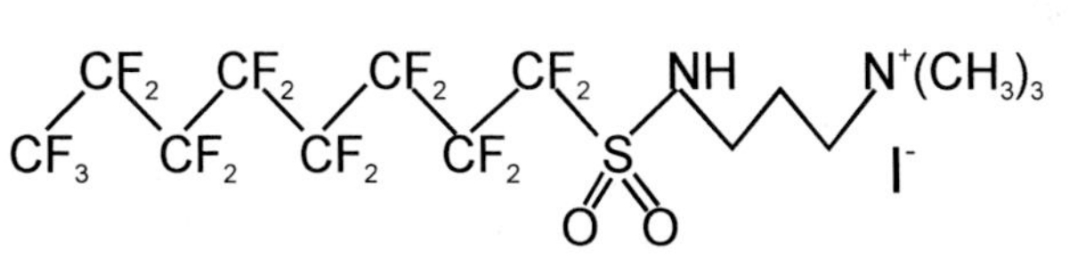

Figure 5. The chemical structure of FC-135 (Fluorad®), a cationic surfactant used to coat the capillary wall.

Addition of non-ionic polymers to the running buffer also can suppress the capillary wall interaction (35). Useful surfactants and polymers are listed in *Table 2*.

ii. Amines and ion-pairing agents

The addition of amines and ion-pairing agents to the running buffer can effectively minimize the wall adsorption problem by either suppressing the inter-

Table 2. Surfactant and polymer buffer additives for CE

Additive	Conditions	Reference
Cationic surfactant		
Fluorad® FC-135[a]	100 mg/ml (0.14 mM) or 50 mg/ml at pH 7	36, 37
Septonex[b]	10 mM at pH 2.5	38
CTAC[c]	50 mM at pH 7 for basic peptides	39
DTAB[d]	DTAB[d]	40
HTAB[e]	50 mM at pH 7	40
Non-ionic surfactant		
Tween 20	0.1% (v/v) at pH 8	41
Triton X-100	3.5 mM at pH 9	42
Polymer		
Poly(vinyl alcohol)	0.05% (w/w) at pH 3	35
HPMC[f]	0.03% (w/v) at pH 2	43, 44
	0.1% (w/v) in 5% (v/v) Pharmalyte 3–10 for cIEF of proteins	45
α-Cyclodextrin	20 mM at pH 9.5 for derivatized peptides	46

[a] Perfluoroalkylammonium iodide.
[b] 1-Ethoxycarbonylpentadecyltrimethylammonium bromide.
[c] Cetyltrimethylammonium chloride.
[d] Dodecyltrimethylammonium bromide.
[e] Hexadecyltrimethylammonium bromide.
[f] Hydroxypropylmethylcellulose.

action of the analytes with silanol groups on the capillary wall surface (47), or by changing the net charge on the analytes (48). Cationic amines, such as spermine (below), suppress analyte–wall interactions in many peptide and protein separations, thus improving the separation efficiency. The unusual use of spermine in the separation of basic proteins (19) probably involves an amine coating on the capillary wall and reduction of the wall interactions due to the Coulombic repulsion between the amine and the basic protein (*Figure 6*).

Anionic ion-pairing reagents can improve separations of hydrophobic and basic synthetic peptides (49). For multisubunit proteins, and proteins which tend to aggregate, ion-pairing reagents can prevent protein–protein interactions from causing peak broadening. Wall adsorption interactions are also lowered if the ion-pairing reagent can coat the region of the peptide or protein which might interact with silanol groups. Amines and ion-pairing agents commonly used for the separation of proteins and peptides are listed in *Table 3*.

iii. Metal ions

The adsorption of proteins and peptides to the capillary wall can be minimized by the addition of metal ions (high salt concentrations) to the running buffer (increasing the ionic strength of the buffer). Metal ions compete with

4: Capillary electrophoresis of peptides and proteins

$$H_2N\text{-}(CH_2)_3\text{-}NH\text{-}(CH_2)_4\text{-}NH\text{-}(CH_2)_3\text{-}NH_2$$
Spermine

$$(CH_3)_3N^+\text{-}(CH_2)_6\text{-}N^+(CH_3)_3 \cdot 2Br^-$$
Hexamethonium bromide

$$(CH_3)_3N^+\text{-}(CH_2)_{10}\text{-}N^+(CH_3)_3 \cdot 2Br^-$$
Decamethonium bromide

Figure 6. The chemical structures of several cationic amines used as buffer additives in CE.

peptides and proteins for the adsorption sites on the capillary wall through ion exchange interactions (56). The uses of K^+ (6, 18, 50, 57, 58), Na^+ (56), Li^+ (56, 59), Zn^{2+} (6), and Cs^+ (56) have been explored with limited success. Potassium sulfate was found to be a more effective additive. Acceptable efficiencies could be achieved by the addition of 250 mM potassium sulfate to the running buffer. In an experiment where 20 mM potassium chloride was added to a running buffer having a pH above the isoelectric points of the protein analytes, very high efficiencies were obtained (50). Alternatively, high ionic strengths/concentrations of the buffer salts themselves may be used to effectively reduce wall adsorption, while simultaneously increasing the buffering capacity (56, 60).

Table 3. Some amines and ion-pairing reagents used in CE

Additive	Condition	Sample	Reference
Amines			
Morpholine	20 mM at pH 8.1	hGH fragments	18, 47
	0.25 M at pH 5.5	Basic proteins	
Spermine	0.2 mM at pH 7	Basic proteins	19
1,4-Diaminobutane	5 mM at pH 8.2	Myoglobins	6, 50, 51
	2 mM at pH 6	Cationic peptides	
Hexamethonium bromide	5 mM at pH 2	Dodecapeptides	52, 53
	1 mM at pH 8.4	hCG isoforms	
Hexamethonium chloride	1 mM at pH 8.4	Glycoproteins	52
Decamethonium bromide	1 mM at pH 8.4	Glycoproteins	52
Ion-pairing reagents			
Pentanesulfonic acid	50 mM at pH 9.9	Synthetic peptides	49
Heptanesulfonic acid	Heptanesulfonic acid	r-HuEPO fragments	48
Hexanesulfonic acid	10 mM at pH 9.2	Dodecapeptides	53
Phytic acid	10 mM at pH 9.2	Peptides and protein	54, 55

High ionic strength buffers will have a high conductivity leading to Joule heating at the high voltages typically used in CE. A lower voltage must then be used for such separations. Lowering the voltage results in a longer analysis time (> 30 min).

2.2.3 CE separations at pH extremes

The manipulation of buffer pH was an early, direct approach used to minimize the adsorption of peptides and proteins on the capillary wall. At pH < 3, the silanol groups on the capillary wall are largely protonated so that the wall adsorption through ionic interaction will be minimal, and the magnitude of EOF is significantly reduced. This elimination of EOF is similar to the effects caused by coating the capillaries. Although the use of low pH buffers has been successful for some proteins (16, 43, 44, 48, 61, 62), when the pH is around 2, most proteins precipitate or will be denatured. None the less, a low pH buffer is satisfactory for the separation of many peptides, and is simple to try. For instance, a 10–100 mM phosphate buffer at pH 2–3 is frequently used in peptide mapping (55, 63).

At pH > 10, both peptide or protein analytes and the silanol groups on the capillary wall are negatively charged. Due to the Coulombic repulsion between the negatively charged groups, the use of high pH running buffers in CE provides low protein adsorption and high separation efficiencies (50, 64), suffers significant drawbacks from the instability of fused-silica at high pH (it dissolves slowly), and the high EOF rates generated (which may reduce resolution). In summary, the use of pH extremes is not a generally useful strategy for peptide and protein separations. In particular, use of pH extremes limits the selection of buffer pH to a narrow pH range which is far from biologically relevant conditions.

3. Buffer

A buffer is used to maintain the pH at a constant value during the CE operation, and ions in the buffer must conduct the current generated when the voltage is applied. Since EOF controls the residence time of the analytes in the capillary and the magnitude of EOF on uncoated capillaries varies with pH, the use of a buffer is required to keep the EOF stable during electrophoresis. Moreover, the charged species to be analysed have different pK_a values and their charges depend on pH. The use of a buffer maintains selectivity by sustaining a constant charge on each analyte during electrophoresis.

3.1 Buffer selection

Generally, the following properties of a buffer should be considered before selecting one for CE separations of peptides and proteins:

(a) pK_a value of a buffer. The desired working pH of the buffer should be within the range of one pH unit around its pK_a.

(b) pK_a variation with temperature. If the buffer pK_a changes dramatically with temperature, the buffer pH may shift during separations.

(c) UV absorption. A buffer should have low UV absorption at the detection wavelength.

(d) Interaction with other species. The buffer may interact with the additives, the peptide and protein samples, or interfere with the interaction of a ligand and substrate.

3.1.1 Useful buffers

Buffers frequently used in the separation of proteins and peptides are listed in *Table 4*. So far, the majority of the buffers adopted for capillary electrophoresis of proteins and peptides are standard buffers such as phosphate, borate, and Tris, while biological 'Good' buffers (65) are used less often. 'Good' buffers have many desirable properties such as a lack of change in pH with changes in temperature, and a lack of interference with biological processes. Most of the 'Good' buffers are zwitterions, having a pK_a averaged between those of their positive and negative ends. Some 'Good' buffers are not transparent to UV light, and this may explain their infrequent use. Standard inorganic buffers are transparent in the UV, are less expensive, and are more readily available.

Phosphate is the most commonly used buffer for peptide and protein CE. The prevalent use of phosphate buffers may be attributed to their high UV transparency, and wide useful pH range. The formation of a complex between silanol and phosphate groups reduces both the EOF and the adsorption of peptides and proteins on the capillary wall (16, 66). Phosphate has three

Table 4. Buffer systems used in protein and peptide separation

Buffer[a]	pK_a	Buffer system
Phosphate	2.15 (25°C) pK_a1[b]	10–150 mM, pH 1.5–3
	6.82 (25°C) pK_a2[b]	10–150 mM, pH 6.8–8
Borate	9.24 (25°C)[b]	20–300 mM, pH 8.2–9.7
Tris	8.08 (25°C)[b] 8.3 (20°C)[c]	10–100 mM, pH 7–8.8
Citrate	3.13 (25°C)[b] pK_a1	10–30 mM, pH 2.1–3
Tricine	8.05 (25°C)[b] 8.15 (20°C)[c]	10–100 mM, pH 7.5–8.2
Ches[d]	9.5 (25°C)[b]	20–100 mM, pH 9–9.2
Acetate	4.76 (25°C)[b]	50–100 mM, pH 3.4–5.5
Mes[e]	6.09 (25°C)[b] 6.15 (20°C)[c]	20–100 mM, pH 6–8
Formate	3.75 (25°C)[b]	20 mM, pH 2.5
Tes[f]	7.4 (25°C)[b] 7.5 (20°C)[c]	50 mM, pH 7

[a] Beginning with the most common one.
[b] Ref. 61.
[c] Ref. 65.
[d] 2-(Cyclohexylamino)ethanesulfonic acid.
[e] 2-(N-morpholino)ethanesulfonic acid.
[f] N-tris(hydroxymethyl)methyl-2-aminoethanesulfonic acid.

ionizeable groups and may be used as a buffer at virtually any pH. Phosphate does form complexes with many proteins, and care must be taken that it does not interact with the analyte (67). Phosphate buffers often have high conductivities, and to avoid high currents, they must be used at low concentrations.

Borate has also been proven a useful buffer in the separation of peptides and proteins. Borate buffer is especially useful in the separation of protein glycoforms, as a result of its ability to complex with *cis*-diols and alter the selectivity of a separation (68). Borate has a higher pK_a than many inorganic buffers, leading to its popularity in separations of basic proteins, such as antibodies. Borate buffers often posses low conductivities, making them excellent for high voltage separations.

In CGE, Tris–Ches is often used as the running buffer, since it has traditionally worked well on slab gels. Also, there are reports that Tris prevents the loss of biological activity of certain proteins, if this is a concern. In general, any low ionic strength buffer used for a given slab gel preparation, may also be used in CGE, provided it fits the buffer criteria listed above.

3.1.2 Choosing a buffer pH and buffer concentration

Buffer pH is the predominant parameter influencing the migration time and selectivity in CE analysis, because it controls the charges on both the capillary wall surface and the charged species. To obtain very high efficiency in CE, the pH of the separations must be optimized. Consider the properties of the substances to be separated, particularly the charge they will carry at a given pH. Sometimes, a biologically relevant pH will be desired. Although high concentration of buffer can suppress both EOF and wall adsorption, this approach is limited by Joule heating. The use of zwitterionic buffers is an attractive alternative, but actually does not improve separation efficiencies (58). Buffer concentrations should also be optimized. The buffer systems used in protein and peptide separation are summarized in *Table 4*.

3.2 Preparing the CE buffer

Before beginning CE, careful buffer selection and preparation will be important to ensure a successful separation. The following general points should be helpful:

(a) Select a buffer with a pK_a close to the desired pH for your separation.
(b) Start work at a low concentration of a buffer. 50 mM is recommended.
(c) Use 18.2 MΩ water to prepare the buffer.
(d) Prepare a buffer at the CE working temperature.
(e) Filter the buffer with a 0.45 μm or 0.2 μm filter.
(f) Degas the buffer in an ultrasonic bath under reduced pressure.

4: Capillary electrophoresis of peptides and proteins

Table 5. Stock ratios for the preparation of 50 mM phosphate buffers[a]

pH	A (ml)	B (ml)	pH	A (ml)	B (ml)	pH	A (ml)	B (ml)
5.8	92	8	6.5	68.5	31.5	7.2	28	72
5.9	90	10	6.6	62.5	37.5	7.3	23	77
6	87.7	12.3	6.7	56.5	43.5	7.4	19	81
6.1	85	15	6.8	51	49	7.5	16	84
6.2	81.5	18.5	6.9	45	55	7.6	13	87
6.3	77.5	22.5	7	39	61	7.7	10.5	89.5
6.4	73.5	26.5	7.1	33	67	7.8	8.5	91.5

[a] By mixing stock solution A (100 mM NaH_2PO_4) and stock solution B (100 mM Na_2HPO_4), and diluting the mixture to a final volume of 200 ml with distilled water.

3.2.1 Phosphate buffer
When a buffer compound and its salt are available, it is convenient to make a series of buffers of different pH by preparing stock solutions containing either pure buffer or salt and then mixing them in calculated volume ratios (see *Table 5*).

3.2.2 Tris buffer
For more details about Tris buffer preparation, see refs 61 and 69, or the Sigma® catalogue. As in the case of phosphate buffers, by mixing Trizma®HCl and Trizma®Base stock solutions, you can obtain any desired pH. Because the pK_a of Tris depends strongly on temperature ($\Delta pK_a/°C = -0.031$) (65), this buffer must be prepared and the pH measured at the CE working temperature.

4. Sample preparation

The following are general guide-lines for sample preparation:

(a) Make the sample solution as concentrated as possible (usually 1 mg/ml) to improve sensitivity.
(b) Match the sample solution and running buffer conductivities (see Section 4.3.4).
(c) Filter the sample solution with a 0.45 μm or 0.2 μm filter to remove any particles which might clog the capillary. Alternatively, transfer the sample solution to a 1.5 ml microcentrifuge tube and centrifuge at > 3000 g for 5 min at room temperature. Remove the supernatant for CE analysis, being careful not to disturb any precipitates which sediment.

4.1 Preparation of samples for CZE, MEKC, and cIEF

(a) Desalt any samples which have a high salt concentration (see Section 4.4).
(b) Dissolve 1 mg of a sample in 1 ml of the running buffer for CZE and MEKC, or in 1 ml of 2% (v/v) ampholyte (for cIEF).
(c) Filter the sample solution with a 0.45 μm filter (or centrifuge the sample).

High amounts of salt in a sample will cause solute defocusing and peak distortion during electrophoresis, thereby leading to a loss of high resolution in CZE and MEKC. In cIEF, the presence of high salt concentrations in a sample resulted in a longer focusing and mobilization times, and a higher risk of precipitation (7). Therefore, sample desalting is also recommended in cIEF.

4.2 Preparation of samples for CGE

(a) It is recommended that you desalt the sample before adding SDS.
(b) Prepare a 1 mg/ml sample solution in 1% (w/v) SDS, 5% (v/v) 2-mercaptoethanol. Heat the sample solution at 100°C in a boiling water-bath for 10 min. Cool to room temperature, and measure 20 μl of the cooled solution into a 0.5 ml sample vial. Add 80 μl of the sample buffer (identical to the running buffer without the 0.1% (w/v) SDS; usually 0.1 M Tris–Ches pH 8.8), and also add 1 μl of 1% (w/v) mellitic acid marker to the sample vial. Vortex for 60 sec.
(c) Filter the sample solution with a 0.45 μm filter (or centrifuge the sample).

The effect of high salt concentrations in the sample varies in SDS–PAGE. High ionic strengths generally reduce the amount of SDS bound to polypeptides. Therefore, sample desalting is generally recommended before electrophoresis. On the other hand, it has been reported that 0.8 M NaCl in a sample had no influence on the results of the molecular weight determination (70). There is yet no comprehensive investigation of the effects of high salt in the sample on CGE results. Keep in mind, however, that high concentrations of K^+, guanidinium, or divalent cations can cause the precipitation of SDS (70).

4.3 Sample concentration

This section describes several methods used for sample pre-concentration. One of the greatest advantages of CE over conventional methods is the small amount of sample required for analysis. However, only nanolitre volumes of a sample solution are injected into the capillary in order to achieve a high resolution separation in CE. The total amount of sample loaded onto the capillary is limited by the sample injection plug length, so that highly concentrated sample solutions should be prepared if their solubility permits. An improvement in detection sensitivity presents a big challenge in instrument design, due to the capillary dimensions and minute sample volumes. In summary, a high solubility of analyte in the sample buffer is required to achieve high resolution separations and reliable, quantitative detection using current instrumentation.

Due to the detection limitations discussed above, the direct injection of dilute samples from biological fluids may not yield detectable peaks and usually, a pre-concentration step is required. However, the use of highly sensitive detectors is another effective approach for analysis of dilute samples.

4: Capillary electrophoresis of peptides and proteins

With fluorescence detection or mass spectrometric detection, the limits of detection can be lowered to the 10^{-18} mol level (4, 71). For more details about detectors, see Section 6. The analyte concentration required to perform a given CE analysis depends mainly on the sensitivity of the detection method.

4.3.1 Lyophilization or vacuum centrifugation

For a large volume of sample solution containing little or no salt, lyophilization and evaporation are the simple and easy methods of choice. For lyophilization, the sample must first be placed in a vacuum-ready vessel, such as a round-bottomed flask or lyophilizer flask (Labconco or VirTis), and then it is frozen by slowly rotating the flask in a dry ice and acetone (or isopropanol) bath. Once the sample is completely frozen, the flask is placed under high vacuum in a freeze-dryer (Labconco or VirTis). Lyophilized samples are fluffy and are usually quite stable during prolonged storage. In the case of samples which cannot be frozen, or when speed is of the essence, vacuum centrifugation concentrates samples quite effectively. Place the sample in centrifuge tubes and balance them, then load the tubes into the SpeedVac® (Savant), close the lid, and turn on the rotor. Turn on the high vacuum pump after the rotor is already spinning, to avoid bumping (the loss of sample which occurs due to bubbling when vacuum is applied). The vacuum is left on until the required volume is achieved or the sample is dried. Both techniques, lyophilization and evaporation under vacuum, leave all buffer salts and impurities behind in the samples, since only the volatile components of a sample are removed.

4.3.2 Ultrafiltration

Choose a stirred ultrafiltration cell (Amicon) with a membrane (M_r cut-off 500) according to the volume of sample solution. Pressurize the cell system using nitrogen pressure (maximum 5.3 kg/cm^2 or 75 psi), and stir until the desired volume is achieved. Salts and small molecular weight impurities will pass through the membrane leading to a selective concentration of the peptide or protein analyte in the solution.

4.3.3 Solid phase extraction

Performing solid phase extraction on the capillary itself (*Protocol 4*) is a convenient way to do both concentration and purification in one step. However, packed-inlet capillaries are expensive or must be prepared in your laboratory. Conditions such as the time and pressure used for sample loading (*Protocol 4*, step 3), the selection of the elution solution (*Protocol 4*, step 4), and the elution volume (time and pressure) should be optimized. At low pHs, such as those frequently used in the separation of peptides, there is a tendency to experience a suppressed or reversed EOF (72) which can be overcome by a simultaneous application of voltage and low pressure (0.5 psi at the inlet of the capillary) during the electrophoresis portion of the protocol (*Protocol 4*, step 5). Off-capillary solid phase extraction is frequently more convenient and may be performed ahead of time. Use a Waters Sep-Pak® C18 or tC18 cartridge (Millipore) according to the volume of sample solution and then follow *Protocol 4*, substituting the word cartridge for capillary. For best results, the acetonitrile eluent containing purified sample should be dried before dissolving the sample in water (or buffer) for injection on the CE (*Protocol 4*, step 5).

Protocol 4. Solid phase extraction[a]

Equipment and reagents
- Capillary electrophoresis system (Beckman)
- AccuSep™C/PRP capillary 75 μm i.d. (Waters) or a self-assembled C18 packed-inlet capillary (72, 73)
- 18.2 MΩ H_2O (Milli-Q water purification system, Millipore)
- Ethanol (HPLC grade)
- Acetonitrile (HPLC grade)

Method
1. Make a detection window on the capillary and install in the CE instrument. Set the detection wavelength to 215 nm.
2. Wash the capillary with ethanol for 10 min, then wash with a 50% (v/v) aqueous solution of acetonitrile for 2 min, wash with pure acetonitrile for 1 min, and then wash with water for 2 min.
3. Load the aqueous sample solution into the capillary by pressure injection (20 psi) for 5 min, then wash the capillary with running buffer for 2 min.
4. Briefly inject a 50% (v/v) aqueous solution of acetonitrile by pressure (10 psi) for 1 min.
5. Start the run by applying a 30 kV voltage.

[a] The procedure is modified from refs 72 and 73.

4.3.4 Stacking the sample

Sample stacking is one of the most widely used on-capillary, sample pre-concentration techniques (74–76). Stacking may allow a ten- to several hundred-fold improvement in detectability to be achieved. Stacking occurs when the capillary is placed under a high voltage and the running buffer has a significantly higher conductivity than that of the sample solution. The required difference in conductivity between the sample plug and the running buffer zone can be obtained either through differences in pH (75) or differences in salt concentration (74).

A simple way of attempting stacking is to create a buffer discontinuity where the sample buffer has a lower concentration (10–100 times) than that of the running buffer (74). First, prepare the sample solution using a low concentration buffer and fill the capillary with a high concentration buffer. Second, inject the sample solution for 2 min by gravity or pressure injection. Then, start the run. The thermal stability of the samples should be considered before attempting this, because the temperature in sample zone will be high during stacking (77). Peptides or proteins in the sample plug will migrate quickly through sample plug and into the running buffer, where they will immediately experience a lower effective electric field and slow down, thus forming a narrow zone of analytes in the running buffer. From this point on, the electric field is homogeneous and electrophoretic separation begins.

Theoretically, optimum stacking will be achieved by dissolving the sample in water and using a high conductivity running buffer. Unfortunately, this large conductivity difference causes the entire sample plug to have a higher electro-osmotic velocity than that of the running buffer zone. The resulting mismatch of electro-osmotic velocities generates a laminar flow which will broaden the peaks and reduce the resolution between peaks (76). In short, both stacking and broadening occur simultaneously if the conductivity differences between sample buffer and run buffers is too large. To balance these effects, the optimal condition required for effective sample stacking is that the concentration of sample buffer should be ten times less than that of running buffer. The sample plug length injected can be up to ten times the diffusion-limited peak width without affecting peak resolution.

4.4 Desalting biological samples

Many buffers containing biologically active proteins have high ionic strengths, yielding high currents in CE. For samples with high salt concentrations (> 100 mM), desalting is recommended. Several methods may need to be tested to find a way to desalt without precipitating the protein. Ultrafiltration (Section 4.3.2) after diluting the sample with 18.2 MΩ water, solid phase extraction (Section 4.3.3), or simply diluting the sample until the sample solution and running buffer conductivities match. If the sample in the resulting solution can be detected, this latter approach is simple and effective.

5. Modes used in CE

CE modes have been classified according to the separation mechanism. The main CE modes, such as capillary zone electrophoresis (CZE), capillary isoelectric focusing (cIEF), capillary gel electrophoresis (CGE), and capillary isotachophoresis (cITP), were adapted from traditional electrophoretic techniques. However, the development of micellar electrokinetic capillary chromatography (MEKC) and capillary gel electrophoresis (CGE) using viscous polymer solutions make CE unique and more powerful than traditional low resolution methods. The various CE modes will be reviewed below. Selection of an appropriate CE mode requires consideration of the properties of the sample and the purpose of CE analysis.

5.1 Capillary zone electrophoresis (CZE)

Capillary zone electrophoresis (CZE) is the most widely used and basic CE mode. Also called 'free-solution' CE, CZE is a fundamental and simple form of CE. In CZE, the separation of analytes occurs based on their different charge-to-mass ratios, and neutral species cannot be resolved from one another.

CZE has been widely used for the separation of native proteins from modified proteins, such as core histones from acetylated derivatives (44), and resolution of the various phosphorylated histone H1 variants (43), recombinant DNA-derived proteins (78), and glycoproteins (51) may also be separated by CZE. Difficult separations may also be separated by CZE. Difficult separations using poorly soluble membrane proteins (79) and collagens (80) have been successfully accomplished by simple CZE. CZE is quite convenient for peptide mapping (48) and for the identification of phosphorylated sites of peptides (81). Protein folding has been explored using CZE, through separations of native and misfolded IGF-1 (36) and by monitoring thermally-induced protein folding (62).

The following factors should be considered to ensure successful separations of proteins and peptides by CZE:

(a) Select an effective strategy to eliminate sample adsorption to the capillary wall (see Section 2.2).

(b) Optimize buffer type, buffer pH, and concentration (see Section 3.1).

(c) Choose an appropriate additive type (Section 2.2.2), and optimize the additive concentration when one is used.

(d) Match the sample solution and buffer conductivities (see Section 4.3.4).

(e) Select an injection mode (hydrodynamic or gravity injection is preferred in most situations), and optimize the sample plug volume for sensitivity, peak shape, and resolution between peaks.

An example of a successful separation of basic and phosphorylated proteins, performed in the CZE mode using a buffer additive, is given in *Protocol 5*.

Protocol 5. Separation of phosphorylated histone H1 variants by CZE[a]

Equipment and reagents
- Capillary electrophoresis system (Beckman)
- Running buffer: 100 mM phosphate buffer pH 2, 0.03% (w/v) HPMC[b]
- Capillary 75 μm i.d. (Polymicro Technologies)
- 0.5 mg/ml protein sample in running buffer, 0.2% (v/v) mesityl oxide[c]

4: Capillary electrophoresis of peptides and proteins

Method

1. Install the capillary (total length 58.8 cm, 50 cm to the detector) on the instrument, set the polarity to (+), and the detection wavelength to 200 nm.
2. Wash the capillary with a 0.1 M NaOH solution for 10 min, rinse with water for 5 min, wash with a 1 M HCl solution for 10 min, rinse with water for 5 min, and then finally rinse with the running buffer for 15 min.
3. Inject the protein sample for 5 sec by pressure injection.
4. Start the run and record the electropherogram.
5. After every five injections, clean the capillary with 0.1 M NaOH for 5 min, rinse with water for 2 min, wash with 1 M HCl for 2 min, rinse with water for 2 min, and finally rinse with the running buffer for 5 min.

[a] Modified from ref. 43.
[b] Low pH and HPMC (hydroxypropylmethylcellulose) eliminate the capillary wall adsorption.
[c] Internal standard.

5.2 cITP

Capillary isotachophoresis (cITP), also called 'displacement electrophoresis', is used for the extremely high resolution separation of charged species, based on the differences in electrophoretic mobilities experienced in different buffers. cITP also finds use as a sample pre-concentration technique similar to stacking (Section 4.3.4), allowing injection of 10–100 times the sample volumes used in CZE (82). The separation mechanism of cITP in a single capillary involves the use of a dual discontinuous buffer system. The leading buffer is chosen to have a higher mobility than that of the analytes and the terminating buffer should have a lower mobility than that of the analytes. This set of buffers leads and trails the sample plug. After a high voltage is applied across the capillary, ions of the leading buffer move fastest, followed by the analyte ions in the order of decreasing mobility. This results in several discrete focused zones containing ionic species which move at the same velocity. The zones form because the effective electric field an ion experiences at any time depends upon the ionic strength of the buffer in that location. The final concentration of ions in each focused zone is determined by that of the leading buffer, whereby the analyte zone is concentrated if its concentration is less than that of leading buffer, or vice versa (82, 83). A regular CZE separation can then be performed on these focused zones, simply by replacing the terminating buffer with more of the leading buffer (now the running buffer). Due to the need for longer capillaries, or an ITP instrument containing two capillaries, and the difficulty of developing appropriate buffer pairs for a given sample, the practical use of cITP as a high resolution technique is less convenient than other methods. cITP gives excellent results when it works.

5.3 Micellar electrokinetic capillary chromatography (MEKC)

Surfactants can form micelles at concentrations above their critical micelle concentrations (CMC). The addition of surfactants to the running buffer at concentrations above their CMC is a fundamental feature of MEKC. This micellar pseudo-phase makes MEKC a powerful and unique technique that acts through the combination of electrophoresis and chromatography. In MEKC, not only charged species but also neutral species can be separated. MEKC separations of charged analytes are based both on differences in the charge-to-mass ratios of ionic species, as well as their different hydrophobicities. Neutral species are resolved solely by partitioning between the micellar pseudo-phase and the buffer aqueous phase in a chromatographic mechanism. MEKC is an effective tool for the separation of closely related peptides and proteins having similar charge-to-mass ratios. The use of MEKC has been successfully used in the separation of angiotensin fragments (30, 40, 84), closely related peptides (85), synthetic peptides (39), protein isoforms and variants (41, 86), and glycoproteins (60).

SDS, the most common surfactant used for the MEKC separation of peptides and proteins, is an anionic surfactant commonly used in concentrations from 20–100 mM (86). Cationic surfactants such as 2.5–50 mM CTAC (39), 50 mM HTAB (40), and 50 mM DTAB (40), non-ionic surfactants such as Tween 20 (41, 84), and zwitterionic surfactants such as 5 mM DAPS (*N*-dodecyl-*N*,*N*-dimethyl-3-ammonio-1-propanesulfonate) (85) have also proved useful for MEKC. The procedure for running MEKC is the same as for CZE, except that surfactants are added to the buffer (see *Protocol 5*). The concentration of the surfactant in running buffer should be optimized, and must be above the surfactant CMC. Moreover, after the addition of a cationic surfactant to running buffer in MEKC (34), it is necessary to change the polarity of the instrument (see *Protocol 3*). In cases where peptides and proteins strongly interact with the micelles, the addition of organic solvents such as 10% acetonitrile (38) or alcohols (85) can reduce hydrophobic interactions, thereby increasing the selectivity of the micelle, and improving peak resolution.

5.4 Capillary isoelectric focusing (cIEF)

Isoelectric focusing gel electrophoresis (IEF) has been used as a high resolution separation technique for proteins, relying upon their lack of migration in an electric field when they have a net neutral charge. Traditional IEF is described in an article containing the detailed procedures (87). Obviously, it is a time-consuming process, which requires practice to master.

By adapting IEF to take advantage of the properties of CE, capillary iso-

4: Capillary electrophoresis of peptides and proteins

electric focusing (cIEF) has proved to be a powerful electrophoretic technique for the high resolution separation of proteins and peptides. IEF separates peptides and proteins based on differences in their isoelectric points (pIs). cIEF has been used for the high resolution separations of rhGH variants (71), and haemoglobin variants (7, 28, 45).

In cIEF, the operation may be divided into three steps: loading, focusing, and mobilization. First, the capillary is filled with a mixture of sample and the ampholytes (Pharmacia). When high voltage is applied at the ends of the capillary using an acidic buffer at the anode and a basic buffer at the cathode, the ampholytes form a stable pH gradient along the length of the capillary. At the same time, the peptides or proteins also migrate along the gradient to a location where the pH is equal to their respective pIs, their charge becomes neutral, and they stop moving. This process is referred to as focusing, which is monitored by a drop in the current during electrophoresis. After the completion of focusing, the entire gradient must be mobilized to pass by the detection window. The mobilization step can be performed in different ways: by addition of salt to one of the electrode reservoirs (7), by application of pressure at the injection end of the capillary, or by application of a vacuum at the exit end of the capillary (88). Alternatively, the EOF itself may be used to elute the proteins, thus combining the focusing and mobilization steps into one (45). The details of a cIEF operation are described in *Protocol 6*. A commercial kit for cIEF is available (Applied Biosystems).

Protocol 6. cIEF separation of proteins and peptides[a]

Equipment and reagents

- Capillary electrophoresis system
- 1 mg/ml protein or peptide sample and 0.2% (w/v) methylcellulose (3000–5000 mPa's Fluka) in 2% (v/v) Pharmalyte 3–10
- Coated capillary 60 cm × 50 μm i.d. (from J & W Scientific or prepared via *Protocol 2*)
- Pharmalyte 3–10 (Pharmacia, narrower pH ranges are available)

Method

1. Make a detection window on the capillary, install the capillary in the CE instrument, set the polarity to (+), and the detection wavelength to 280 nm.

2. Programme the system to fill the capillary with a 20 mM sodium hydroxide solution.

3. Load the sample mixture into the capillary up to the detection point of the capillary by pressure injection.

4. Place the injection end of the capillary into 10 mM phosphoric acid anolyte and place the outlet end into 20 mM sodium hydroxide catholyte.

5. Apply high voltage (30 kV) and observe the current.

Protocol 6. *Continued*

6. After the current has dropped to a constant value, turn off the power supply. Exchange the sodium hydroxide catholyte for a buffer containing 0.1 M sodium chloride dissolved in 20 mM sodium hydroxide.[b]
7. Reapply the high voltage, and record the electropherogram.

[a] Modified from refs 7 and 88.
[b] Step 6 may be replaced by applying pressure at the injection end of the capillary or vacuum at the exit end of the capillary while maintaining a high voltage (88).

Theoretically, the elimination of EOF is not necessary in cIEF as long as the EOF is slow enough so that focusing can be completed before the analytes pass the detection window. However, the use of a hydrolytically stable coated capillary, with a negligible EOF, is recommended for high resolution cIEF, since complete elimination of the EOF is necessary to permit longer focusing times (7, 28). Coatings also hinder protein adsorption to the capillary wall and can thus prevent variation in the EOF during a run.

After focusing, the peptides or proteins typically become concentrated 200- to 300-fold from a large injected volume into narrow focused zones. Precipitation has been a problem during IEF in both gels and capillaries, due to the increase in protein concentration and the low solubility of peptides and proteins at their pI. Frequently, the addition of non-ionic detergents to the sample and ampholyte mixture can circumvent this difficulty (7).

In cIEF, there is a short distance between the point of detection and the outlet of the capillary. If the peptides or proteins focus in that region of the capillary, they will not be detected during the mobilization step. The use of an extended pH gradient generated by addition of TEMED to the ampholyte mixture is one way to solve this problem (7, 45). TEMED fills the cathodic region during focusing, acting as a blocking agent. Another approach is to fill the region past the detection window with catholyte when the cathode is at the detector end (88). So far, on-line UV detection of cIEF runs has been limited to wavelengths greater than 280 nm because of the high UV absorption of ampholytes which occurs at wavelengths below 260 nm.

5.5 Capillary gel electrophoresis (CGE)

Cross-linked polyacrylamide gels were first prepared in the capillary using the same methods as for traditional SDS–PAGE, a technique here called 'permanently gel-filled capillaries'. The use of cross-linked polyacrylamide gels has been highly successful for the separation of negatively charged complexes of peptides and proteins with SDS according to their molecular weight (89–91). Due to the mechanical instability of cross-linked polyacrylamide gels, linear polyacrylamide gels were later used (91). Although separations have been improved, permanently filled gels suffer from a limited lifetime, and detection

4: Capillary electrophoresis of peptides and proteins

sensitivity due to the high UV absorption of polyacrylamide at 200–260 nm (91). Because of their expense and limited lifetime, the permanently gel-filled capillaries are gradually being replaced in CGE by the use of viscous polymer solutions. These polymer solutions are made from pre-polymerized monomers which are then introduced into the capillary, where they serve as a sieving matrix (92).

5.5.1 Viscous polymer solutions

The efficacy of viscous polymer solutions as sieving matrices is related to the concentration of the polymer solution. Theoretically, at a concentration of polymer solution above its entanglement threshold, the polymer chains become entangled, forming a transient network (92). The use of viscous polymer solutions as dynamic sieving matrices has the advantage of easy preparation, low UV absorption, good reproducibility, and easy replacement of the sieving matrix.

Preparation of a viscous polymer solution is achieved either by simply dissolving a preformed polymer in running buffer or by first polymerizing the desired monomer. Then, the prepared polymer solution is loaded into the capillary; it can be replaced after analysis of each sample. Commercial CE kits using polymer solutions for both analysis and molecular weight determinations of proteins are available from ABI (89, 93), Beckman (89, 94), and Bio-Rad (95). *Protocol 7* describes the CGE procedure for separating protein–SDS complexes using a viscous polymer solution. The various types of polymer solutions commonly used for protein separations are listed in *Table 6*.

Protocol 7. Separation of protein–SDS complexes by CE[a]

Equipment and reagents

- Capillary electrophoresis system
- Capillary 60 cm × 50 µm i.d. (Polymicro Technologies)
- Mellitic acid (Aldrich)
- Standard proteins: α-lactalbumin (14 kDa), carbonic anhydrase (29 kDa), ovalbumin (45 kDa), bovine albumin (66 kDa), phosphorylase b (97 kDa), β-galactosidase (116 kDa), myosin (205 kDa) (Sigma)
- ProSort™ SDS–protein analysis reagent (ABI)
- 18.2 MΩ water (Milli-Q, Millipore)
- 200 µg/ml protein standard mixture, with 10 µg/ml mellitic acid[b] added, in 0.2% (w/v) SDS, and 1% (v/v) 2-mercaptoethanol
- 200 µg/ml protein sample, with 10 µg/ml mellitic acid added, in 0.2% (w/v) SDS, and 1% (v/v) 2-mercaptoethanol

Method

1. Make a detection window on the capillary, install the capillary on the CE instrument, set the polarity to (–), and set the detection wavelength to 215 nm.

2. Clean the capillary with a 0.1 M NaOH solution for 10 min, rinse with water for 5 min, and coat the capillary with ProSort™ reagent for 60 min.[c]

Protocol 7. *Continued*

3. Load the protein standard mixture into the capillary by electrokinetic injection (10 sec, 5 kV). First run the protein standard mixture using ProSort™ reagent as the running buffer (15 min, 12 kV), and record the electropherogram.

4. Flush the capillary with ProSort™ reagent for 12 min,[d] then load your protein sample by electrokinetic injection (10 sec, 5 kV),[e] and run the sample in ProSort™ reagent as the running buffer. Record the electropherogram (15 min, 12 kV).

[a] Modified from ref. 93.
[b] Reference marker has a low molecular weight and six negative charges.
[c] ProSort™ reagent acts as a combination coating agent and sieving matrix.
[d] Replace ProSort™ reagent after analysis of each sample.
[e] Electrokinetic injection is advantageous in CGE.

Uncoated capillaries can be used in CGE with polymer solutions if the polymer solutions can also act as a coating agent. Otherwise, a coated capillary should be used to minimize the EOF and eliminate protein adsorption to the capillary wall.

Table 6. Viscous polymer solutions used for CGE separations of protein–SDS complexes

Polymer solution	Running buffer[a]	Reference
Linear polyacrylamide		
10% (w/v) gel	50 mM phosphate pH 5.5[b]	5
10% (w/v), M_r 2 × 10^6	60 mM CACO/AMPD[c] pH 8.8	91, 96
10% (w/v), M_r 2 × 10^6	0.1 M Tris–Ches pH 8.6	15
10% (w/v), M_r varied	0.1 M Tris–Ches pH 8.8	97
15% (w/v), M_r 72 × 10^3	0.1 M Tris–Ches pH 8.8	98
Poly(ethylene glycol)		
3% (w/v), M_r 10^5	0.1 M Tris–Ches pH 8.8	91
Poly(ethylene oxide)		
3% (w/v), M_r 10^5	0.1 M Tris–Ches pH 8.8	98
	0.1 M Tris–Ches pH 8.5	99
Poly(vinyl alcohol)		
4–6% (w/v) in buffer, M_r 133 000	60 mM CACO/AMPD[c] pH 8.8	96
Pullulan		
1–10% (w/v), M_r 5–10 × 10^4	0.1 M Tris–Ches pH 8.7	13

[a] Contains 0.1% (w/v) SDS.
[b] Contains 0.5% (w/v) SDS.
[c] Cacodylic acid/2-amino-2-methyl-1,3-propanediol.

5.5.2 Permanently gel-filled capillaries

In early attempts to use CE for the separation of protein–SDS complexes, the gel in the capillary was prepared in the same way as in traditional SDS–PAGE. The capillary was first prepared by coating it with a bifunctional reagent so that the polymer would be anchored to the capillary surface. Secondly, ammonium persulfate and TEMED were added to initiate polymerization of an acrylamide and *bis*acrylamide monomer mixture, which was quickly loaded into the prepared capillary. Polymerization then occurred within the capillary (90), forming a permanently gel-filled capillary.

To prepare permanently gel-filled capillaries, use *Protocol 1*, as for a coated capillary, but in step 3, substitute a 4–12% (w/v) solution of acrylamide monomer, and if cross-linked polyacrylamide gels are desired, add 0–4% (w/w) *bis*acrylamide (weight *bis*acrylamide/weight acrylamide + *bis*acrylamide). In step 4, after filling the capillary with the monomer solution, allow the capillary to stand for one hour to complete the polymerization. After one hour, the capillary is ready for use. Caution is needed to avoid gel breakage during the preparation and storage of this type of capillary. Capillaries should be examined under a microscope before use to confirm that no bubbles have formed during polymerization. During both cutting and storage, prevent the ends of the capillary from drying out by keeping them in buffer at all times. Do not heat the capillary to form a detector window.

6. CE detection strategies

Detection in CE is often a big challenge due to the nanolitre volumes of samples present in the solute zones, and the small dimensions of the capillary. On-capillary detection has been used to meet this challenge. Although a variety of detection methods have been attempted in CE (100), UV, fluorescence, and mass spectrometric detection are the most commonly used for peptides and proteins.

6.1 UV detection

Despite a relatively low sensitivity, UV detection is the most widely used detection method and has proven extremely useful for the on-line quantitation of peptides and proteins in a mixture. Detection is based on the strong UV absorption of the amide groups present in peptides and proteins at wavelengths from 190–215 nm. The significant absorbance of tryptophan-containing peptides and proteins around 280 nm allows detection in a second wavelength range. It is possible to calculate the extinction coefficient of a protein from its amino acid sequence (101) and predict how sensitive UV detection will be in a given case. It is best to keep in mind, when quantitating your peaks, that for most CE detectors, the UV path length is simply the diameter of the capillary (usually 50 μm). Thus, absorbances of even concentrated solutions are quite small. The UV detectors available in commercial CE instruments provide routine and reliable detection of absorbances as low as 0.001 AU.

6.2 Fluorescence detection

Fluorescence detection in CE has the inherent advantage of a higher sensitivity than UV and also greater selectivity. With the use of lasers as excitation sources, laser-induced fluorescence detection can lower the limit of detection to the 10^{-18} mol level (46) or below (71) for peptides and proteins. The emission wavelength monitored by the detector may be chosen to enhance the natural selectivity of fluorescence by specific detection of one fluorophore among many.

6.2.1 Native fluorescence

In cases where derivatization of the peptide or protein is inconvenient or impossible, examination of the protein sequence may reveal the presence of tryptophan or tyrosine amino acids. In these cases, the native fluorescence of tryptophan- and/or tyrosine-containing peptides and proteins can be detected with laser-induced excitation at 280 nm (102) or 275 nm. Because only a few peptides and proteins produce adequate fluorescence in the excitation region available, the derivatization of peptides and proteins with a fluorescent compound or tag is usually performed prior to fluorescence detection.

6.2.2 Derivatization with fluorescent tags

Peptides and proteins are usually derivatized with amine-reactive fluorogenic compounds. Pre-capillary derivatization has been widely used for the analysis of peptides and proteins in CE, but the electrophoretic properties of the peptides and proteins are changed by derivatization. Post-capillary derivatization is usually not practical for CE because it requires a post-capillary reactor (57, 102) and the detection limits are not as low as those obtained using pre-capillary approaches (103). In summary, post-capillary derivatization is useful only when the preformed derivatives are not stable, or a protein has multiple derivatization sites that may produce a mixture of peaks due to incomplete derivatization of that protein (57). A useful procedure, detailed in *Protocol 8*, entails pre-capillary derivatization of proteins and peptides with fluorescamine (59, 104).

Protocol 8. Derivatization of peptides and proteins with fluorescamine[a]

Reagents
- Derivatization reagent (fluorescamine, Fluram® solution): 3 mg/ml fluorescamine in acetone containing 20 μl pyridine (Fluka)
- 0.1 M sodium borate buffer, either pH 8–9 (peptides) or pH 9–9.5 (proteins)

4: Capillary electrophoresis of peptides and proteins

> *Method*
> 1. Prepare a sample solution by dissolving the sample (approx. 0.1 mg) in 100 µl of the appropriate 0.1 M sodium borate buffer.
> 2. Measure 70 µl of the sample solution into a 500 µl microcentrifuge tube, then add 30 µl of the derivatization reagent to the tube.
> 3. Vortex continuously and vigorously for 2 min.
> 4. Transfer the solution into a CE sample vial for analysis (first, centrifuge or filter the sample).
>
> [a] Modified from refs 59 and 104.

Fluorescamine (Fluram®) is an initially non-fluorescent reagent that reacts readily with primary amino groups in peptides and proteins to give stable derivatives which fluoresce following excitation at 390 nm. A list of other derivatization methods commonly used so that peptides and proteins may be detected via fluorescence during CE separations is presented in *Table 7*.

6.3 CE-mass spectrometry

Mass spectrometric (MS) detection has long been the favoured method for confirming the identity of peaks separated by GC and HPLC. With the high

Table 7. Fluorescent derivatization methods for peptides and proteins

Derivatizing reagent	Sample	λ_{ex} (nm)	λ_{em} (nm)	Reference
Tetramethylrhodamine-iodoacetamide	Antibody fragment	488	580	71
CBQCA[a]	Peptides	442	550	40, 46
o-Phthalic dicarboxaldehyde	Peptides Proteins	350	400 LP[b]	95, 102
Fluorescamine (Fluram®)	Peptides Proteins	390	480	95, 102
FITC[c]	Proteins	490	519	102 105
FMOC-Cl[d]	Peptides	260	305 LP[b]	102
Benzoin	Arginine-containing peptides	327	440	38
1. CHCl₃/OH⁻ 2. 4-Methoxy-1,2-phenylenediamine	Tyrosine-containing peptides	330	438	38
AQC[e]	Peptides	250	395	106

[a] 3-(4-carboxybenzoyl)-2-quinolinecarboxaldehyde.
[b] Long pass filter.
[c] Fluorescein isothiocyanate.
[d] 9-Fluorenylmethyl chloroformate.
[e] 6-Amino-quinolyl-N-hydroxysuccinimidyl carbamate.

resolution separation capability of CE, the combination of CE and MS provides a very powerful system for the characterization of mixtures of peptides and proteins (107). CE/fast atom bombardment-mass spectrometry (FAB-MS) (14), CE/plasma desorption-mass spectrometry (PDMS) (24), and CE/electrospray-mass spectrometry (ESI-MS) (4), all provide high resolution separations of analyte mixtures in combination with accurate measurement of their molecular weights. CE/ESI-MS/MS (108) yields both molecular weights and some additional structural information about the peptides and proteins from each CE peak.

When combining a CE instrument with a mass spectrometer, the following requirements must be satisfied:

(a) An interface must couple the capillary with the mass spectrometer, so that the separated sample ions are transferred from the CE running buffer into the high vacuum chamber in a mass spectrometer.
(b) Use of volatile buffers in CE.
(c) Minimal amounts of inorganic salts and detergents in the CE buffer.

Several companies now manufacture interfaces for CE instruments compatible with the different commonly-found types of mass spectrometer. More information about companies which sell interfaces is listed in the suppliers section, they include Analytica of Branford, Bio-Rad, Finnigan, LC packings, Micromass, and PE Sciex.

7. Applications of CE to peptide and protein research

7.1 Evaluation of sample purity

The use of any of the CE modes discussed above to check the purity of a protein or peptide sample has the following advantages over conventional methods:

(a) With the highest efficiency and selectivity of any readily available analytical method, CE may indicate the presence of impurities that cannot be detected by other analytical methods.
(b) A short analysis time (10 min) and a small amount of sample (nl) are needed.
(c) CE operates on an orthogonal separation principle to HPLC and provides additional information not obtained by chromatography alone.
(d) CE avoids the low resolution separations and peptide staining difficulties encountered in slab gel electrophoresis.

7.2 Protein mapping

Protein or peptide mapping (fingerprinting) has been used for the fundamental characterization of proteins. The mapping process involves enzymatic or chemical hydrolysis of a peptide or protein, followed by separation of the small peptide fragments found in the hydrolysate. Analysis of changes in the peak pattern, or map, formed by separation of these small peptides can be used to detect subtle differences between large proteins. Mapping has also allowed a comparison of the protease susceptibility between different conformations of related proteins, and has helped to identify the location and extent of post-translational modifications in proteins. A detailed and practical description of peptide mapping can be found in ref. 109.

As a separation technique, the use of CE has been quite successful for peptide mapping of proteins and glycoproteins (48, 108). The dual advantage of CE is that it generates a peptide map using less sample, and gives a higher resolution of the peptide fragments than competing methods. Also, CE is quite useful for the second-dimension analysis of a peptide map done first by HPLC, because it relies on a different separation principle than HPLC (48). For instance fragment peaks which overlap in the chromatogram may be resolved in the electropherogram.

7.3 Binding constant determinations

Affinity capillary electrophoresis (ACE) is a newly developed technique in CE for studying protein–ligand interactions (110–112). ACE began with the adaptation of the widely used electrophoretic mobility shift assay (EMSA) to CZE. When a protein interacts with other molecules to form a complex, the electrophoretic mobility of the complex may be measurably different from that of the free protein. Based on the differences between the electrophoretic mobilities of the ligand-complexed and free protein, ACE has been used to measure protein–ligand binding constants (111, 113), and to determine the binding stoichiometry (67, 114) of these equilibria. *Table 8* lists examples of

Table 8. Biological interactions studied by affinity capillary electrophoresis (ACE)

Interaction	Sample solution	Reference
r-hGH[a] and anti-hGH fragment	r-hGH and anti-hGH fragment	71
Streptavidin and biotin	Streptavidin and biotin	114
HSA[b] and anti-HSA	HSA and anti-HSA	114
DNA and protein	DNA and protein	115
Proteins and verapamil	Proteins and verapamil	12
Lectin and carbohydrate	Proteins	116
Protein and Ca^{2+}, Zn^{2+}	Proteins	117
Protein and deoxyspergualin	Proteins	118
Peptide and vancomycin	Vancomycin	119
Peptide and vancomycin	Peptide	120
Peptide and liposome	Peptides	121
Antibody and antigen	Antibody and antigen	67, 111

[a] Recombinant human growth hormone.
[b] Human serum albumin.

the types of biomolecular interactions that have been studied through use of ACE. The advantages of ACE include the rapid analysis of protein–ligand interactions using small amounts of samples, and the simultaneous determination of multiple binding constants (110, 112).

7.3.1 ACE case 1: equilibrium in a tightly-binding system

In the case of a high affinity system (having slow dissociation rates), such as is found in many antibody–antigen interactions, the protein–ligand complex will not dissociate during the time of the CE separation (107). The procedure for measurement of such a protein–ligand binding constant is outlined as follows:

(a) Prepare the sample solution by mixing the ligand and protein in the running buffer.
(b) Inject the sample solution (see *Protocol 5*), and run the sample.
(c) Measure the peak areas of the protein–ligand complex, free protein, and free ligand.
(d) Calculate the binding constant K_b.

7.3.2 ACE case 2: equilibrium in a weakly-binding system

In the case of a low affinity system (having fast dissociation rates), the protein–ligand complex is not stable during the CE separation time. The thermodynamic binding constant may still be obtained by measuring changes in the electrophoretic mobility of a protein as a function of the concentration of the ligand present in the running buffer. The data is then analysed via Scatchard analysis (122). The procedure for measurement of such a binding constant is described below:

(a) Prepare a series of solutions having different concentrations of the ligand in the running buffer.
(b) Prepare the sample solution by dissolving the protein and mesityl oxide (a neutral marker) in the running buffer.
(c) Inject the sample solution and electrophorese in the running buffer without any ligand, following *Protocol 5*. Record the relative migration time (t_0) of the protein to mesityl oxide (which migrates at the EOF velocity).
(d) Repeat step (c) using a running buffer containing some concentration of the ligand [L]. Change the buffer after each run, keeping the amount of protein sample used constant. Run until all of the buffers (n) containing a series of concentrations of the ligand [L] have been used. Record the relative migration times (t_n) of the protein in the running buffers containing different concentrations of the ligand [L]. Note the relative migration time of the protein (t_{max}) in the running buffer containing the highest concentration of ligand $[L]_{max}$.
(e) Calculate the R_fs of the protein in the running buffer containing a series of concentrations of the ligand [L], using the equation $R_f = (t_n - t_0)/(t_{max} - t_0)$.

4: Capillary electrophoresis of peptides and proteins

(f) For Scatchard analysis, plot $R_f/[L]$ versus R_f for all buffers. Perform a linear regression analysis of your data: the negative slope is equal to the binding constant, K_b (122).

7.4 Peptide and protein molecular weight determinations

CE can be used for the measurement of the molecular weights of peptides and proteins using CE-ESI-MS or CGE of protein–SDS complexes. As mentioned in Section 6.3, an interface for a mass spectrometer may be purchased for the most accurate determinations of peptide and protein molecular weights. However, CGE of protein–SDS complexes using polymer sieving solutions gives rapid and inexpensive molecular weight determinations using simple equipment, particularly when the $M_r > 100$ kDa. The measurement of molecular weights of polypeptides and proteins by CGE of protein–SDS complexes is described here.

7.4.1 Molecular weights of globular peptides and proteins using SDS-CGE

For normally behaving proteins (proteins which bind 1.4 g SDS/g of protein when saturated), the procedure for molecular weight measurement is relatively simple (93):

(a) Run the protein M_r standard mixture according to *Protocol 7*. The peak of the reference marker, mellitic acid, comes out first (t_{rm}), followed by proteins in the order of increasing mass. Then run your unknown protein sample.

(b) Calculate the relative migration times (R_f) of the standard proteins by dividing the migration time of the reference marker (t_{rm}) by the migration times of the standard proteins $R_f = t_{rm}/t$. Plot log M_r versus R_f for the standard proteins, and perform a linear regression to generate the M_r standard curve.

(c) Calculate the relative migration time (R_f) of any unknown protein(s) by dividing the migration time of reference marker (t_{rm}) by the migration time of the unknown(s), then locate the log M_r of the unknown(s) on the graph using the standard curve generated in step (b).

7.4.2 Molecular weights of glycoproteins

Some proteins, such as glycoproteins, contain chemical moieties that do not bind to SDS, resulting in a change in the constant charge-to-mass ratio of the SDS–protein complex found in normal proteins (123). Consequently, the CGE migration time of these proteins will not behave normally, and an inaccurately higher or lower apparent molecular weight is obtained in Section 7.4.1. The Ferguson method is used to avoid such errors in M_r determinations through the construction of Ferguson plots (99). This procedure is readily performed in CGE using buffers containing viscous polymers.

Ferguson plots may be prepared in two steps: first, measure the migration times of protein M_r standards in different concentrations [P] of the sieving polymer solution, plotting the logarithms of relative migration times (R_f) for the protein M_r standards versus the concentrations of polymer solution [P] used in each run. The slope of each line will be the negative retardation coefficient ($-K_r$) of that protein. Secondly, plotting the logarithm of the molecular weight versus the square root of the retardation coefficients ($K_r^{1/2}$) for the protein M_r standards generates a universal standard curve. This new curve is used for the reliable estimation of the molecular weights of unknown proteins, and this method is not affected by the presence of attached chemical moieties. This procedure for the molecular weight measurement of glycoproteins is modified from refs 93 and 99. To modify *Protocol 7* and prepare Ferguson plots, the following procedure is recommended:

(a) Make 50%, 67%, and 83% solutions of ProSort™ reagent by diluting the ProSort™ reagent with a buffer. Run the protein M_r standard mixture (see *Protocol 7*) using 50%, 67%, 83%, and 100% ProSort™ reagent, respectively. The peak of the reference marker, mellitic acid, elutes first, followed by the standard proteins in the order of increasing mass.

(b) Calculate the relative migration times of the protein M_r standards by dividing the migration time of the reference marker by the migration times of the standard proteins ($R_f = t_{rm}/t$). Plot log (R_f) versus the concentration of ProSort™ reagent for each standard protein, and perform a linear regression to find the slope (equal to $-K_r$) for each standard.

(c) Plot log M_r versus the square roots of K_r for the protein M_r standards. Perform a linear regression to generate universal standard curve.

(d) Perform steps (a), (b), and (c) using your unknown protein(s). Find the K_r of the unknown(s), and then calculate the M_r of the unknown(s) using the universal M_r standard curve generated in step (c).

7.5 Micropreparative CE

The high efficiency of CE enables it to separate peaks that may not be resolved by any other analytical method. The ability of the user to collect these peaks is an important complement to CE analysis. In practical terms, CE is most valuable as a micropreparative technique for peptides and proteins. CE can facilitate protein structure determinations performed by amino acid sequencing, carbohydrate analysis, and mass spectrometry. The preparation of sufficient quantities of a protein using CE for bioassays, NMR, and X-ray crystallographic analysis remains tricky at best. To perform micropreparative CE on a larger scale, there are two major challenges:

(a) The volume of sample injected into the capillary is on the scale of nanolitres, in order to maintain the high resolution. The amount of sample which can be dissolved in such a volume remains small, so a single injection

4: Capillary electrophoresis of peptides and proteins

usually cannot provide enough purified substance for subsequent analyses. Therefore, highly reproducible separations and fraction collection are required to perform multiple sequential collections of the peaks.

(b) To maintain high resolution during fraction collection, a sufficient volume (microlitres) of buffer must be present in collection vials to maintain an electric circuit between the electrodes. Buffer in the outlet reservoir can easily dilute the collected peaks (nanolitres diluted into microlitres).

To date, there is not a generalized collection method suitable for all purposes, although peptides, and glycopeptides (124), as well as proteins (125) have been successfully purified using micropreparative CE. Currently, attempts at fraction collection following CE have adopted the following strategies:

(a) After achieving the separation in the capillary by capillary electrophoresis, the power supply was turned off and the desired peak was collected; it was eluted by pressure rather than by electrophoresis (126).

(b) From the beginning of electrophoresis, the separated peaks were collected continuously onto a wetted and moving membrane suitable for sequencing and mass spectrometry (127, 128).

(c) The desired peaks were collected into vials containing small volumes of running buffer under field-programmed or fixed voltages with designated interrupted collection times (124, 129).

(d) The separated peaks were continuously collected into empty collection vials by using a coaxial capillary interface (used in both CE/MS and CE/HPLC) with an auxiliary sheath flow (125, 130), or a conductive capillary interface (125) to maintain a constant electric field.

8. CE of peptides and proteins: future prospects

Much progress has been made in the last decade toward standardizing the capillary electrophoresis of peptides and proteins. There is no doubt that CE will continue to grow as an important analytical technique in this area. Although CE is still in its developing stages, it will eventually become a routine tool for the separation of peptides and proteins, due to the wealth of electrophoretic methods now used in chemistry, biology, and medicine. The development and improvement of CE techniques in novel areas will be the key to the future success of research in laboratories which perform the analysis of peptides and proteins. These frontier areas include:

(a) Minimization of sample adsorption and creation of new types of capillaries.

(b) The development of UV transparent ampholytes for cIEF.

(c) Performance of electrochromatography (HPLC) inside the capillaries.

(d) Increasingly sensitive detection so that biological fluids may be directly analysed.
(e) New methodologies for micropreparation of larger quantities of protein.

CE-MS will become a major tool of investigators who analyse and screen biological fluids, due to their familiarity with electrophoresis and CE-MS's ability to provide quantitative and structural information in a rapid fashion. The micropreparative aspect of CE will yield enough pure protein for molecular weight determinations, microsequencing, and sensitive bioassays. Rudimentary CE separations have been performed on microchips, so miniaturization of CE instruments may develop in the near future. More on-capillary microreactors will be used to monitor biochemical reactions, and already several laboratories have probed single cells with a capillary to measure metabolite levels *in vivo*. Clearly, convenient small scale, high resolution electrophoresis is an idea whose time has come.

References

1. Låås, T. (1989). In *Protein purification: principles, high resolution methods, and applications* (ed. J.-C. Janson and L. Rydén), p. 349. VCH Publishers, Inc., New York.
2. Karger, B.L. (1989). *Nature*, **339**, 641.
3. Giddings, J.C. (1969). *Sep. Sci.*, **4**, 181.
4. Valaskovic, G.A., Kelleher, N.L., and McLafferty, F.W. (1996). *Science*, **273**, 1199.
5. Widhalm, A., Blaas, C., and Kenndler, E. (1991). *J. Chromatogr.*, **549**, 446.
6. Stover, F.S., Haymore, B.L., and McBeath, R.J. (1989). *J. Chromatogr.*, **470**, 241.
7. Zhu, M., Rodriguez, R., and Wehr, T. (1991). *J. Chromatogr.*, **559**, 479.
8. Grossman, P.D., (1990). *Am. Biotechnol. Lab.*, **8**, 35.
9. Hjertén, S. (1985). *J. Chromatogr.*, **347**, 191.
10. Chiari, M., *et al.* (1995). *J. Chromatogr. A*, **717**, 1.
11. Cobb, K.A., Dolnik, V., and Novotny, M. (1990). *Anal. Chem.*, **62**, 2478.
12. Shibukawa, A., (1994). *J. Pharm. Sci.*, **83**, 616.
13. Nakatani, M., Shibukawa, A., and Nakagawa, T. (1994). *J. Chromatogr. A*, **672**, 213.
14. Moseley, M.A., (1991). *Anal. Chem.*, **63**, 109.
15. Lausch, R., *et al.* (1993). *J. Chromatogr. A*, **654**, 190.
16. McCormick, R.M. (1988). *Anal. Chem.*, **60**, 2322.
17. Bruin, G.J.M., (1989). *J. Chromatogr.*, **471**, 429.
18. Cifuentes, A., *et al.* (1993). *J. Chromatogr. A*, **655**, 63.
19. Schmalzing, D., *et al.* (1993). *J. Chromatogr. A*, **652**, 149.
20. Towns, J.K., Bao, J., and Regnier, F.E. (1992). *J. Chromatogr.*, **599**, 227.
21. Bruin, G.J.M., (1989). *J. Chromatogr.*, **480**, 339.
22. Swedberg, S.A. (1990). *Anal. Biochem.*, **185**, 51.
23. Maa, Y.-F., Hyver, K.J., and Swedberg, S.A. (1991). *J. High Resol. Chromatogr.*, **14**, 65.
24. Weinmann, W., (1994). *Electrophoresis*, **15**, 228.
25. Erim, F.B., *et al.* (1995). *J. Chromatogr. A*, **708**, 356.
26. Iki, N. and Yeung, E.S. (1996). *J. Chromatogr. A*, **731**, 273.
27. Towns, J.K. and Regnier, F.E. (1991). *Anal. Chem.*, **63**, 1126.
28. Yao, X.-W., Wu, D., and Regnier, F.E. (1993). *J. Chromatogr. A*, **636**, 21.
29. Wiktorowicz, J.E. and Colburn, J.C. (1990). *Electrophoresis*, **11**, 769.
30. Kelly, J.F., (1996). *J. Chromatogr. A*, **720**, 409.
31. Towns, J.K. and Regnier, F.E. (1990). *J. Chromatogr.*, **516**, 69.
32. Ng, C.L., Lee, H.K., and Li, S.F.Y. (1994). *J. Chromatogr. A*, **659**, 427.
33. Emmer, Å., Jansson, M., and Roeraade, J. (1991). *J. Chromatogr.*, **547**, 544.

4: Capillary electrophoresis of peptides and proteins

34. Lucy, C.A. and Underhill, R.S. (1996). *Anal. Chem.*, **68**, 300.
35. Gilges, M., *et al.* (1992). *J. High Resol. Chromatogr.*, **15**, 452.
36. Emmer, Å., Jansson, M., and Roeraade, J. (1991). *J. High Resol. Chromatogr.*, **14**, 738.
37. Reubsaet, J.L.E., (1994). *Anal. Biochem.*, **220**, 98.
38. Cobb, K.A. and Novotny, M.V. (1992). *Anal. Biochem.*, **200**, 149.
39. Gaus, H.-J., Beck-Sickinger, A.G., and Bayer, E. (1993). *Anal. Chem.*, **65**, 1399.
40. Liu, J., Cobb, K.A., and Novotny, M. (1990). *J. Chromatogr.*, **519**, 189.
41. Paterson, G.R., Hill, J.P., and Otter, D.E. (1995). *J. Chromatogr. A*, **700**, 105.
42. Tadey, T. and Purdy, W.C. (1993). *J. Chromatogr. A*, **652**, 131.
43. Lindner, H., (1992). *J. Chromatogr.*, **608**, 211.
44. Lindner, H., (1992). *Biochem. J.*, **283**, 467.
45. Mazzeo, J.R. and Krull, I.S. (1991). *Anal. Chem.*, **63**, 2852.
46. Liu, J., (1991). *Anal. Chem.*, **63**, 408.
47. Nielsen, R.G. and Rickard, E.C. (1990). *J. Chromatogr.*, **516**, 99.
48. Rush, R.S., (1993). *Anal. Chem.*, **65**, 1834.
49. Martin, L.M. (1995). In *Peptides: chemistry, structure and biology* (ed. P.T.P. Kaumaya and R.S. Hodges), Vol. 14, p. 144. Mayflower Scientific Ltd., England.
50. Lauer, H.H. and McManigill, D. (1986). *Anal. Chem.*, **58**, 166.
51. Watson, E. and Yao, F. (1993). *Anal. Biochem.*, **210**, 389.
52. Oda, R.P., (1994). *J. Chromatogr. A*, **680**, 85.
53. Oda, R.P., (1994). *J. Chromatogr. A*, **680**, 341.
54. Okafo, G.N., (1994). *Anal. Biochem.*, **219**, 201.
55. Kornfelt, T., (1996). *J. Chromatogr. A*, **726**, 223.
56. Green, J.S. and Jorgenson, J.W. (1989). *J. Chromatogr.*, **478**, 63.
57. Nickerson, B. and Jorgenson, J.W. (1989). *J. Chromatogr.*, **480**, 157.
58. Bushey, M.M. and Jorgenson, J.W. (1989). *J. Chromatogr.*, **480**, 301.
59. Guzman, N.A., (1992). *J. Chromatogr.*, **598**, 123.
60. James, D.C., (1994). *Anal. Biochem.*, **222**, 315.
61. Dawson, R.M.C., *et al.* (1986). *Data for biochemical research*, 3rd edn, p. 418. Oxford University Press, Oxford.
62. Hilser, V.J., Worosila, G.D., and Freire, E. (1993). *Anal. Biochem.*, **208**, 125.
63. Perrett, D., Birch, A., and Ross, G. (1994). *Biochem. Soc. Trans.*, **22**, 127.
64. Chen, F.-T.A. (1991). *J. Chromatogr.*, **559**, 445.
65. Good, N.E., *et al.* (1966). *Biochemistry*, **5**, 467.
66. Zhu, M., (1990). *J. Chromatogr.*, **516**, 123.
67. Martin, L.M. (1996). *J. Chromatogr. B*, **675**, 17.
68. Landers, J.P. (1993). *Trends Biochem. Sci.*, **18**, 409.
69. Gomori, G. (1955). In *Methods in enzymology* (ed. S.P. Colowick and N.O. Kaplan), Vol. 1, p. 138. Academic Press, Inc., New York.
70. See, Y.P. and Jackowski, G. (1989). In *Protein structure: a practical approach* (ed. T.E. Creighton), p. 1. IRL Press, Oxford.
71. Shimura, K. and Karger, B. (1994). *Anal. Chem.*, **66**, 9.
72. Strausbauch, M.A., Landers, J.P., and Wettstein, P.J. (1996). *Anal. Chem.*, **68**, 306.
73. Beattie, J.H., Self, R., and Richards, M.P. (1995). *Electrophoresis*, **16**, 322.
74. Burgi, D.S. and Chien, R.-L. (1991). *Anal. Chem.*, **63**, 2042.
75. Aebersold, R. and Morrison, H.D. (1990). *J. Chromatogr.*, **516**, 79.
76. Chien, R.-L. and Burgi, D.S. (1992). *Anal. Chem.*, **64**, 1046.
77. Vinther, A. and Soeberg, H. (1991). *J. Chromatogr.*, **559**, 27.
78. Wu, S.-L., (1990). *J. Chromatogr.*, **516**, 115.
79. Josic, D., Zeilinger, K., and Reutter, W. (1990). *J. Chromatogr.*, **516**, 89.
80. Deyl, Z., Rohlicek, V., and Adam, M. (1989). *J. Chromatogr.*, **480**, 371.
81. Dawson, J.F., Boland, M.P., and Holmes, C.F.B. (1994). *Anal. Biochem.*, **220**, 340.
82. Stegehuis, D.S., (1991). *J. Chromatogr.*, **538**, 393.
83. McDonnell, T. (1991). *J. Chromatogr.*, **559**, 489.
84. Matsubara, N. and Terabe, S. (1992). *Chromatographia*, **34**, 493.

85. Greve, K.F., Nashabeh, W., and Karger, B.L. (1994). *J. Chromatogr. A*, **680**, 15.
86. Beattie, J.H. and Richards, M.P. (1995). *J. Chromatogr. A*, **700**, 95.
87. Righetti, P.G. (1989). In *Protein structure: a practical approach* (ed. T.E. Creighton), p. 23. IRL Press, Oxford.
88. Chen, S.-M. and Wiktorowicz, J.E. (1992). *Anal. Biochem.*, **206**, 84.
89. Tsuji, K. (1994). *J. Chromatogr. B*, **662**, 291.
90. Cohen, A.S. and Karger, B.L. (1987). *J. Chromatogr.*, **397**, 409.
91. Ganzler, K., (1992). *Anal. Chem.*, **64**, 2665.
92. Heller, C. (1995). *J. Chromatogr. A*, **698**, 19.
93. Werner, W.E., (1993). *Anal. Biochem.*, **212**, 253.
94. Shieh, P.C.H., (1994). *J. Chromatogr. A*, **676**, 219.
95. Gump, E.L. and Monnig, C.A. (1995). *J. Chromatogr. A*, **715**, 167.
96. Simo-Alfonso, E., *et al.* (1995). *J. Chromatogr. A*, **689**, 85.
97. Karim, M.R., Janson, J.-C., and Takagi, T. (1994). *Electrophoresis*, **15**, 1531.
98. Guttman, A., Horvath, J., and Cooke, N. (1993). *Anal. Chem.*, **65**, 199.
99. Benedek, K. and Thiede, S. (1994). *J. Chromatogr. A*, **676**, 209.
100. Wallingford, R.A. and Ewing, A.G. (1989). In *Advances in chromatography* (ed. J.C. Giddings, E. Grushka, and P.R. Brown), Vol. 29, p. 1. Marcel Dekker, Inc., New York.
101. Gill, S.C. and von Hippel, P.H. (1989). *Anal. Biochem.*, **182**, 319.
102. Albin, M., (1991). *Anal. Chem.*, **63**, 417.
103. Szulc, M.E. and Krull, I.S. (1994). *J. Chromatogr. A*, **659**, 231.
104. Findlay, J.B.C. (1987). In *Biological membranes: a practical approach* (ed. J.B.C. Findlay and W.H. Evans), p. 179. IRL Press, Oxford.
105. Zhao, J.Y., (1992). *J. Chromatogr.*, **608**, 239.
106. Antonis, K.M.D., *et al.* (1994). *J. Chromatogr. A*, **661**, 279.
107. Karger, B.L., Chu, Y.-H., and Foret, F. (1995). *Annu. Rev. Biophys. Biomol. Struct.*, **24**, 579.
108. Figeys, D., (1996). *Anal. Chem.*, **68**, 1822.
109. Carrey, E.A. (1989). In *Protein structure: a practical approach* (ed. T.E. Creighton), p. 117. IRL Press, Oxford.
110. Heegard, N.H.H. and Robey, F.A. (1992). *Anal. Chem.*, **64**, 2479.
111. Heegard, N.H.H. (1994). *J. Chromatogr. A*, **680**, 405.
112. Chu, Y.-H., (1995). *Acc. Chem. Res.*, **28**, 461.
113. Martin, L.M., Rotondi, K.S., and Merrifield, R.B. (1994). In *Peptides: chemistry, structure and biology* (ed. R.S. Hodges and J.A. Smith), p. 249. ESCOM, Leiden.
114. Chu, Y.-H., (1994). *Biochemistry*, **33**, 10616.
115. Xian, J., Harrington, M.G., and Davidson, E.H. (1996). *Proc. Natl. Acad. Sci. USA*, **93**, 86.
116. Honda, S., *et al.* (1992). *J. Chromatogr.*, **597**, 377.
117. Kajiwara, H. (1991). *J. Chromatogr.*, **559**, 345.
118. Nadeau, K., (1994). *Biochemistry*, **33**, 2561.
119. Goodall, D.M. (1993). *Biochem. Soc. Trans.*, **21**, 125.
120. Liu, J., (1994). *Anal. Chem.*, **66**, 2412.
121. Zhang, Y., (1995). *Electrophoresis*, **16**, 1519.
122. Avila, L.Z., (1993). *J. Med. Chem.*, **36**, 126.
123. Segrest, J.P. and Jackson, R.L. (1972). In *Methods in enzymology* (ed. S.P. Colowick and N.O. Kaplan), Vol. 28, p. 54. Academic Press, Inc., New York.
124. Boss, H.J., Rhode, M.F., and Rush, R.S. (1995). *Anal. Biochem.*, **230**, 123.
125. Chiu, R.W., (1995). *Anal. Chem.*, **67**, 4190.
126. Camilleri, P., Okafo, G.N., and Southan, C. (1991). *Anal. Biochem.*, **196**, 178.
127. Eriksson, K.-O., Palm, A., and Hjerten, S. (1992). *Anal. Biochem.*, **201**, 211.
128. Cheng, Y.-F., (1992). *J. Chromatogr.*, **608**, 109.
129. Lee, H.G. and Desiderio, D.M. (1994). *J. Chromatogr. A*, **686**, 309.
130. Weinmann, W., (1994). *J. Chromatogr. A*, **680**, 353.

Section B
Selected affinity approaches

5

Lectins as affinity probes

PHILIP S. SHELDON

1. Properties and uses of lectins—a historical perspective

In 1888, Hermann Stillmark described the agglutination of erythrocytes by an extract from castor beans. Soon after this initial observation, the proteins which we now know as lectins, began to be used as tools in biomolecular research (1). Paul Ehrlich carried out pioneering studies in immunology using plant agglutinins and toxins as model antigens. Concanavalin A (Con A) from jackbeans was the first lectin to be purified and crystallized. The observations that Con A precipitated glycogen and that sucrose inhibited haemagglutination led to the proposal that haemagglutination was a consequence of interactions between Con A and carbohydrate on the cell surface. As it emerged that haemagglutination was species- and blood type-specific, the term lectin became used to refer to plant agglutinins which were blood group-specific.

In the early 1950s, understanding of the chemical nature of lectin specificity was advanced when it was shown that *Phaseolus limensis* lectin haemagglutination of type A erythrocytes was inhibited by *N*-acetylgalactosamine (GalNAc) whilst *Lotus tetragonolobus* haemagglutination of type O cells was inhibited by fucose. In the next decade the mitogenic activity of lectins was discovered. The finding that lectins could interact with malignant cells and show a degree of specificity led to the proposition that abnormal glycosylation may be a property associated with the surface of cancerous cells. More recently there has been extensive research focusing on the structure and molecular biology of lectins (2–7). The definition of a lectin now includes all carbohydrate-binding proteins of non-immune origin which agglutinate cells, regardless of origin and blood group specificity (8).

In comparison with the wealth of information on the physico-chemical properties of plant lectins, knowledge of their biological functions is only just beginning to emerge. Evidence that plant lectins can act as defence molecules is reviewed in ref. 9. A number of plant toxins, for example ricin, contain a lectin domain which is combined with a cytotoxic enzymic domain. Plant root lectins may have a role in conferring species specificity in symbiotic interactions with *Rhizobium* bacteria (10). The biological function of animal lectins is reviewed in ref. 11. Example of functions include the hepatic asialoglycoprotein receptor (glycoprotein internalization), the mannose phosphate receptor (part of the mechanism for lysosomal targeting), mannose-binding proteins (involved in defence against pathogens), galectins (associated with higher growth rates and resistance to apoptosis in Jurkat leukaemia T cells) (12), and selectins (recruitment of cells from the circulatory system in response to wounding or infection) (13). This chapter describes the use of lectins as affinity ligands for chromatographic purification of glycans and glycoconjugates.

2. The types of oligosaccharides found in glycoproteins

Lectins can interact with a wide variety of glycoconjugates. However chromatographic applications are usually applied to separation of glycoproteins, glycopeptides, and oligosaccharides. Oligosaccharides are usually attached to proteins either via asparagine side chains (*N*-linked) or serine/threonine side chains (*O*-linked). Linkage can also occur to tyrosine side chains (e.g. in glycogenin) and hydroxyproline (e.g. in collagen and plant arabinogalactans) (14).

2.1 *N*-linked oligosaccharides

N-linked oligosaccharides are found in glycoproteins attached to the side chains of asparagine residues residing within the amino acid sequence motif Asn–X–Ser/Thr. However it has been estimated that only a third of potential *N*-glycosylation sites in this sequence actually are glycosylated. Some typical glycan structures are illustrated in *Figure 1*.

All *N*-glycans possess a common core Manα1,3(Manα1,6)Manβ1,4GlcNAcβ1,4GlcNAcβ1Asn, and the huge variety of *N*-glycans results from different branches which can be attached to the core. The three basic classes of *N*-glycan are: high mannose type (*Figure 1a*), complex type (*Figure 1b–d*), and hybrid type (*Figure 1e*). Complex oligosaccharides can be classified further according to the number of branches they contain and are described as being of biantennary (*Figure 1b*), triantennary (*Figure 1c*), or tetraantennary type (*Figure 1d*). Hybrid and some complex structures may also contain a bisecting *N*-acetylglucosamine residue linked β1,4 to the β-mannose of the core (*Figure 1e, f*). Another potential variation is the presence of fucose in α1,6 (animals) (*Figure 1f–h*) or α1,3 (plants and insects) (*Figure 1i*) linkage to the innermost core GlcNAc. In plants complex oligosaccharides frequently contain xylose but sialic acids are absent (*Figure 1j*) (15–17). Yeast glycans may contain linear chains of ~ 50–100 mannose residues (18).

The assembly of *N*-linked oligosaccharides is reviewed in Kornfeld and

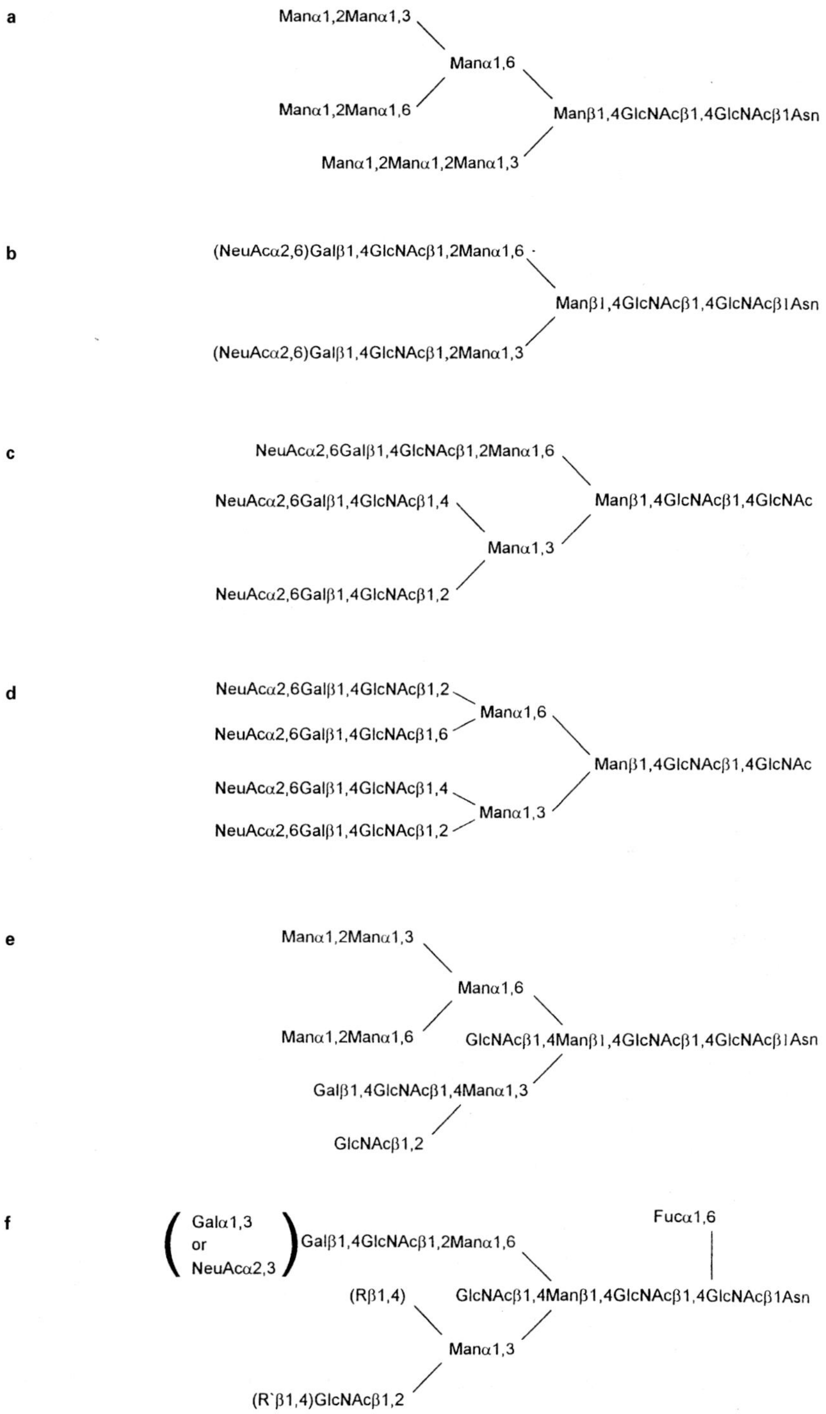

5: Lectins as affinity probes

a

Manα1,2Manα1,3
　　　　　　　　＼
　　　　　　　　　Manα1,6
Manα1,2Manα1,6 ／　　　　＼
　　　　　　　　　　　　　　Manβ1,4GlcNAcβ1,4GlcNAcβ1Asn
Manα1,2Manα1,2Manα1,3 ／

b

(NeuAcα2,6)Galβ1,4GlcNAcβ1,2Manα1,6
　　　　　　　　　　　　　　　　　　＼
　　　　　　　　　　　　　　　　　　　Manβ1,4GlcNAcβ1,4GlcNAcβ1Asn
(NeuAcα2,6)Galβ1,4GlcNAcβ1,2Manα1,3 ／

c

NeuAcα2,6Galβ1,4GlcNAcβ1,2Manα1,6
　　　　　　　　　　　　　　　　　＼
NeuAcα2,6Galβ1,4GlcNAcβ1,4　　　　　Manβ1,4GlcNAcβ1,4GlcNAc
　　　　　　　　　　　　　＼　　　 ／
　　　　　　　　　　　　　　Manα1,3
NeuAcα2,6Galβ1,4GlcNAcβ1,2 ／

d

NeuAcα2,6Galβ1,4GlcNAcβ1,2
　　　　　　　　　　　　　＼
　　　　　　　　　　　　　　Manα1,6
NeuAcα2,6Galβ1,4GlcNAcβ1,6 ／　　　＼
　　　　　　　　　　　　　　　　　　　Manβ1,4GlcNAcβ1,4GlcNAc
NeuAcα2,6Galβ1,4GlcNAcβ1,4
　　　　　　　　　　　　　＼　　　 ／
　　　　　　　　　　　　　　Manα1,3
NeuAcα2,6Galβ1,4GlcNAcβ1,2 ／

e

Manα1,2Manα1,3
　　　　　　　　＼
　　　　　　　　　Manα1,6
Manα1,2Manα1,6 ／　　　＼
　　　　　　　　　GlcNAcβ1,4Manβ1,4GlcNAcβ1,4GlcNAcβ1Asn
Galβ1,4GlcNAcβ1,4Manα1,3 ／
　　　　　　　／
GlcNAcβ1,2

f

$\begin{pmatrix} \text{Galα1,3} \\ \text{or} \\ \text{NeuAcα2,3} \end{pmatrix}$ Galβ1,4GlcNAcβ1,2Manα1,6   　　　Fucα1,6
　　　　　　　　　　　　　　　　　　　　　　＼　　　　　　 |
　　　　　　　　　　　　　(Rβ1,4)　　　GlcNAcβ1,4Manβ1,4GlcNAcβ1,4GlcNAcβ1Asn
　　　　　　　　　　　　　　　＼　　 ／
　　　　　　　　　　　　　　　　Manα1,3
(R'β1,4)GlcNAcβ1,2 ／

g

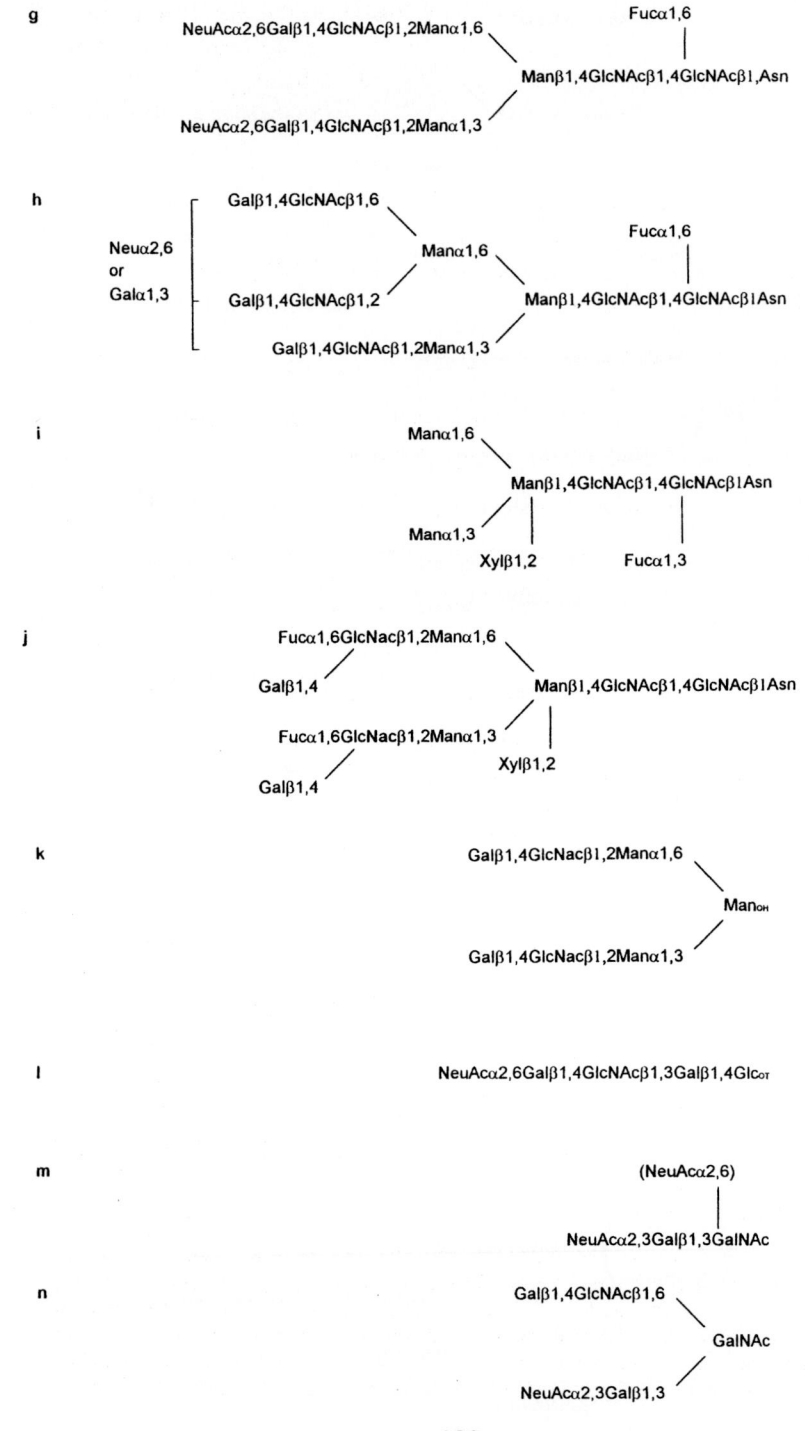

```
NeuAcα2,6Galβ1,4GlcNAcβ1,2Manα1,6                Fucα1,6
                                   \                |
                                    Manβ1,4GlcNAcβ1,4GlcNAcβ1,Asn
                                   /
NeuAcα2,6Galβ1,4GlcNAcβ1,2Manα1,3
```

h

```
              ⎡ Galβ1,4GlcNAcβ1,6
              ⎢                   \
  Neuα2,6     ⎢                    Manα1,6              Fucα1,6
    or        ⎢                           \               |
  Galα1,3     ⎢ Galβ1,4GlcNAcβ1,2           Manβ1,4GlcNAcβ1,4GlcNAcβ1Asn
              ⎢                           /
              ⎣ Galβ1,4GlcNAcβ1,2Manα1,3
```

i

```
              Manα1,6
                    \
                     Manβ1,4GlcNAcβ1,4GlcNAcβ1Asn
                    /           |              |
              Manα1,3         Xylβ1,2        Fucα1,3
```

j

```
         Fucα1,6GlcNacβ1,2Manα1,6
        /                        \
  Galβ1,4                         Manβ1,4GlcNAcβ1,4GlcNAcβ1Asn
                                 /           |
         Fucα1,6GlcNacβ1,2Manα1,3          Xylβ1,2
        /
  Galβ1,4
```

k

```
         Galβ1,4GlcNacβ1,2Manα1,6
                                 \
                                  Man_OH
                                 /
         Galβ1,4GlcNacβ1,2Manα1,3
```

l

NeuAcα2,6Galβ1,4GlcNAcβ1,3Galβ1,4Glc_OT

m

```
                    (NeuAcα2,6)
                         |
              NeuAcα2,3Galβ1,3GalNAc
```

n

```
         Galβ1,4GlcNAcβ1,6
                          \
                           GalNAc
                          /
         NeuAcα2,3Galβ1,3
```

5: Lectins as affinity probes

o
```
              NeuAcα2,3Galβ1,4GlcNAcβ1,6
                                        \
                                         GalNAc
                                        /
                                  Galβ1,3
```

p
```
                   — 3Galβ1,4GlcNAcβ1,6
                                       |
GalNAcα1,3Galβ1,4GlcNAcβ1,[3Galβ14GlcNAcβ1,3Galβ1,4GlcNAcβ1,]2Manα1,6
       |                                                 n        \
     Fucα1,2                                                       Manβ1,4 —
                                                                  /
                                                          — 2Manα1
```

q
```
        Galβ1,4GlcNAcβ1,6
                         \
                          Manα1,6
                         /        \
        Galβ1,4GlcNAcβ1,2           Manβ1,4GlcNAc_OT
                                   /
            Galβ1,4GlcNAcβ1,2Manα1,3
```

r
```
        Galβ1,4GlcNAcβ1,2Manα1,6
                                \
           GlcNAcβ1,4            Manβ1,4GlcNAc_OT
                     \          /
                      Manα1,3
                     /
        Galβ1,4GlcNAcβ1,2
```

s
```
        Galβ1,4GlcNAcβ1,2
                         \
                          GlcNAcβ1,4Man
                         /
        Galβ1,4GlcNAcβ1,2
```

t
```
     NeuAcα2,3Galβ1,4GlcNAcβ1,2Manα1,6
                                      \
     NeuAcα2,6Galβ1,4GlcNAcβ1,2         Manβ1,4GlcNAcβ1,4GlcNAcβ1 Asn
                               \       /
                                Manα1,3
                               /
     NeuAcα2,3Galβ1,4GlcNAcβ1,4
```

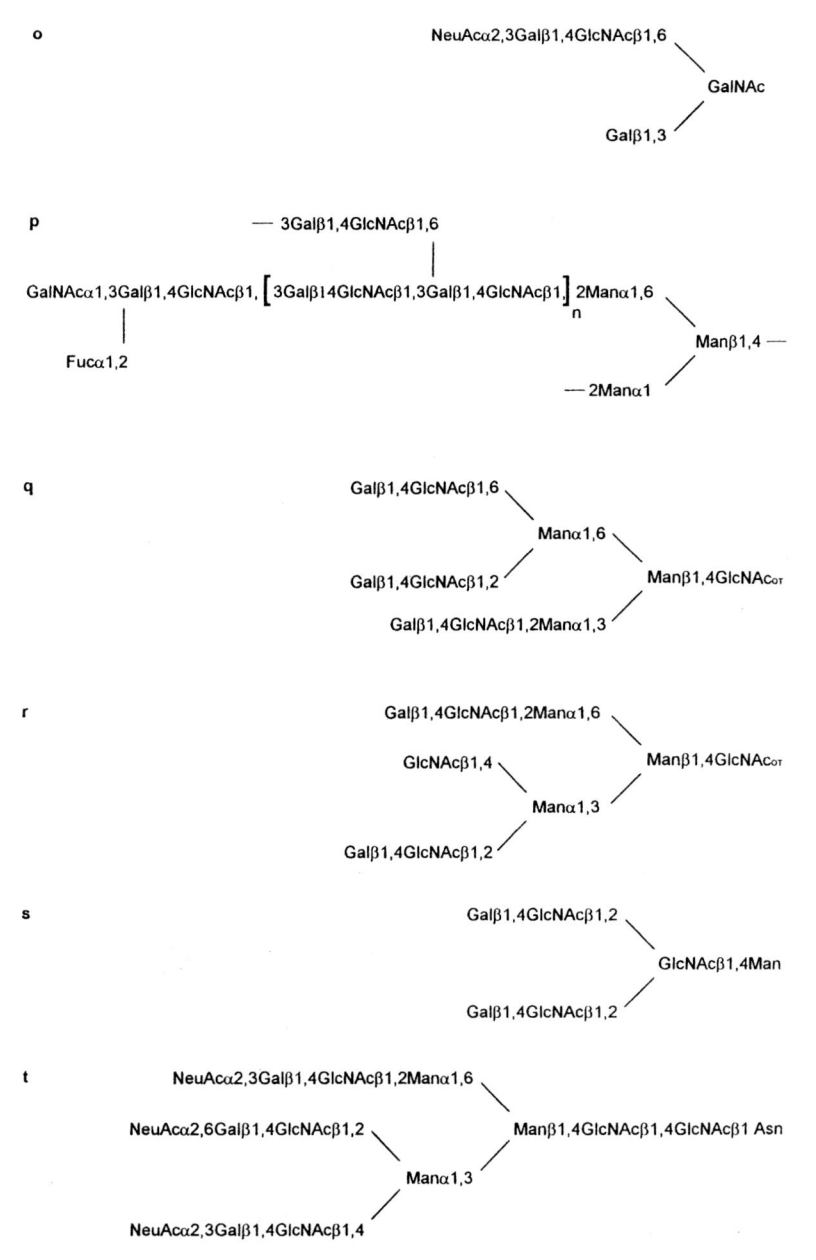

Figure 1. Glycan structures recognized by lectins. Abbreviations used are: Man, D-mannose; GlcNAc, N-acetyl-D-glucosamine; NeuAc, N-acetyl-D-neuraminic acid; Gal, D-galactose; Fuc, L-fucose; Xyl, D-xylose; GalNAc, N-acetyl-D-galactosamine. The linkage between sugars describes the anomeric form of the sugar at the non-reducing side of the bond and the numbered atoms in the two sugars which are linked, e.g. GlcNAcβ1,4Man describes N-acetyl-β-D-glucosamine linked through C1 to C4 of mannose.

Kornfeld (19). Proteins are *N*-glycosylated co-translationally within the endoplasmic reticulum by transfer of $Glc_3Man_9GlcNAc_2$ from the glycolipid $Glc_3Man_9GlcNAc_2$–P–P–dolichol. The first steps of oligosaccharide processing, removal of the three terminal glucose residues and possible removal of one or two α1,2-linked mannose residues can also take place co-translationally within the endoplasmic reticulum. Further processing of the oligosaccharide is dependent on transport to the Golgi. Therefore, proteins which are resident in the endoplasmic reticulum and precursors destined for the secretory pathway which have not passed through the Golgi, can only be expected to possess high mannose structures. As they enter the *cis*, *trans*, and *medial* Golgi, glycoproteins are exposed to sequential arrays of glycosidases and transferases which can convert oligosaccharides into complex and hybrid structures.

In the *cis* Golgi, phosphorylation of mannose residues of specific glycoproteins enables binding to the mannose-6-phosphate receptor (itself a lectin) and transport to the lysosomes. In the *medial* Golgi, addition of GlcNAc to $GlcNAc_2Man_5$ is a key step in committing oligosaccharide processing to the formation of complex and hybrid structures. Addition of sugars such as galactose and sialic acid to the non-reducing termini is thought to take place in the *trans* Golgi. Whilst *N*-linked oligosaccharides have diverse structures, their formation is far from random since the processing enzymes in the endoplasmic reticulum have quite specific structural requirements. Oligosaccharide side chains of glycoproteins have many different roles including protein stabilization, solubilization, modification of activity, intracellular and extracellular recognition (20).

2.2 *O*-linked oligosaccharides (21)

O-linked oligosaccharides, attached to the hydroxyl group of serine or threonine side chains can range in size from a single sugar, GalNAc, to over 20 sugars. The more commonly encountered structural motifs are shown in *Figure 2*. The longer oligosaccharide structures can be thought of as consisting of three sections:

- a core (linked to serine or threonine)
- a backbone consisting of potentially repeated motifs
- non-reducing termini, α-linked sugars which may be antigenic determinants

O-glycan biosynthesis, unlike that of *N*-glycans, does not involve transfer of an oligosaccharide *en bloc* from a lipid precursor. The first step is addition of GalNAc from UDP-*N*-acetylgalactosamine. Whilst *O*-glycans are not associated with a precise amino acid sequence, they tend to occur near proline residues. The subcellular localization of *O*-glycan biosynthesis, whilst probably involving the endoplasmic reticulum and Golgi compartments, has not been precisely defined.

5: Lectins as affinity probes

3. Carbohydrate recognition and lectin specificity

3.1 Abrus precatorius

The seeds contain a toxic lectin and a non-toxic agglutinin (22). Abrin A (APA) recognizes internal β1-linked Gal (*Figure 1k*) sequences and has a high affinity for the Thomsen–Friedenreich antigen (T, Galβ1,3GalNAc). It also recognizes the I/II antigen (Galβ1,3/4GlcNAc) although with less specificity than *Ricinus communis* agglutinin (RCA1) and the E antigen (Galα1,4Gal). Its relatively weak affinity for sialylated glycopeptides is increased on desialylation (23, 24).

3.2 Agaricus bisporus

This mushroom lectin binds to Galβ1,3GalNAcα1Ser/Thr but unlike *Arachis hypogea* lectin, it can recognize sialylated glycans (25–27). It also binds to sulfated lipids, phosphatidic acid, and phosphatidylglycerol (28).

3.3 Aleuria aurantia

This is a lectin (AAL) which binds to fucose residues linked α1,6 to the innermost Glc of the *N*-linked oligosaccharide core. Unlike pea, lentil, and *Vicia faba* lectins, it does not recognize the trimannosyl part of the core and it can also bind weakly to outer fucose residues such as those found in the X-antigenic determinant (*Figure 2*, type 2). The affinity is increased if the oligosaccharide contains two antigens (29). AAL has been used to detect the abnormal occurrence of α1,2-linked fucose in membrane glycoproteins of hepatoma cells (30).

3.4 Allomyrina dichotoma

Two lectins having similar binding properties have been purified from beetle extracts by adsorption onto acid treated Sepharose and elution with lactose. They bind strongly to an *O*-linked oligosaccharide from human milk (*Figure 1l*). Weak binding to *N*-linked oligosaccharides is correlated in strength with increasing numbers of Galβ1,4GlcNAc sequences (31).

3.5 Amaranthus caudatus

This lectin binds to the T antigen, Galβ1,3GalNAcα1Ser and has a specificity distinguished from that of *Arachis hypogea* agglutinin by the ability to tolerate substitution at C3 of Gal in addition to C6 of GalNAc (32).

3.6 Anguilla anguilla

This eel lectin is blood group O-specific and recognizes fucose in the sequence Fucα1,2Gal (33, 34).

Core Classes:

1: Galβ1,3GalNAcα1Ser/Thr

2: GlcNAcβ1,6GalNAcα1Ser/Thr
  |
  Galβ1,3

3: GlcNAcβ1,3GalNAcα1Ser/Thr

4: GlcNAcβ1,6GalNAcα1Ser/Thr
  |
  GlcNAcβ1,3

5: GalNAcα1,3GalNAcα1Ser/Thr

6: GlcNAcα1,6GalNAcα1Ser/Thr

Backbone classes:

1: Galβ1,3GlcNAcβ1- (Type 1)

2: Galβ1,4GlcNAcβ1- (Type 2)

3: (Galβ1,4GlcNAcβ1,3-)$_n$

Repeating type 2 disaccharide, polylactosaminoglycan, recognised by anti blood group i antibodies.

4:
Galβ1,4GlcNAcβ1,6
            \
             Galβ1-
            /
Galβ1,4GlcNAcβ1,3

(can be blood group I antigen)

5: GlcNAcβ1,6Galβ1-

(found only in human glycoproteins to date)

Non reducing terminal antigenic determinants

Type 1

Fucα1,2Galβ1,3GlcNAc-   Blood group H, type 1

GalNAcα1,3Galβ1,3GlcNAc-
  |
  Fucα1,2
                         Blood group A, type 1

Galα1,3Galβ1,3GlcNAc-
  |
  Fucα1,2
                         Blood group B, type 1

Fucα1,4GlcNAc-
  |
  Galβ1,3
                         Blood group Lewis*, Le* type 1

Fucα1,4GlcNAc
  |
  Fucα1,2Galβ1,3
                         Blood group Lewis^b, Le^b type 1

5: Lectins as affinity probes

Type 2

Fucα1,2Galβ1,4GlcNAc- Blood group H, type 2

GalNAcα1,3Galβ1,4GlcNAc- Blood group A, type 2
|
Fucα1,2

Galα1,3Galβ1,4GlcNAc- Blood group B, type 2
|
Fucα1,2

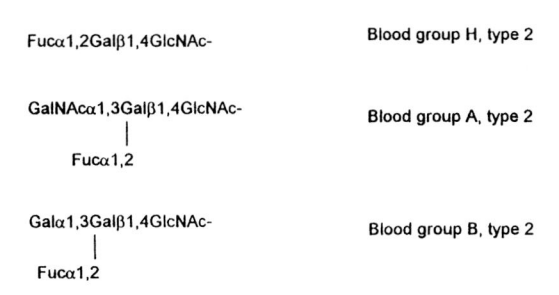

Non reducing terminal antigenic determinants (continued)

Type 2 (continued)

Galβ1,4GlcNAc- Determinant X*(Le*), Stage
| specific embryonic antigen 1
Fucα1,3   (SSEA-1)

Fucα1,2Galβ1,4GlcNAc- Determinant Y (Le^y)
|
Fucα1,3

Polylactosaminoglycans

Galβ1,4GlcNAcβ1,3Galβ1,4GlcNAcβ1,3- Blood group i

Galβ1,4GlcNAcβ1,6   blood group I
            \
             Galβ1,4GlcNAc-
            /
Galβ1,4GlcNAcβ1,3

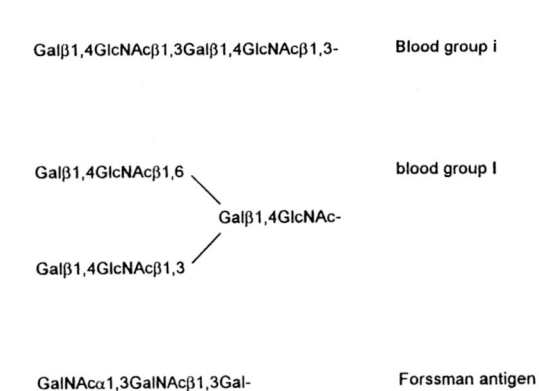

GalNAcα1,3GalNAcβ1,3Gal- Forssman antigen

Figure 2. *O*-glycan structures. As described in the text, *O*-glycans can vary from very simple to complex repeated structures. The figure shows the main motifs which are found. The same convention is used to describe the structure as is in *Figure 1*.

127

3.7 Arachis hypogea
Peanut agglutinin (PNA) recognizes primarily the sequence Galβ1,3Gal-NAcSer/Thr (27, 35, 36). It does not recognize α2,3 sialylated glycans but can bind glycans in which sialic acid is α2,6-linked (37, 38).

3.8 Artocarpus integrifolia
Jacalin is a lectin which binds to Thomsen–Friedenreich (T) antigen, (NeuAcα2,3)Galβ1,3GalNAc, and is used to isolate *O*-glycan containing structures (*Figure 1m–o*) such as those found in IgA (serum and secretory forms), plasminogen, protein z, and insulin-like growth factor (39–43). Sialylation does not appear to have a significant effect on the lectin affinity. It has been shown to interact with a P selectin ligand (44). Recently another lectin with a high affinity for a trimannosyl core containing Xylβ1,2-linked has been isolated from jackfruit seeds (45).

3.9 Bauhinia pururea
This agglutinin has similar specificity to *Arachis hypogea* agglutinin (27, 46, 47).

3.10 Canavalia einsiformis
Concanavalin A (Con A) is one of the most widely used and characterized lectins. It binds to α-mannose and α-glucose residues which are unsubstituted at the C3, C4, and C6 positions (48) in polymers such as starch. It has a very high affinity for the trimannosyl core of *N*-linked high mannose and complex biantennary structures (*Figure 1a, b*) (49–53). The presence of fucose linked α1,6 to the innermost βGlcNAc residue has little effect on affinity. Nor do sialylation of complex structures (51) and phosphorylation of high mannose structures (54). The presence of a bisecting GlcNAc residue on a complex biantennary structure (*Figure 1f*) significantly lowers the affinity (51, 55). Since so many glycan structures will bind to it, immobilized Con A is used to purify a wide range of proteins (56–60). Divalent derivatives, prepared by chemical modification of Con A, have slightly altered binding characteristics (61, 62). Weakly bound (complex biantennary) components are eluted from immobilized Con A with 10 mM methyl α-glucoside and those which are strongly bound (high mannose and hybrid types) are eluted with up to 0.5 M methyl α-mannoside which may be warmed to 60°C (63).

3.11 Codium fragile
The lectin from this alga recognizes GalNAcα1,3(Fucα1,2)Gal (A_h structure) and Galβ1,3GalNAc (T antigen) (64).

5: Lectins as affinity probes

3.12 Datura stramonium
This agglutinin (DSA) binds oligosaccharides containing poly N-acetyllactosamine [Galβ1,4GlcNAcβ1,3]$_n$ repeats (*Figure 1p*). Triantennary complex structures containing outer mannose residues substituted by N-acetyllactosamine have variable affinity depending on the branching pattern. Structure q (*Figure 1*) is strongly bound whereas r (*Figure 1*) is merely retarded on an affinity column. A tetraantennary structure (*Figure 1e*) is strongly bound. Fucosylation and sialylation of the outer branches strongly reduces the affinity. A long thin column is used to reduce the amount of gel and hapten eluent required (63). Elution from the column is carried out using chitobiose, chitotriose, and chitotetraose. The presence of fucose α1,6-linked to the core or bisecting GlcNAc have little effect on affinity (52, 65, 66).

3.13 Dolichos biflorus
This lectin binds to terminal N-acetyllactosamine, for example in blood group A (67–69).

3.14 Erythrina species
E. corralodendron lectin binds N-acetyllactosamine and has been used to identify N-acetyllactosamine terminated glycosphingolipids (24, 70–72). The presence of fucose α1,2-linked either to the galactose or GlcNAc of N-acetyllactosamine has no effect on binding since the lectin binds to the H determinant of type 2 chain (*Figure 2*, type 2). *E. crystagalli* lectin has been used to remove glycosylated CHIP 28 water channels from enzymatically deglycosylated CHIP 28 (73).

3.15 Galanthus nivalis
Snowdrop lectin binds to non-reducing terminal mannose residues and has greatest affinity for those which are α1,3-linked (74–77). It neutralizes human immunodeficiency virus and binds to virus envelope glycoproteins (78, 79).

3.16 Glycine max
Soybean agglutinin (SBA) has a high affinity for N-acetylgalactosamine which can be in α1-*O* linkage to serine or threonine (27, 52). It can also bind to galactose in β1,4 linkage in N-linked glycans (*Figure 1s*) (24).

3.17 Griffonia (Bandeiraea) simplicifolia
The seed contains several lectins. GS1 is made from two polypeptide subunits having different binding specificities which associate to form either single-species or hybrid tetramers (80). The A$_4$ species recognizes αGalNAc (Tn antigen) and GalNAcα1,3GalNAcβ (Forssman antigen). Sialylation and

fucosylation abolish or reduce the affinity of the lectin for oligosaccharide (81, 82). The B_4 recognizes Galα1,4Gal (blood group B) and galabiose (83). A long thin (0.3 × 14 cm) column is used to resolve weakly interacting oligosaccharide. Raffinose (0.1–10 mM) can be used as an eluent (63). GS-II recognizes non-reducing GlcNAc and its oligomers (84). GS-IV has a specificity for difucosyl oligosaccharides such as the Le^b antigen (*Figure 2*, type 1) (85, 86).

3.18 Helix pomatia
This agglutinin binds to *N*-acetylgalactosamine and has been used to purify a hemomucin from *Drosophila* (87), a lymphocyte cell surface protein (88), and a poly *N*-acetyllactosamine containing cell surface glycoprotein from mouse teratocarcinoma cells (89). *Helix pomatia* agglutinin will also bind an *O*-glycosylated form of human laccase and separate it from non-*O*-glycosylated glycoforms (90).

3.19 Lens culinaris
Lentil lectin has a similar specificity to *Pisum sativum* lectin (see below). Uses include the separation of glycoforms of thymocyte Thy 1 protein (91), purification of a mutant insulin receptor from COS 7 cells (92), and physophilin, a synaptic vesicle-binding protein (93).

3.20 Limax flavus
This lectin reacts with sialic acid regardless of the type of linkage (94, 95).

3.21 Lotus tetragonolobus
The lectin from winged pea is blood group O-specific, recognizing α-linked fucose residues (52, 96).

3.22 Lycopersicon esculentum
Tomato lectin has a high affinity for oligosaccharides containing three or more linear repeats of [Galβ1,4GlcNAcβ1] (64) and has been used to purify H$^+$/K$^+$ ATPase from porcine gastric cells (97, 98). Chitotriose may be used as an eluent.

3.23 Maackia amurensis
The seeds contain two lectins, a haemagglutinin and a leukoagglutinin (95, 99). The leukoagglutinin has an affinity for α2,3-linked sialic acid in the structural determinant NeuAcα2,3Galβ1,4GlcNAc found in bovine glycopeptide 1 and laminin (*Figure 1t*). The simple sugars with highest hapten activity are *N*-acetyllactosamine and lactose. The lectin also interacts with *O*-glycans from human glycophorin (100). A long column (0.6 × 17 cm) may be used to

resolve weakly interacting components. Elution is carried out with 100 mM lactose.

3.24 Phaseolus limensis
Lima bean lectin recognizes the blood group A trisaccharide (*Figure 2*, types 1 and 2) (101). It contains a cysteine thiol which must be in reduced form for activity (102, 103).

3.25 Phaseolus vulgaris
Red kidney bean lectins (PHA) are tetrameric proteins composed of varying proportions of two subunits having differing specificities, E (erythroagglutinatin) and L (leukoagglutin). The E_4-PHA species binds to non-sialylated complex bisected biantennary structures (*Figure 1f*) (64, 104). The equivalent non-bisected structure is merely retarded as it passes through the column at room temperature but binding is stronger at lower temperature (105). A long thin column is usually used to resolve weakly interacting components. Elution from the matrix may be carried out with 0.4 M *N*-acetylgalactosamine or 50 mM glycine pH 3.5. Outer chain fucosylation can prevent binding (106). E_4-PHA has been used to isolate cell surface glycoproteins from fibroblasts (107).

The L_4-PHA species retards tri- and tetraantennary Asn-linked oligosaccharides containing out α-mannose residues substituted at C2 and C6 by Galβ1,4GalNAcβ1 such as those found in thyroglobulin (*Figure 1h*) (64, 104, 108). Removal of sialic acid or the α1,6-linked core fucose has no effect on binding but subsequent removal of galactose abolishes the interaction. L_4-PHA agarose is used in long columns to resolve retarded oligosaccharides from those which do not interact with the matrix and if necessary GalNAc is used to elute strongly interacting components.

3.26 Phytolacca americana
Pokeweed contains a number of lectins (109). Pa-1, Pa-2, and Pa-4 bind oligosaccharides containing poly *N*-acetyllactosamine from erythrocyte band 3 glycoprotein (*Figure 1p*) (110).

3.27 Pisum sativum
Pea lectin binds to the fucosylated trimannosyl core of high mannose, complex biantennary, and certain complex triantennary structures (*Figure 1g*) (111–113). A complex bisected biantennary structure has reduced affinity and is retarded as it passes through the column. The requirement for α1,6 fucose to the innermost GlcNAc of the core distinguishes the specificity of pea lectin from that of concanavalin A. Weakly binding oligosaccharides may be eluted with 10 mM methyl α-glucoside whilst those which are strongly bound may be

eluted with up to 0.5 M methyl α-mannoside. The lectins from *Lens culinaris* and *Vicia faba* have similar specificities which may differ in their requirements for attachment to asparagine and exposure of terminal mannose residues at the non-reducing termini (52, 64).

3.28 Ricinus communis

Castor bean seeds contain two galactose binding proteins, RCA-I, a haemo-agglutinin with subunit composition A_2B_2 and RCA-II (ricin), a highly potent non-agglutinating toxin which must be handled with utmost caution, with subunit composition A'B' (114). Although the two proteins share extensive sequence similarities, they have slightly differing carbohydrate binding specificities. RCA-I binding is directed mainly towards the sequence NeuAcα-2,6Galβ1,4GlcNAc in which the galactose residue is unsubstituted at C2, C3, and C4 (*Figure 1u*). The affinity is lower if sialic acid is in α2,3 linkage and is higher if sialic acid is removed (52, 114). Whilst the two proteins have indistinguishable affinities for complex triantennary structures, RCA-I binds more tightly to biantennary structures than RCA-II. The presence of a bisecting *N*-acetylglucosamine can affect the strength of the interaction (55). RCA-I can also bind weakly to *O*-linked glycans in which case there is an absolute requirement for removal of α2,3-linked sialic acid. Binding is strengthened when two *O*-linked oligosaccharides are in close proximity. The haptenic sugar most commonly used for elution from RCA-I matrices is lactose.

RCA-II has a higher affinity than RCA-I for the sequence NeuAcα-2,3Galβ1,4GlcNAc and for *O*-linked structures. Galβ1,3GalNAc containing structures can be irreversibly bound and are not eluted with lactose (114). RCA-II also has a higher affinity for *N*-acetylgalactosamine than RCA-I.

3.29 Sambucus nigra

The lectin from elder bark has a strong affinity for the sequence NeuAcα-2,6Galβ1,4GlcNAc (115). Affinity is significantly lowered when sialic acid is present in α2,3 linkage. The strongest simple sugar haptens in order of decreasing strength are Galβ1,6Gal > Galβ1,4GlcNAc > GalNAcα1,6Gal > galactose. *N*-acetylneuraminic acid is not a hapten.

3.30 Solanum tuberosum

Potato lectin binds to glycopeptides containing a high density of non-reducing *N*-acetylglucosamine and like wheat germ agglutinin is also able to interact with sialic acid (52).

3.31 Triticum vulgare

Wheat germ agglutinin (WGA) recognizes a variety of structural determinants (52, 116–119). It binds to chitobiose and to larger oligomers of β1,4-linked

N-acetylglucosamine. High mannose, complex biantennary, and triantennary glycans from porcine thyroglobulin did not bind to the affinity matrix. However, a hybrid structure did bind and the essential requirement here was the motif containing a bisecting N-acetylglucosamine, GlcNAcβ1,4Manβ1,4Glc-NAcβ1,4GlcNAcβ1Asn (*Figure 1e*). The presence of fucose linked α1,6 to the innermost core N-acetylglucosamine (*Figure 1f*) can intefere with binding (116). WGA can interact weakly with sialylated oligosaccharides (118) and binds strongly to oligosaccharides containing poly N-acetyllactosamine repeats. Examples of use include isolation of the cystic fibrosis transmembrane conductance regulator (120), purification of a P selectin ligand (44), and a hepatoma protein phosphotyrosine phosphatase (121).

3.32 Ulex europaeus
Gorse seeds contain at least three lectins which have differing specificities (100). Lectin 1 binds to fucose, for example when present in the blood group H oligosaccharide (*Figure 2*, types 1 and 2) (96, 122). It may be used to isolate human von Willebrand factor (123).

3.33 Vicia faba
Fava bean lectin has a similar specificity to that of *Pisum sativum* lectin (see Section 3.27).

3.34 Vicia villosa
Hairy vetch contains several tetrameric lectins made up of varying proportions of A and B subunits. The B_4 species has an affinity for N-acetylgalactosamine which is increased when two of these sugars are in close proximity (27, 124, 125).

3.35 Wisteria floribunda
This lectin interacts principally with N-acetylgalactosamine residues (126).

The lectins detailed above, proteins that have been resolved, and typical elution conditions for these are summarized in *Table 1*.

4. Use of lectins: practical aspects
Before the structures of carbohydrates can be determined, they must first be purified. The wide range of specificities of lectins for particular carbohydrate epitopes makes them ideal tools for purification of oligosaccharides and glycoconjugates.

The formation of complexes between glycolipids and lectins has been utilized in the elucidation of the spatial orientation within the endoplasmic reticulum of dolichol phosphate–oligosaccharide intermediates (127, 128). However, there are few examples to date where lectins have been used for the

Table 1. Summary of lectins, proteins isolated, and elution conditions

Species	Common name	Proteins/glycans isolated	Eluent
Abrus precatorius	Abrin A	Thomsen–Friedenreich antigen (T).	200 mM galactose.
Artocarpus integrifolia	Jacalin	Thomsen–Friedenreich antigen (T). IgA (serum and secretory forms, plasminogen, protein z, insulin-like growth factor.	100 mM methy α-galactoside.
Canavalia ensiformis	Concanavalin A, Con A	Insulin receptor.	10 mM methyl α-glucoside followed by up to 0.5 M methyl α-mannoside.
Datura stramonium		Poly N-acetyllactosamine repeats.	10 mM mixture of chitobiose, chitotriose, and chitotetraose.
Dolichos biflorus		Terminal N-acetyllactosamine (blood group A).	100 mM N-acetylgalactosamine.
Galanthus nivalis	Snowdrop lectin	Human immunodeficiency virus envelope glycoproteins.	200 mM methyl α-mannoside.
Glycine max	Soybean agglutinin		N-acetylgalactosamine, galactose.
Griffonia (Bandeiraea) simplicifolia	GS1 A$_4$	TN and Forssmann antigens.	N-acetylgalactosamine.
Griffonia (Bandeiraea) simplicifolia	GS1 B$_4$	Blood group B, galabiose.	0.1–10 mM raffinose.
Helix pomatia		N-acetylgalactosamine Drosophila hemomucin.	50 mM GalNAc.
Lens culinaris	Lentil	Thy 1, physophilin.	200 mM methyl α-mannoside.
Lotus tetragonolobus	Winged pea lectin	Blood group O.	25 mM fucose.
Lycopersicon esculentum	Tomato lectin	H^+/K^+ ATPase.	10 mM mixture of chitobiose, chitotriose, and chitotetraose.
Maackia amurensis		Bovine glycopeptide 1, laminin.	100 mM N-acetyllactosamine, lactose.
Phaseolus vulgaris	Red kidney bean erythroagglutin, E$_4$	Fibroblast cell surface glycoproteins.	400 mM N-acetylgalactosamine or 50 mM glycine pH 3.5.
Phaseolus vulgaris	Red kidney bean leukoagglutinin, L$_4$	Fibroblast cell surface glycoproteins.	400 mM N-acetylgalactosamine.
Phytolacca americana	Pokeweed	Erythrocyte band-3 glycoprotein.	10 mM chitotriose.
Pisum sativum	Garden pea		200 mM methyl α-mannoside.
Ricinus communis	Castor bean haemoagglutinin	LDL complexes, gastric mucus glycoprotein.	100 mM lactose.
Sambucus nigra	Elder lectin	Fetuin glycopeptides.	100 mM N-acetyllactosamine or lactose.
Triticum vulgare	Wheat germ agglutinin (WGA)	Cystic fibrosis conductance regulator hepatoma protein phosphotyrosine phosphatase.	100 mM GlcNAc.
Ulex europaeus	Gorse lectin 1	Human von Willebrand factor.	25 mM fucose.

5: Lectins as affinity probes

purification of glycolipids. More commonly, they are used for purification of glycoproteins, glycopeptides, and glycans.

4.1 Immobilization of lectins

For use in chromatographic purification lectins are usually coupled to activated matrices to form affinity adsorbents. Two frequently used matrices for immobilization are CNBr–Sepharose 4B (Pharmacia) and Affi-Gel 10 (Bio-Rad). Coupling to CNBr–Sepharose can be expected to occur with higher efficiency (~ 90%) than coupling to Affi-Gel 10 (~ 60%) under conditions which maintain lectin activity. However, the lectin coupled to Affi-Gel 10 is less susceptible to leaching through hydrolysis of the bond between lectin and matrix than when the lectin is coupled using CNBr–Sepharose. This may be a particularly important consideration if chromatography is to be carried out at high pH. Lectins are coupled to these matrices using standard procedures, in the presence, if possible, of haptenic sugars to protect the carbohydrate binding site. The more commonly used lectins are available ready-coupled (e.g. from Sigma, Pharmacia, and Bio-Rad).

Protocol 1. Lectin coupling to CNBr–Sepharose (56, 58, 99, 129)

Reagents
- CNBr-activated Sepharose 4B (Pharmacia)
- Blocking buffer: 0.2 M ethanolamine–HCl pH 8.3, or 0.2 M glycine/Na$^+$ pH 8.3
- Buffer A: 100 mM NaHCO$_3$ pH 8.3, 0.5 M NaCl
- Buffer B: 100 mM sodium acetate pH 4.5, 0.5 M NaCl

Method

1. Resuspend cyanogen bromide–Sepharose in ice-cold 1 mM HCl for 15 min and wash with 20 vol. 1 mM HCl on a sintered glass funnel.

2. Transfer the moist gel to an equal volume of buffer A, haptenic sugar,[a] and lectin (5–10 mg/ml moist gel). Tumble end-over-end overnight[b] at 4°C. Amine buffers such as Tris should *not* be included at this stage.[c]

3. Remove uncoupled lectin by washing the matrix on a sintered glass funnel alternately with coupling buffer (haptenic sugar not required) and buffer B, frequently switching between these two buffers.

4. Block unreacted groups by resuspending the gel in blocking buffer and tumbling overnight[b] at 4°C.

5. Remove the blocking reagent by washing with buffer A. The affinity matrix may then be stored at 4°C, typically in PBS or TBS containing 0.02% (w/v) sodium azide to prevent bacterial growth.

[a] Examples of haptenic sugars are 200 mM methyl α-mannoside for Con A or pea lectin, 200 mM lactose for RCA-1, and 50 mM GalNAc for *Maackia amurensis* agglutinin.
[b] Alternatively tumble for 2 h at room temperature.
[c] A lectin which is unstable under alkaline conditions can be coupled at a lower pH. For example, Con A will aggregate at high pH and may therefore be coupled in sodium phosphate pH 7.

> **Protocol 2.** Lectin coupling to *N*-hydroxysuccinimide-activated matrices (104)
>
> *Reagents*
> - Affi-Gel 10 (Bio-Rad)
> - Buffer A: 100 mM NaHCO$_3$ pH 8.3, 0.5 M NaCl
> - Blocking buffer: 100 mM ethanolamine–HCl pH 8
>
> *Method*
> 1. Quickly wash the slurry of activated matrix, supplied as a suspension in isopropyl alcohol, on a sintered glass funnel with ice-cold water making sure that the gel does not dry out. Immediately add the moist gel to 0.6 vol. of lectin solution in buffer A containing hapten sugar (see *Protocol 1*).[a] Tumble the mixture for 4 h at 4°C.
> 2. Remove uncoupled protein by washing with coupling buffer and block any remaining active groups on the matrix with blocking buffer, tumbling the mixture for 2 h at 4°C.
> 3. Wash the affinity matrix on a sintered glass funnel with storage buffer and store at 4°C in PBS or TBS containing 0.02% (w/v) sodium azide.
>
> [a] Whilst an alkaline buffer such as NaHCO$_3$ pH 8.5–9 is likely to give a high efficiency of coupling, maintenance of lectin activity may be favoured by use of a lower pH buffer, e.g. Mops pH 7 was used for *Allomyrina dichotoma* agglutinin whilst NaHCO$_3$ pH 8 used for PHA. Do not use a buffer such as Tris which contains an amine group.

4.2 Purification of glycoproteins by lectin affinity chromatography

Glycoproteins bind to lectin affinity columns through recognition by the lectins of specific carbohydrate epitopes as described in Section 3. However, the behaviour of a glycoprotein on a lectin affinity column is more difficult to predict than that of a single glycan or glycopeptide. If a glycoprotein is to bind to a lectin affinity matrix, the carbohydrate epitope recognized by the lectin must have sufficient exposure and freedom of movement to allow formation of a lectin–carbohydrate complex. This may not always be the case. On the other hand, clustering of glycans on a polypeptide can increase the avidity of interaction with a lectin (130–132). Sometimes it may be difficult to elute a glycoprotein from a lectin affinity column with a haptenic sugar and harsh conditions such as glycine–HCl pH 2.5 may be required for elution.

4.2.1 Conditions for chromatography

Most of the legume lectins described require divalent cations such as Ca^{2+}, Mn^{2+}, or Mg^{2+} as cofactors and if possible, the appropriate combination of these should be included in chromatography buffers in the concentration range 0.1–1 mM. However, under mild conditions, for example in PBS, the ions remain bound to the lectins and therefore their inclusion is not absolutely necessary. Chelating agents such as EDTA should be avoided at the binding stage. If phosphate buffers are used, divalent ions should not be used at concentrations higher than 0.1 mM in order to avoid precipitate formation. Mn^{2+} should not be allowed to come into contact with alkaline solutions because it is susceptible to oxidation.

Interaction of a glycan with a lectin can usually occur under a wide range of conditions and is unlikely to be greatly affected by ionic strength or pH in the ranges normally required for maintenance of native protein structure. However, non-specific ionic interactions can take place under conditions of very low ionic strength and hydrophobic interactions may occur at very high ionic strength. Typically the glycoprotein is applied to the affinity matrix in PBS or TBS. The column is then extensively washed with the same buffer to remove non-binding components. Glycoproteins which bind to the matrix are eluted with haptenic sugar. If an appropriate haptenic sugar is not available, EDTA may be used as eluent.

4.2.2 Membrane proteins

Most membrane proteins are glycosylated. The glycan chains help maintain the correct orientation of the protein in the membrane lipid bilayer and may also act as ligands for endogenous or exogenous lectins. Lectin affinity chromatography is therefore often a useful technique in membrane protein purification. Detergents may be included in buffers if required to solubilize membrane glycoproteins (44, 57, 58, 87, 89, 93, 131). The stability of selected lectins in various detergents is reviewed by Lotan and Nicholson (133). Glucoside detergents should be avoided in cases where they could interact with the lectins.

4.2.3 Choice of lectin

The selection of an appropriate lectin for use will depend on what is known about a glycoprotein. If the glycoprotein is identified as a particular polypeptide on SDS–PAGE, it can be transferred to nitrocellulose and screened with various lectin probes. A lectin may have been shown to be an interesting histological marker and the objective would then be to identify and purify polypeptides carrying the glycan recognized by the lectin. In this case it would be logical to identify polypeptides containing the glycan recognized by the lectin by two strategies:

(a) Separation of cellular polypeptides by SDS–PAGE followed by affinity blotting using the lectin.

(b) Use of the immobilized lectin to directly purify binding glycoproteins (44, 134).

In some cases there may be no information about the actual nature of glycosylation of a protein. In these circumstances, lectin affinity could be one of a number of possible techniques to be screened in preliminary investigations of purification methods. Consideration should be made to the biosynthesis and subcellular location of the protein of interest. Glycosylation can only take place on polypeptides in which biosynthesis has involved segregation into the endoplasmic reticulum. The formation of hybrid and complex glycans takes place in the *medial* and *trans* Golgi compartments. Therefore proteins residing in the endoplasmic reticulum or precursors which have not passed through the Golgi cannot possess hybrid or complex *N*-linked structures. They must be of the high mannose type. Another modification thought to take place in the *medial* Golgi is α1,6 linkage of fucose to the innermost core GlcNAc. For example, it is possible to use lentil lectin chromatography to separate UDP galactose:ceramide galactosyltransferase (an endoplasmic reticulum protein containing a non-fucosylated high mannose type oligosaccharide) from the glutamate/aspartate transporter (a plasma membrane protein with an oligosaccharide containing α1,6-linked fucose) (131).

As mentioned above, whilst there may be differences in the behaviour towards lectins of a glycoprotein and the glycans derived from it, many principles remain in common between the two in selection of suitable lectins. These principles are outlined in Sections 3 and 5. Some of the most widely used lectins in protein purification are Con A, WGA, RCA-1, and lentil agglutinin. Jacalin is often used in conjunction with *O*-linked glycans. If the protein of interest does bind to any of these lectins, it may then be worthwhile investigating other lectins which have narrower specificity ranges. Whilst lectin affinity chromatography of glycoproteins will often provide a very significant degree of purification, usually it needs to be used in conjunction with other protein purification techniques in order to provide a homogeneous preparation of the protein of interest.

5. Lectin affinity chromatography of oligosaccharides, glycosylasparagines, and glycopeptides

The specificities of the lectins described in Section 2 were generally determined using well characterized glycans. Since the interaction of these ligands with lectins is relatively well understood in terms of the presence of certain structural determinants, lectin affinity chromatography can be a powerful tool for purification of oligosaccharides and glycopeptides which may be used with other techniques such as HPLC, paper chromatography, and electrophoresis. At the same time, it provides suggestions which can be used in conjunction with techniques which directly determine glycan structure (135).

5.1 Cleavage of glycans from glycoproteins

Glycans may be released from glycoproteins by a number of methods:
(a) Protease digestion, if carried out with Pronase will usually give rise to products containing very little of the polypeptide backbone. Glycosylas-

paragines are the most likely products but longer peptides may be produced where several glycosylation sites are clustered into a short region. Indeed one of the functions of clustered glycosylation sites in mucins may be to render them resistant to proteases. The presence of several glycans on a glycopeptide may affect its behaviour towards lectin affinity columns (114).

(b) Oligosaccharides may also be released by digestion with N-glycanase from *Flavobacterium meningosepticum* (PNGase F). Deglycosylation will often occur more efficiently when the glycoprotein substrate has first been denatured. Whilst PNGase F activity can be resistant to low concentrations of SDS, the enzyme is not able to cleave glycans containing fucose α1,3-linked to the innermost core GlcNAc (136, 137). Whilst almond N-glycanase (PNGase A) has broader specificity, its activity is not resistant to denaturants (138). The product of N-glycanase digestion is an oligosaccharide with an amino group at the reducing terminus which is slowly hydrolysed to give a normal reducing terminal. The asparagine side chain to which the glycan was attached is converted to aspartic acid.

(c) Oligosaccharides can be released by chemical methods. Anhydrous hydrazine treatment of a glycoprotein releases de-N-acetylated glycans as their hydrazones. A radiolabel can be introduced either as these are re-N-acetylated or as they are reduced to alditols. This method is commonly used when only a small amount of glycoprotein is available. Alkaline borohydride is usually used to specifically release O-glycans.

Protocol 3. Preparation of glycopeptides and glycosylasparagines by Pronase digestion of glycoproteins (50, 139–141)

Reagents
- Digestion buffer: 100 mM Tris–HCl pH 8, 2 mM CaCl$_2$
- Pronase E (Merck)
- Sephadex G25 (Pharmacia)
- Dowex AG-50W × 2 (200–400 mesh) and BioGel P2 (Bio-Rad)
- [^{14}C]acetic anhydride (Amersham)

Method
1. Dissolve the glycoprotein in digestion buffer to give a 50 mg/ml solution. Add Pronase E at a glycoprotein:Pronase E ratio of 50:1 (w/w) and incubate the mixture at 37°C for 24 h under a small layer of toluene. Adjust to pH 8 with Tris at frequent intervals.
2. Add another portion of Pronase E using a glycoprotein:Pronase E ratio of 100:1 (w/w) and incubate the mixture for a further 24 h. If maximal digestion is required, a third portion of Pronase E can be added and the mixture incubated for a further period, the pH being maintained at 8 by the addition of Tris.

Protocol 3. *Continued*

3. Dry the digest by rotary evaporation or lyophilization, redissolve in a small amount of water, and separate by Sephadex G25 chromatography using a column which has been pre-equilibrated in 7% (v/v) propan-1-ol or 5% (v/v) acetic acid. A 5 × 87 cm gel filtration column should be sufficient to process 20 g glycoprotein starting material. Glycan containing molecules, detected by the phenol–sulfuric acid assay (142), elute near the void volume and may be lyophilized.

4. Optionally, if further purification is desired before lectin affinity chromatography, the glycan containing fraction can be passed through a cation exchange column. A 2 × 30 cm column of AG-50W × 2 of 200–400 mesh which has been previously equilibrated in 1 mM $Na^+/CH_2CO_2^-$ pH 2.6 would be suitable. Glycosylasparagines are mainly contained in the non-binding and retarded fractions. Glycopeptides with additional amino acids may be eluted with 50 mM $Na^+/CH_2CO_2^-$ pH 6.

5. The glycopeptides can be radiolabelled by [^{14}C] or [^{3}H]acetylation (87, 141, 143). To the glycopeptide (10 nmol) dissolved in water, add 0.2 ml 1 M $NaHCO_3$. Add [1-^{14}C]acetic anhydride (370 kBq) and then incubate at room temperature for 30 min.

6. Add a further 0.2 ml 1 M $NaHCO_3$ followed by 0.2 ml 2% (v/v) unlabelled acetic anhydride in acetone. Add 0.2 ml 4 M acetic acid, then remove free radiolabel by three or four repeated cycles of evaporation from 0.1 M pyridine:acetic acid pH 5. This is followed by chromatography on Sephadex G25 or BioGel P2, equilibrated in 7% propan-1-ol. Alternatively the glycopeptide can be purified by paper chromatography in butan-1-ol:acetic acid:water (12:3:5). It is eluted from the origin with water.

Protocol 4. Release of oligosaccharides from *N*-glycosylated glycoproteins by *N*-glycanase digestion (108)

Reagents
- *N*-glycanase from *Flavobacterium meningosepticum* (PNGase F) (Genzyme or Boehringer Mannheim)
- NP-40: 10% (v/v) NP-40 in water
- Borate buffer: 200 mM sodium borate pH 9.8
- Phosphate buffer: 1 M sodium phosphate pH 8.6
- 1,10 phenanthrolene stock: 100 mM 1,10 phenanthrolene in methanol

Method

1. The glycoprotein may be first treated with 2 mg/ml unlabelled $NaBH_4$ in borate buffer, at 30°C for 5 h, to eliminate potential reducing groups. Desalt the treated protein by Sephadex G25 chromatography,

5: Lectins as affinity probes

or dialysis, and if necessary concentrate to a volume of < 5 μl by lyophilization.

2. Resuspend the glycoprotein (20 mg) in 10 ml 0.5% (w/v) SDS and heat to 95°C for 2 min. After cooling, add 4 μl phosphate buffer followed by 2 μl 1,10 phenanthrolene stock, 2.5 μl NP-40, and 1.5 μl (0.375 U) N-glycanase. Incubate the sample at 37°C for 16 h.

3. Desalt the sample by Sephadex G25 gel filtration in a column pre-equilibrated in 7% propan-1-ol, and lyophilize the desalted oligosaccharides. The oligosaccharides are made up to a concentration of 2 mg/ml in borate buffer and can now be radiolabelled by reduction with NaB^{3}H$_4$ (0.7–1.7 Ci/mol) at 30°C for 2–3 h.

4. Desalt the oligosaccharides by gel filtration, pass through a column of Dowex AG-50 × 12 (H$^+$ form) (see *Protocol 3*, step 4), and separate by paper chromatography in butanol:ethanol:water (4:1:1). Elute the labelled oligosaccharides, located at the origin, with water.

Protocol 5. Release of glycans by hydrazinolysis (131, 144)

Reagents
- Anhydrous hydrazine (< 1% water, freshly distilled or from a fresh vial) (Sigma)
- Dowex 50 × 8 (25–50 mesh; H$^+$ form) (Bio-Rad)
- NaB^{3}H$_4$ (Amersham)
- Toluene
- 50 mM NaOH

Method

1. Desalt the glycoprotein by dialysis against a volatile solution (water, 0.1% (v/v) trifluoroacetic acid, 1% (v/v) acetic acid, NH$_4$HCO$_3$, and NH$_4$CH$_3$CO$_2^-$ are acceptable). Lyophilize in an acid washed glass vessel.

2. Add anhydrous hydrazine at a protein:hydrazine ratio of 5–25 mg/ml. A larger excess of hydrazine is acceptable. Seal the reaction vessel.

3. For cleavage of both *N*- and *O*- linked glycans, incubate the mixture at 95°C for 4 h. For release of *O*-linked glycans only, incubate at 60°C for 5 h.

4. Remove excess hydrazine using a high vacuum pump fitted with a charcoal/alumina trap, with the sample maintained at > 25°C. Remove the final trace of hydrazine by the addition and evaporation of toluene.

5. The glycan is purified by gel filtration and *N*-acetylated as described in *Protocol 3*.

6. Dissolve the glycan at 1 mg/ml in 50 mM NaOH and carry out re-

Protocol 5. *Continued*

    duction with NaBH$_4$ using 5 mg/mg sugar for 16 h at 20°C. Stop the reaction by the addition of Dowex 50 × 8 resin (25–50 mesh; H$^+$ form).

7. After filtration and washing of the Dowex resin with water, lyophilize the combined filtrates. Remove boric acid by repeated addition and evaporation of methanol under vacuum.
8. The glycan may be purified by gel filtration as described in *Protocol 3*.

Protocol 6. Reductive cleavage of *O*-glycosidic linkages (140)

Reagents
- Sephadex G50 fine (Pharmacia)
- BioGel P4 (Bio-Rad)
- 0.1 M NaOH
- Borohydride solution: 0.1 M NaOH in 2 M NaBH$_4$
- 50% (v/v) acetic acid

Method

1. Adjust the glycoprotein solution (10–20 mg/ml) to pH 10 with 0.1 M NaOH and add an equal volume of cold freshly prepared borohydride solution.
2. Incubate at 45°C for 16 h.
3. Neutralize the cooled solution with 50% acetic acid to pH 6. Separate *O*-glycans (released as oligosaccharide alditols) from *N*-glyco-oligopeptides by gel filtration on a 75 × 2 cm column of Sephadex G50 fine or BioGel P4.

5.2 Purification of oligosaccharides, glycosylasparagines, and glycopeptides by serial lectin chromatography

Glycans and glycopeptides in a complex mixture can be purified by chromatography on a series of lectin columns with differing ranges of specificity. In general, between successive lectin affinity chromatography steps, glycans are separated from haptenic sugars by size exclusion chromatography using columns equilibrated in 7% (v/v) propan-2-ol, followed by lyophilization. This systematic methodology was originally applied to the asparagine-linked oligosaccharides from a mouse lymphoma cell line (145). It is largely based on the principle of successively selecting progressively more specific structural determinants, starting from the core structure and working out through the branches. The original method is outlined in *Figure 3* whilst the glycans resolved (C-1 to C-8, HM and H-1) are presented in *Figure 4*. The lectins utilized within the separation scheme are given in Section 3.

5: Lectins as affinity probes

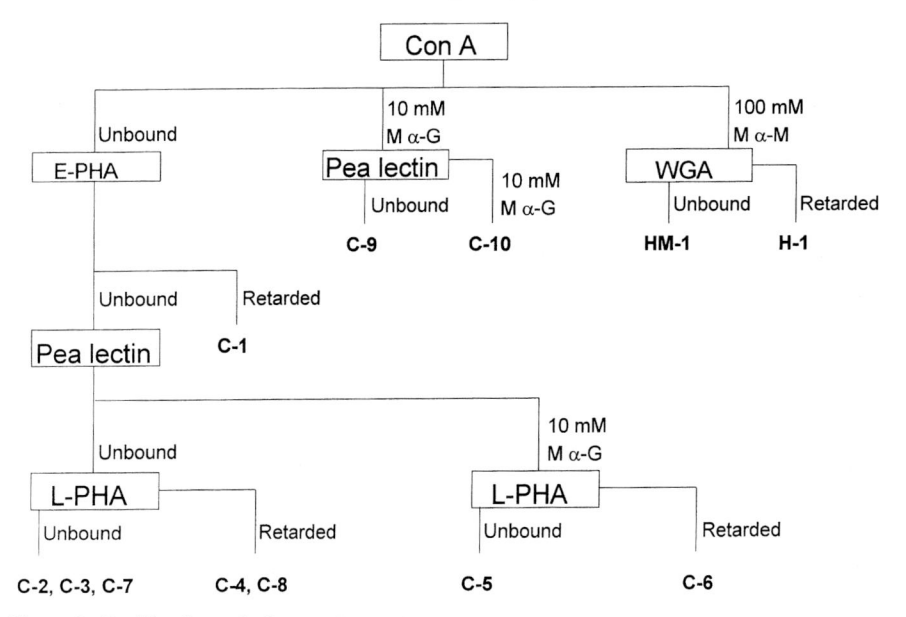

```
                          ┌─────────┐
                          │  Con A  │
                          └─────────┘
         ┌───────────────────┼────────────────────┐
         │              10 mM│               100 mM
    Unbound             M α-G                M α-M
   ┌───────┐          ┌──────────┐          ┌──────┐
   │ E-PHA │          │Pea lectin│   10 mM  │ WGA  │
   └───────┘          └──────────┘   M α-G  └──────┘
      │              Unbound                Unbound   Retarded
      │                C-9         C-10       HM-1      H-1
      │
   ┌──┴──────┐
 Unbound    Retarded
┌──────────┐   C-1
│Pea lectin│
└──────────┘
      │
   ┌──┴─────────────────────────┐
 Unbound                   10 mM
┌───────┐                   M α-G
│ L-PHA │                  ┌───────┐
└───────┘                  │ L-PHA │
 Unbound  Retarded         └───────┘
                            Unbound   Retarded
C-2, C-3, C-7  C-4, C-8      C-5       C-6
```

Figure 3. Purification of glycans by serial lectin chromatography. Asn-linked oligosaccharides, labelled *in vivo*, were released from BW5147 mouse lymphoma cells by Pronase digestion and resolved by sequential lectin chromatography as indicated in the flow chart. The structures of the individual oligosaccharides are shown in *Figure 4*.

Con A Sepharose chromatography separates the Asn-linked oligosaccharides into three fractions:

(a) The 100 mM (60°C) methyl α-mannoside eluate contains high mannose and hybrid structures. These are separated on wheat germ agglutinin Sepharose in which high mannose structures are unbound and hybrid structures are retarded because of the bisecting GlcNAc sugar.

(b) The 10 mM methyl α-glucoside eluate contains non-bisected biantennary complex glycans. Pea lectin Sepharose is used to bind and separate structures containing fucose α1,3-linked to the innermost GlcNAc from those in which it is absent.

(c) The unbound fraction contains triantennary, tetraantennary, and bisected biantennary complex glycans. On E-PHA Sepharose, the bisected biantennary structure (C-1) is retarded and resolved from the other glycans. When these are applied to pea lectin Sepharose, certain triantennary structures (C-5 and C-6) containing fucose α1,3-linked to the innermost GlcNAc are bound. Pea (*Pisum sativum*) and fava (*Vicia faba*) (Section 3) (27, 33) lectins require asparagine attached to the glycans and are suitable for use when the glycans have been released by proteolytic digestion. Lentil (*Lens culinaris*) lectin requires an intact non-reduced inner core GlcNAc and can be used in conjunction with *N*-glycanase released

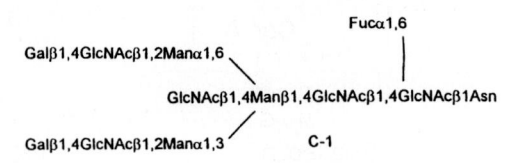

Galβ1,4GlcNAcβ1,2Manα1,6
　　　　　　　　　　　　　　Fucα1,6
　　　　　　　　　　　　　　　|
　　　　　　　GlcNAcβ1,4Manβ1,4GlcNAcβ1,4GlcNAcβ1Asn
Galβ1,4GlcNAcβ1,2Manα1,3　　C-1

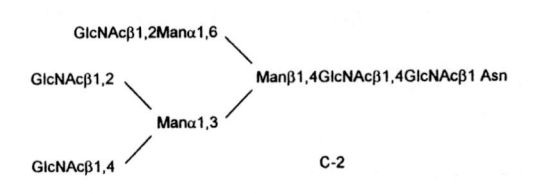

GlcNAcβ1,2Manα1,6
GlcNAcβ1,2　　　　　　Manβ1,4GlcNAcβ1,4GlcNAcβ1 Asn
　　　　Manα1,3
GlcNAcβ1,4　　　　　　C-2

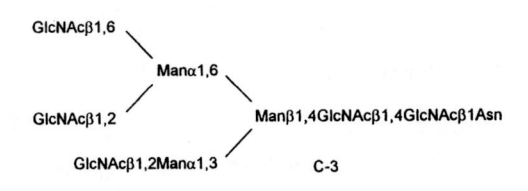

GlcNAcβ1,6
　　　　Manα1,6
GlcNAcβ1,2　　　　　　Manβ1,4GlcNAcβ1,4GlcNAcβ1Asn
　　GlcNAcβ1,2Manα1,3　　C-3

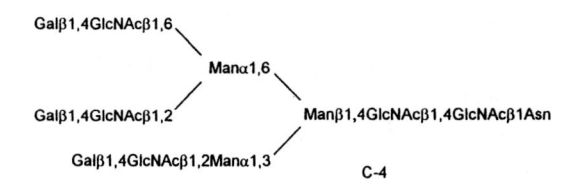

Galβ1,4GlcNAcβ1,6
　　　　Manα1,6
Galβ1,4GlcNAcβ1,2　　　Manβ1,4GlcNAcβ1,4GlcNAcβ1Asn
　Galβ1,4GlcNAcβ1,2Manα1,3
　　　　　　　　　　　C-4

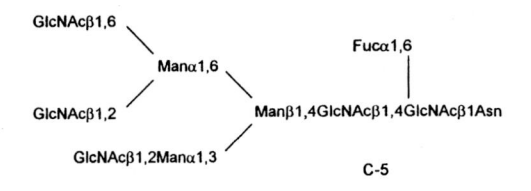

GlcNAcβ1,6
　　　　Manα1,6　　　Fucα1,6
GlcNAcβ1,2　　　　　　|
　　　　　　　　Manβ1,4GlcNAcβ1,4GlcNAcβ1Asn
　GlcNAcβ1,2Manα1,3
　　　　　　　　　　C-5

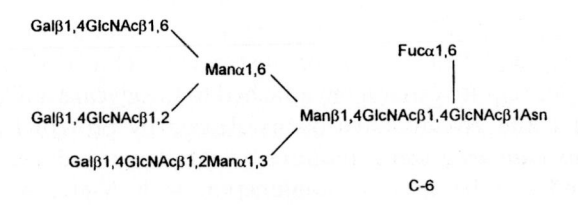

Galβ1,4GlcNAcβ1,6
　　　　　　　　　　　Fucα1,6
　　　　Manα1,6　　　　|
Galβ1,4GlcNAcβ1,2　　Manβ1,4GlcNAcβ1,4GlcNAcβ1Asn
　Galβ1,4GlcNAcβ1,2Manα1,3
　　　　　　　　　　C-6

5: Lectins as affinity probes

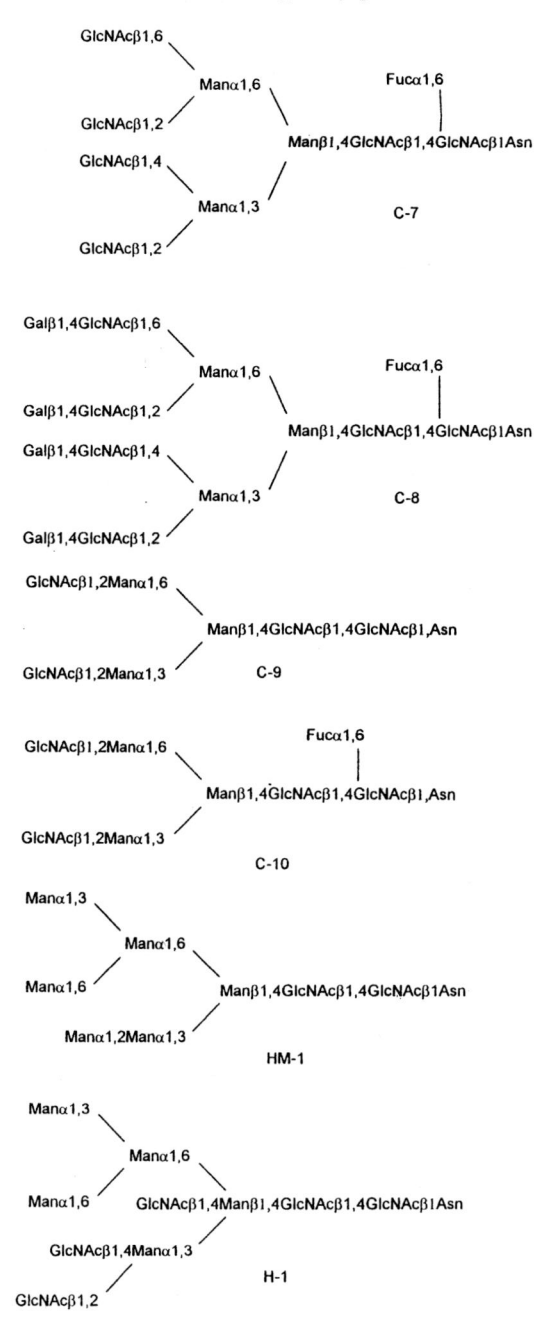

```
GlcNAcβ1,6
         \
          Manα1,6              Fucα1,6
GlcNAcβ1,2/      \                |
GlcNAcβ1,4        Manβ1,4GlcNAcβ1,4GlcNAcβ1Asn
         \       /
          Manα1,3            C-7
         /
GlcNAcβ1,2
```

```
Galβ1,4GlcNAcβ1,6
                \
                 Manα1,6            Fucα1,6
Galβ1,4GlcNAcβ1,2/      \              |
Galβ1,4GlcNAcβ1,4        Manβ1,4GlcNAcβ1,4GlcNAcβ1Asn
                \       /
                 Manα1,3          C-8
                /
Galβ1,4GlcNAcβ1,2
```

```
GlcNAcβ1,2Manα1,6
                \
                 Manβ1,4GlcNAcβ1,4GlcNAcβ1,Asn
                /
GlcNAcβ1,2Manα1,3     C-9
```

```
GlcNAcβ1,2Manα1,6         Fucα1,6
                \            |
                 Manβ1,4GlcNAcβ1,4GlcNAcβ1,Asn
                /
GlcNAcβ1,2Manα1,3
                      C-10
```

```
Manα1,3
       \
        Manα1,6
       /       \
Manα1,6         Manβ1,4GlcNAcβ1,4GlcNAcβ1Asn
               /
Manα1,2Manα1,3
                HM-1
```

```
Manα1,3
       \
        Manα1,6
       /       \
Manα1,6         GlcNAcβ1,4Manβ1,4GlcNAcβ1,4GlcNAcβ1Asn
               /
GlcNAcβ1,4Manα1,3
       /        H-1
GlcNAcβ1,2
```

Figure 4. Structure of glycans purified by serial lectin chromatography. The figure shows the structures of the BW5147 mouse lymphoma glycans purified as indicated in *Figure 3*. The glycan mixture contained complex (C-1, C-2.....C-10), high mannose (HM-1), and hybrid (H-1) structures.

glycans. An alternative to pea, lentil, and fava lectins at this stage is the use of *Aleuria aurantia* agglutinin Sepharose (47). This will also bind certain inner core fucosylated complex triantennary and tetraantennary structures in which branches connecting to the α-linked mannose sugar prevent interaction with pea lectin. *Aleuria aurantia* agglutinin will also interact with structures containing linked fucose in outer non-reducing branches. The final step is chromatography on L-PHA Sepharose which retards triantennary and tetraantennary glycans containing galactose at the non-reducing termini. This step separates C-5 from C-6 and C-2, C-3, and C-7 from C-4 and C-8.

There are various other steps which can be used in a serial lectin fractionation scheme. Poly *N*-acetyllactosamine containing oligosaccharides interact with lectins from *Phytolacca americana* or *Datura stramonium* (47, 64). RCA-1 is used to bind complex oligosaccharides containing galactose on non-reducing branches (123, 146). Other lectins used in fractionation schemes include those from *Griffonia simplicifolia* (GSA-II), *Wisteria floribunda* (123), RCA II, *Vicia villosa* agglutinin (147), and *Ulex europaeus* agglutinin 1 (148).

References

1. Sharon, N. and Lis, H. (1987). *Trends Biochem. Sci.*, **12**, 488.
2. Goldstein, I.J. and Hayes, C.E. (1978). In *Advances in carbohydrate chemistry and biochemistry* (ed. R.S. Tipson and D. Horton), Vol. 35, p. 127. Academic Press, London.
3. Liener, I.E., Sharon, N., and Goldstein, I.J. (ed.) (1986). *The lectins*. Academic Press, London.
4. Rini, J.M. (1995). *Annu. Rev. Biophys. Biomol. Struct.*, **24**, 551.
5. Sharon, N. (1993). *Trends Biochem. Sci.*, **18**, 221.
6. Weiss, W.I. and Drickamer, K.D. (1996). *Annu. Rev. Biochem.*, **65**, 441.
7. Lis, H. and Sharon, N. (1986). *Annu. Rev. Biochem.*, **55**, 35.
8. Sharon, N. and Lis, H. (1972). *Science*, **177**, 949.
9. Chrispeels, M.J. and Raikhel, N.V. (1991). *Plant Cell*, **3**, 1.
10. Diaz, C., Melchers, L.S., Hooykas, P.J.J., Lugtenberg, B.J.J., and Kijne, J.W. (1989). *Nature*, **338**, 589.
11. Drickamer, K. and Taylor, M.E. (1993). *Annu. Rev. Cell Biol.*, **9**, 237.
12. Yang, R.-Y., Hsu, D.K., and Liu, F.-T. (1996). *Proc. Natl. Acad. Sci. USA*, **93**, 6737.
13. Nelson, R.M., Venot, A., Bevilaqua, M.P., Linhardt, R.J., and Stamenkovic, I. (1995). *Annu. Rev. Cell. Dev. Biol.*, **11**, 601.
14. Fincher, G.B., Stone, B.A., and Clarke, A.E. (1983). *Annu. Rev. Plant Physiol.*, **34**, 47.
15. Kaushal, G.P., Szumilo, T., and Elbein, A.D. (1988). In *The biochemistry of plants* (ed. J. Preiss), Vol. 14, p. 421. Academic Press, London.
16. Faye, L., Johnson, K.D., and Chrispeels, M.J. (1989). *Physiol. Plant.*, **75**, 309.
17. Kaushal, G.P. and Elbein, A.D. (1989). In *Methods in enzymology* (ed. V. Ginsburg), Vol. 179, p. 452. Academic Press, London.

18. Kukuruzinska, M.A., Bergh, M.L.E., and Jackson, B.J. (1987). *Annu. Rev. Biochem.*, **56**, 915.
19. Kornfeld, R. and Kornfeld, S. (1985). *Annu. Rev. Biochem.*, **54**, 631.
20. Varki, A. (1993). *Glycobiology*, **3**, 97.
21. Schachter, H. and Brockhausen, I. (1992). In *Glycoconjugates* (ed. H.J. Allen and E.C. Kisailus), p. 263. Marcel Dekker Inc., New York.
22. Olsnes, S. (1978). In *Methods in enzymology* (ed. V. Ginsburg), Vol. 50, p. 323. Academic Press, London.
23. Bhattacharyya, L., Haraldsson, M., and Brewer, C.F. (1988). *Biochemistry*, **27**, 1034.
24. Wu, A.M., Lin, S.R., Chin, L.K., Chow, L.P., and Lin, L.Y. (1992). *J. Biol. Chem.*, **267**, 19130.
25. Presant, C.A. and Kornfeld, S. (1972). *J. Biol. Chem.*, **247**, 6937.
26. Sueyoshi, S., Tsuji, T., and Osawa, T. (1985). *Biol. Chem. Hoppe-Seyler*, **366**, 213.
27. Sueyoshi, S., Tsuji, T., and Osawa, T. (1988). *Carbohydr. Res.*, **178**, 213.
28. Rosén, S., Bergström, J., Karlsson, K.-A., and Tunlid, A. (1996). *Eur. J. Biochem.*, **238**, 830.
29. Yamashita, K., Kochibe, N., Ohkura, T., Ueda, I., and Kobata, A. (1985). *J. Biol. Chem.*, **260**, 4688.
30. Nuck, R., Orthen, B., and Reutter, W. (1992). *Eur. J. Biochem.*, **208**, 669.
31. Yamashita, K., Kobata, A., Suzuki, T., and Umetsu, K. (1989). In *Methods in enzymology* (ed. V. Ginsburg), Vol. 179, p. 331. Academic Press, London.
32. Boland, C.R., Chen, Y.-F., Rinderle, S.J., Resav, J.H., Luk, G.D., Lynch, H.T., et al. (1991). *Cancer Res.*, **51**, 657.
33. Kelly, C. (1984). *Biochem. J.*, **220**, 221.
34. Baldus, S.E., Thiele, J., Park, Y.O., Hanisch, F.G., Bara, J., and Fischer, R. (1996). *Glycoconjugate J.*, **13**, 585.
35. Lotan, R. and Sharon, N. (1978). In *Methods in enzymology* (ed. V. Ginsburg), Vol. 50, p. 361. Academic Press, London.
36. Swamy, M.J., Gupta, D., Mahanta, S.K., and Surolia, A. (1991). *Carbohydr. Res.*, **213**, 59.
37. Springer, G.F. and Desai, P.R. (1982). *J. Biol. Chem.*, **257**, 2744.
38. Hemmerich, S., Leffler, H., and Rosen, S.D. (1995). *J. Biol. Chem.*, **270**, 12035.
39. Hortin, G.L. (1990). *Anal. Biochem.*, **191**, 262.
40. Hortin, G.L. and Trimpe, B.L. (1990). *Anal. Biochem.*, **188**, 271.
41. Daughaday, W.H., Trivedi, B., and Baxter, R.C. (1993). *Proc. Natl. Acad. Sci. USA*, **90**, 5823.
42. Maemura, K. and Fukuda, M. (1992). *J. Biol. Chem.*, **267**, 24379.
43. Roque-Barreira, M.C. and Campos-Neto, A. (1985). *J. Immunol.*, **134**, 1740.
44. Norgard, K.E., Moore, K.L., Diaz, S., Stults, L., Ushiyama, S., McEver, R.P., et al. (1993). *J. Biol. Chem.*, **268**, 12764.
45. Misquith, S., Rani, P.G., and Surolia, A. (1994). *J. Biol. Chem.*, **269**, 30393.
46. Osawa, T., Irimura, T., and Kawaguchi, T. (1978). In *Methods in enzymology* (ed. V. Ginsburg), Vol. 50, p. 367. Academic Press, London.
47. Osawa, T. and Tsuji, T. (1987). *Annu. Rev. Biochem.*, **56**, 21.
48. Goldstein, I.J., Hollerman, C.E., and Smith, E.E. (1965). *Biochemistry*, **4**, 876.
49. Krusius, T., Finne, J., and Rauvala, H. (1976). *FEBS Lett.*, **71**, 117.
50. Narasimhan, S., Wilson, J.R., Martin, E., and Schachter, H. (1979). *Can. J. Biochem.*, **57**, 83.

51. Baenziger, J.U. and Fiete, D. (1979). *J. Biol. Chem.*, **254**, 2400.
52. Debray, H., Decout, D., Strecker, G., Spik, G., and Montreuìl, J. (1981). *Eur. J. Biochem.*, **117**, 41.
53. Naismith, J.H. and Field, R.A. (1996). *J. Biol. Chem.*, **271**, 972.
54. Gabel and Kornfeld, S. (1982). *J. Biol. Chem.*, **257**, 10605.
55. Narasimhan, S., Freed, J.C., and Schachter, H. (1986). *Carbohydr. Res.*, **149**, 65.
56. Asberg, K. and Porath, J. (1970). *Acta Chem. Scand.*, **24**, 1839.
57. Steinemann, A. and Stryer, L. (1973). *Biochemistry*, **12**, 1499.
58. Cuatrecasas, P. and Parikh, I. (1974). In *Methods in enzymology* (ed. W.B. Jakoby and M. Wilchek), Vol. 34, p. 653. Academic Press, London.
59. Liener, I.E., Garrison, O.R., and Pravda, Z. (1973). *Biochem. Biophys. Res. Commun.*, **51**, 436.
60. Sølling, H. and Wang, P. (1973). *Biochem. Biophys. Res. Commun.*, **53**, 1234.
61. Young, N.M. (1974). *Biochim. Biophys. Acta*, **336**, 46.
62. Mandal, D.K. and Brewer, C.F. (1993). *Biochemistry*, **32**, 5116.
63. Cummings, R.D. (1994). In *Methods in enzymology* (ed. J. Lennarz and G.W. Hart), Vol. 230, p. 66. Academic Press, London.
64. Wu, A.M., Song, S.-C., Hwang, P.-Y., Wu, J.-H., and Chang, K.S.S. (1995). *Eur. J. Biochem.*, **233**, 145.
65. Merkle, R.K. and Cummings, R.D. (1987). In *Methods in enzymology* (ed. V. Ginsburg), Vol. 138, p. 232. Academic Press, London.
66. Yamashita, K., Totani, K., Ohkura, T., Takasaki, S., Goldstein, I.J., and Kobata, A. (1987). *J. Biol. Chem.*, **262**, 1602.
67. Etzler, M.E. (1981). *J. Biol. Chem.*, **256**, 2367.
68. Etzler, M.E. (1989). In *Methods in enzymology* (ed. V. Ginsburg), Vol. 179, p. 341. Academic Press, London.
69. Etzler, M.E. (1994). *Glycoconjugate J.*, **11**, 395.
70. Lis, H. and Sharon, N. (1987). In *Methods in enzymology* (ed. V. Ginsburg), Vol. 138, p. 544. Academic Press, London.
71. Teneberg, S., Ångström, J., Jovall, P.-Å., and Karlsson, K.A. (1994). *J. Biol. Chem.*, **269**, 8554.
72. Erlich-Rogozinski, S., De Maio, A., Lis, H., and Sharon, N. (1987). *Glycoconjugate J.*, **4**, 379.
73. Van Hoek, A.N., Wiener, M.C., Verbavatz, J.M., Brown, D., Lipniunas, P.H., Townsend, R.R., *et al.* (1995). *Biochemistry*, **34**, 2212.
74. Shibuya, N., Goldstein, I.J., Van Damme, E.J.M., and Peumans, W.J. (1988). *J. Biol. Chem.*, **263**, 728.
75. Kaku, H. and Goldstein, I.J. (1989). In *Methods in enzymology* (ed. V. Ginsburg), Vol. 179, p. 327. Academic Press, London.
76. Vandamme, E.J.M., Allen, A.K., and Peumans, W.J. (1987). *FEBS Lett.*, **215**, 140.
77. Hester, G., Kaku, H., Goldstein, I.J., and Wright, C.S. (1995). *Nature Struct. Biol.*, **2**, 472.
78. Hammar, L., Hirsch, I., Machado, A.A., Demareuil, J., Baillon, J.G., Bolmont, C., *et al.* (1995). *AIDS Res. Hum. Retroviruses*, **11**, 87.
79. Gilljam, G. (1993). *AIDS Res. Hum. Retroviruses*, **9**, 431.
80. Murphy, L.A. and Goldstein, I.J. (1978). In *Methods in enzymology* (ed. V. Ginsburg), Vol. 50, p. 345. Academic Press, London.

81. Chen, Y.-F., Boland, C.R., Kraus, E.R., and Goldstein, I.J. (1994). *Int. J. Cancer*, **57**, 561.
82. Piller, V., Piller, F., and Cartron, J.-P. (1990). *Eur. J. Biochem.*, **191**, 461.
83. Wu, A.M., Song, S.C., Wu, J.H., and Kabat, E.A. (1995). *Biochem. Biophys. Res. Commun.*, **216**, 814.
84. Ebisu, S. and Goldstein, I.J. (1978). In *Methods in enzymology* (ed. V. Ginsburg), Vol. 50, p. 350. Academic Press, London.
85. Shibata, S., Goldstein, I.J., and Baker, D.A. (1982). *J. Biol. Chem.*, **257**, 9324.
86. Kaladas, P.M., Kabat, E.A., Shibata, S., and Goldstein, I.J. (1983). *Arch. Biochem. Biophys.*, **223**, 309.
87. Theopold, U., Samakovlis, C., Erdjument-Bromage, H., Dillon, N., Axelsson, B., Schmidt, O., *et al.* (1996). *J. Biol. Chem.*, **271**, 12708.
88. Axelsson, B., Kimura, A., Hammarström, S., Wigzell, H., Nilsson, K., and Mellstedt, H. (1978). *Eur. J. Immunol.*, **8**, 757.
89. Spillmann, D. and Finne, J. (1994). *Eur. J. Biochem.*, **220**, 385.
90. Naim, H.Y. and Lentze, M.J. (1992). *J. Biol. Chem.*, **267**, 25494.
91. Parekh, R.B., Tse, A.G.D., Dwek, R., Williams, A.F., and Rademacher, T.W. (1987). *EMBO J.*, **6**, 1233.
92. Haruta, T., Sawa, T., Takata, Y., Imamura, T., Takada, Y., Morioka, H., *et al.* (1995). *Biochem. J.*, **305**, 599.
93. Siebert, A., Lottspeich, F., Nelson, N., and Betz, H. (1994). *J. Biol. Chem.*, **269**, 28329.
94. Miller, R.L., Collaun, J.F., and Fish, W.W. (1982). *J. Biol. Chem.*, **257**, 7574.
95. Knibbs, R.N., Goldstein, I.J., Ratcliffe, R.M., and Shibuya, N. (1991). *J. Biol. Chem.*, **266**, 83.
96. Sugii, S., Kabat, E.A., and Baer, H.H. (1982). *Carbohydr. Res.*, **99**, 99.
97. Callaghan, J.M., Toh, B.-H., Simpson, R.J., Baldwin, G., and Gleeson, P.A. (1992). *Biochem. J.*, **283**, 63.
98. Jones, C.M., Toh, B.H., Pettitt, J.M., Martinelli, T.M., Humphris, D.C., Callaghan, J.M., *et al.* (1991). *Eur. J. Biochem.*, **197**, 49.
99. Wang, W.-C. and Cummings, R.D. (1988). *J. Biol. Chem.*, **263**, 4576.
100. Konami, Y., Yamamoto, K., and Osawa, T. (1991). *Biol. Chem. Hoppe-Seyler*, **372**, 95.
101. Jordan, E.T. and Goldstein, I.J. (1995). *Eur. J. Biochem.*, **230**, 958.
102. Roberts, D.D. and Goldstein, I.J. (1984). *J. Biol. Chem.*, **259**, 903.
103. Roberts, D.D. and Goldstein, I.J. (1984). *J. Biol. Chem.*, **259**, 909.
104. Cummings, R.D. and Kornfeld, S. (1982). *J. Biol. Chem.*, **257**, 11230.
105. Kobata, A. and Yamashita, K. (1989). In *Methods in enzymology* (ed. V. Ginsburg), Vol. 179, p. 46. Academic Press, London.
106. Yamashita, K., Hitoi, A., and Kobata, A. (1983). *J. Biol. Chem.*, **258**, 14753.
107. Asaga, H. and Yoshizato, K. (1992). *J. Cell Sci.*, **101**, 625.
108. Green, E.D. and Baenziger, J.U. (1987). *J. Biol. Chem.*, **262**, 12018.
109. Waxdal, M.J. (1978). In *Methods in enzymology* (ed. V. Ginsburg), Vol. 50, p. 354. Academic Press, London.
110. Irimura, T. and Nicolson, G.L. (1983). *Carbohydr. Res.*, **120**, 187.
111. Kornfeld, K., Reitman, M.L., and Kornfeld, R. (1981). *J. Biol. Chem.*, **256**, 6633.
112. Narasimhan, S. (1982). *J. Biol. Chem.*, **257**, 235.

113. Rini, J.M., Hardman, K.D., Einspahr, H., Suddath, F.L., and Carver, J.P. (1993). *J. Biol. Chem.*, **268**, 10126.
114. Baenziger, J.U. and Fiete, D. (1979). *J. Biol. Chem.*, **254**, 9795.
115. Shibuya, N., Goldstein, I.J., Broekoert, W.F., Nsimba-Lubaki, M., Peeters, B., and Peumans, W.J. (1987). *J. Biol. Chem.*, **262**, 1569.
116. Yamamoto, K., Tsuji, T., Matsumoto, I., and Osawa, T. (1981). *Biochemistry*, **20**, 5894.
117. Goldstein, I.J., Hammarström, S., and Sundblad, G. (1975). *Biochim. Biophys. Acta*, **405**, 53.
118. Gallagher, J.T., Morris, A., and Dexter, T.M. (1985). *Biochem. J.*, **231**, 115.
119. Monsigny, M., Roche, A.-C., Sene, C., Maget-Dana, R., and Delmotte, F. (1980). *Eur. J. Biochem.*, **104**, 147.
120. Ostedgaard, L.S. and Welsh, M.J. (1992). *J. Biol. Chem.*, **267**, 26142.
121. Meyerovitch, J., Backer, M.J., Csermely, P., Shoelson, S.E., and Kahn, R.C. (1992). *Biochemistry*, **31**, 10338.
122. Pereira, M.E., Kisailus, E.C., Gruezo, F., and Kabat, E.A. (1978). *Arch. Biochem. Biophys.*, **185**, 108.
123. Matsui, T., Titani, K., and Mizuochi, T. (1992). *J. Biol. Chem.*, **267**, 8723.
124. Tollefsen, S.E. and Kornfeld, R. (1983). *J. Biol. Chem.*, **258**, 5165.
125. Tollefsen, S.E. and Kornfeld, R. (1983). *J. Biol. Chem.*, **258**, 5172.
126. Torres, B.V., McCrumb, D.K., and Smith, D.F. (1988). *Arch. Biochem. Biophys.*, **262**, 1.
127. Snider, M.D. and Robbins, P.W. (1982). *J. Biol. Chem.*, **257**, 6796.
128. Snider, M.D. and Rogers, O.C. (1984). *Cell*, **36**, 753.
129. Lloyd, K.O. (1970). *Arch. Biochem. Biophys.*, **137**, 460.
130. Moore, K.L., Stults, N.L., Diaz, S., Smith, D.F., Cummings, R.D., Varki, A., et al. (1992). *J. Cell Biol.*, **118**, 445.
131. Schulte, S. and Stoffel, W. (1995). *Eur. J. Biochem.*, **233**, 947.
132. Hayes, B.K., Greis, K.D., and Hart, G.W. (1995). *Anal. Biochem.*, **228**, 115.
133. Lotan, R. and Nicolson, G.L. (1979). *Biochim. Biophys. Acta*, **559**, 329.
134. Damjanov, A. and Damjanov, I. (1992). *J. Reprod. Fertil.*, **95**, 679.
135. Dwek, R.A., Edge, C.J., Harvey, D.J., Wormald, M.R., and Parekh, R.B. (1993). *Annu. Rev. Biochem.*, **62**, 65.
136. Chu, F.K. (1986). *J. Biol. Chem.*, **261**, 172.
137. Tretter, V., Altmann, F., and März, L. (1991). *Eur. J. Biochem.*, **199**, 647.
138. Taga, E.M., Waheed, A., and Van Etten, R.L. (1984). *Biochemistry*, **23**, 815.
139. Huang, C.-C., Mayer, H.E., and Montgomery, R. (1970). *Carbohydr. Res.*, **13**, 127.
140. Montreuil, J., Bouquilet, S., Debray, H., Fournet, B., Spik, G., and Strecker, G. (1986). In *Carbohydrate analysis: a practical approach* (ed. M.F. Chaplin and J.F. Kennedy), p. 143. IRL Press, Oxford.
141. Finne, J. and Krusius, T. (1982). In *Methods in enzymology* (ed. V. Ginsburg), Vol. 83, p. 269. Academic Press, London.
142. Dubois, M., Gilles, K.A., Hamilton, J.K., Rebers, P.A., and Smith, F. (1956). *Anal. Chem.*, **28**, 350.
143. Narasimhan, S., Harpaz, N., Longmore, G., Carver, J.P., Grey, A.A., and Schachter, H. (1980). *J. Biol. Chem.*, **255**, 4876.
144. Patel, T.P. and Parekh, R.B. (1994). In *Methods in enzymology* (ed. W.J. Lennarz and G.W. Hart), Vol. 230, p. 57. Academic Press, London.

5: Lectins as affinity probes

145. Cummings, R.D. and Kornfeld, S. (1982). *J. Biol. Chem.*, **257**, 11235.
146. Matsui, T., Mizuochi, T., Titani, K., Okinaga, T., Hoshi, M., Bousfield, G.R., *et al.* (1994). *Biochemistry*, **33**, 14039.
147. Green, E.D., Brodbeck, R.M., and Baenziger, J.U. (1987). *J. Biol. Chem.*, **262**, 12030.
148. Chochula, J., Fabre, C., Bellan, C., Lui, J., Bourgerie, S., Abadie, B., *et al.* (1993). *J. Biol. Chem.*, **268**, 2312.

6

Nucleotide– and dye–ligand chromatography

BO MATTIASSON and IGOR YU. GALAEV

1. Introduction

As the level of competition between the manufacturers of rival biotechnological products increases, downstream processing costs, which can account for up to 80% of the overall production costs, will become more and more critical in determining product competitiveness (1). Since the late 1960s biospecific separation methods have come into focus and have been continuously improved. A fundamental advantage of affinity techniques is their predictive and rational character, since the ligand selected is designed to interact specifically with the protein to be purified. Among main important developments in affinity techniques is the use of more robust adsorbents and coupling chemistry that are not prone to leaking immunogenic degradation products. One such group of robust adsorbents is presented by matrices with nucleotide–ligands or dye–ligands, the latter exhibit a remarkable propensity to bind, sometimes specifically, to a plethora of proteins and enzymes.

The aim of the present chapter is to give an overview of the recent developments in laboratory scale applications of nucleotide– and dye–ligand chromatography of proteins and to give some practical advices in preparation of nucleotide– and dye–ligand containing agarose matrices as well as in tackling some problems encountered during nucleotide– and dye–ligand chromatography.

2. Nucleotide–ligand chromatography

Nearly one-third of the known enzymes require a nucleotide coenzyme. Thus, the idea of coupling nucleotide to beads of cross-linked agarose matrix (the most popular matrix in protein chromatography) and using the latter for affinity purification of nucleotide-binding enzyme is very attractive. The first problem to be tackled is that nucleotides cannot be covalently bound directly to agarose. Since early 1970s many papers were published on different syntheses of nucleotide derivatives and their immobilization on agarose. The first feature to be recognized in ligand coupling is that a correct orientation of ligand to protein ligate has to be found. For instance AMP derivatives immobilized to beaded agarose by the adenine N^6-amino group or the phosphate group display different affinities towards a number of AMP-binding proteins (2). NAD derivatives modified at the N^6-position of the adenine moiety usually interact better with dehydrogenases than derivatives modified at the C^8-position, as structurally a more favourable access to the adenine binding site is maintained (3). One can imagine three main ways of modifying nucleotide for immobilization: via nitrogenous base, via sugar, or via terminal phosphate group (*Figure 1*),

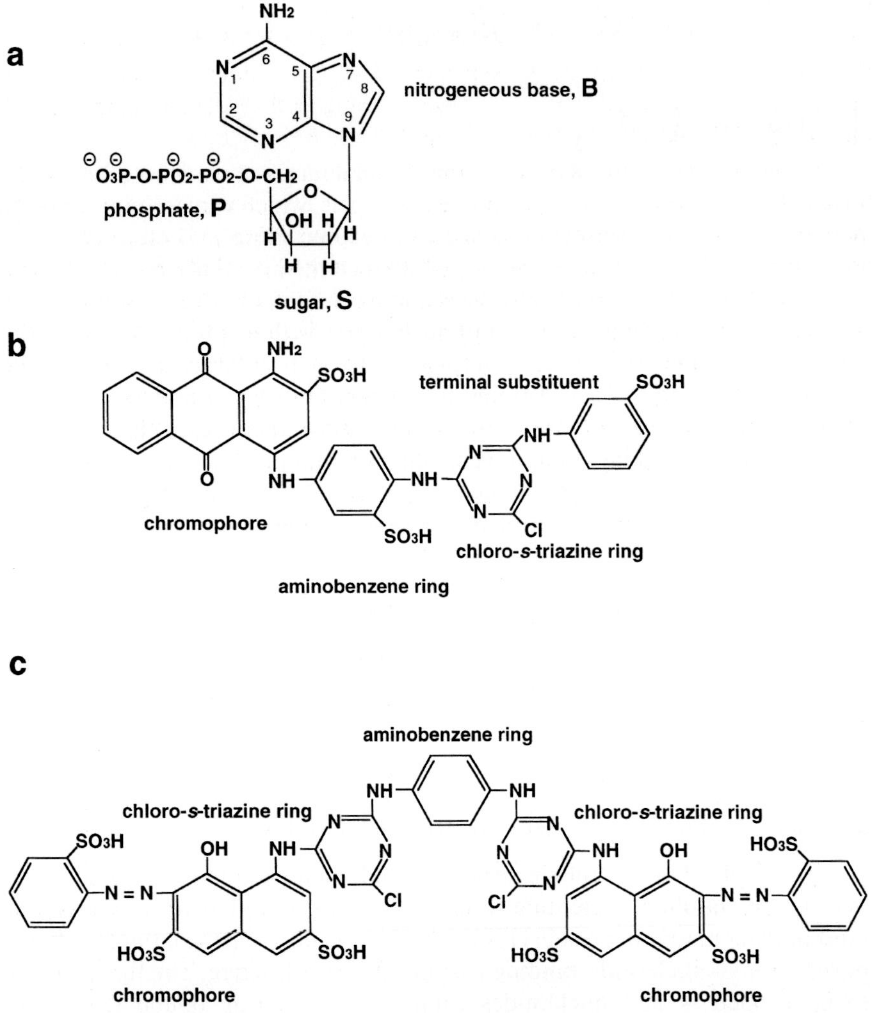

Figure 1. Chemical structure of (a) nucleotide ATP, (b) dye Cibacron Blue F3GA, and (c) dye Procion Red HE-3B.

6: Nucleotide- and dye-ligand chromatography

each of the ways to be discussed in its turn below. The abbreviations used are **N** for nitrogenous base, **S** for sugar, and **P** for phosphate whenever these groups are not involved in chemical modification.

Different nucleotide–ligand matrices are available on the market with various points of nucleotide attachment and various length of the spacer between the nucleotide and the matrix (*Table 1*). The chemical modification of nucleotides requires both expensive chemicals, i.e. nucleotides, and an elaborate chemistry. Hence, the prices of nucleotide–ligand matrices are relatively high, about 15–20 US $/ml which is at least an order of magnitude higher than the prices for dye–ligand matrices (0.5–1 US $/ml). The latter have also proved to be an efficient means for purification of nucleotide-binding enzymes (discussed in Section 3). Thus, the use of nucleotide–ligand matrices is restricted mainly to those cases where they prove to be much superior to dye–ligand matrices. An extensive list of protein purification using immobilized nucleotides is presented by Månsson and Mosbach (4).

Nucleotides are complex organic molecules with different types of covalent bonds. They participate in a plethora of reactions in living organisms, hence numerous enzymes are capable of chemical modification and degradation of nucleotides. Thus, one can expect a reduced stability of nucleotide–ligand matrices as compared to other small ligand-containing affinity matrices in general and dye–ligand matrices in particular. Alongside with general problems of partial matrix degradation and splitting the bond between the matrix and the ligand (discussed below for dye–ligand matrices), the leachate from the nucleotide–ligand matrices may contain the products of nucleotide decomposition.

Table 1. Examples of commercial nucleotide–ligand matrices

Nucleotide–ligand	Point of attachment	Spacer, number of atoms	Ligand content, µmoles/ml	Company
Adenosine 3′,5′-cyclic monophosphate	Base, *C*-8	9	2–10	Sigma
Adenosine 2′,5′- diphosphate	Base, *N*-6	8	2–10	Sigma
Adenosine 2′,5′- diphosphate	Base, *N*-6	6	≈ 2	Pharmacia
Adenosine 5′- triphosphate	Base, *C*-8	9	2–10	Sigma
Adenosine 5′- monophosphate	Base, *N*-6	6	≈ 2	Pharmacia
Cytidine 5′-diphosphate	Sugar	11	1–3	Sigma
Guanosine 5′-triphosphate	Sugar	11	1–3	Sigma
7-Methyl-guanosine 5′-triphosphate	Phosphate	6	0.2	Pharmacia
Uridine 5′-diphosphate	Sugar	11	1–3	Sigma
Flavine adenine dinucleotide	Sugar	11	0.5–1.0	Sigma
β-NAD	*C*-8	9	2–10	Sigma
β-NADP	Sugar	11	2–4	Sigma

2.1 Nucleotides coupled via base

The nitrogenous base moiety of nucleotides provides different sites for covalent binding of spacers terminated with amino groups followed by coupling the synthesized derivative to an agarose matrix. Two sites are mainly used for modification, namely C^6- and C^8-positions of the adenine ring (*Figure 1*). Introduction of an amino terminated spacer into the C^6-position is achieved by direct coupling of alkyldiamines to adenine derivatives substituted with chlorine (5) or mercapto groups (6) at the C^6-position (*Figure 2a*). C^6-mercapto substituted derivatives can also be coupled directly to bromoacetamidohexyl–Sepharose 4B by incubation for five days at room temperature (*Figure 2b*)

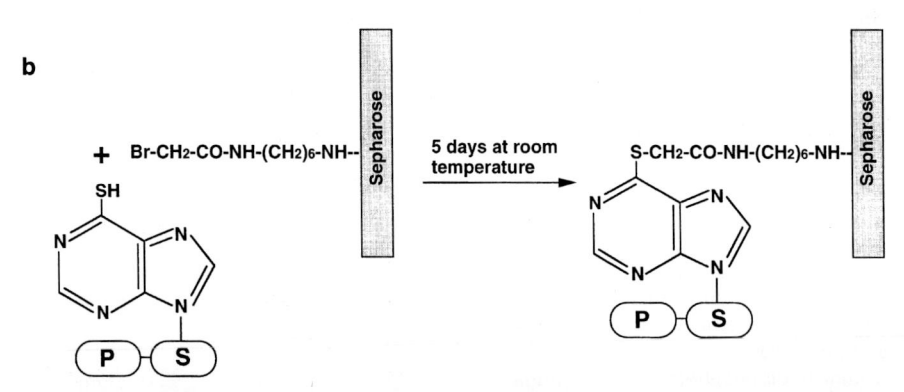

Figure 2. Modification of nitogeneous base in nucleotides with amino group terminated spacer. (a) Binding of alkyldiamine spacer to C^6-Cl- or C^6-SH- derivatives. (b) Direct coupling of C^6-SH- derivatives to bromoacetamidohexyl–Sepharose 4B. (c *see overleaf*) Modification at C^6-position via Dimroth rearrangement. (d) Modification at C^8-position.

(7). Bromoacetamidohexyl–Sepharose 4B is synthesized by treating aminohexyl–Sepharose 4B with *O*-bromacetyl-*N*-hydroxysuccinimide in mild aqueous conditions (8).

Despite the simplicity of coupling procedures of amino or mercapto derivatives of nucleotides which consist mainly of incubation at room or elevated temperature, this approach is applicable only in the cases when such deriva-

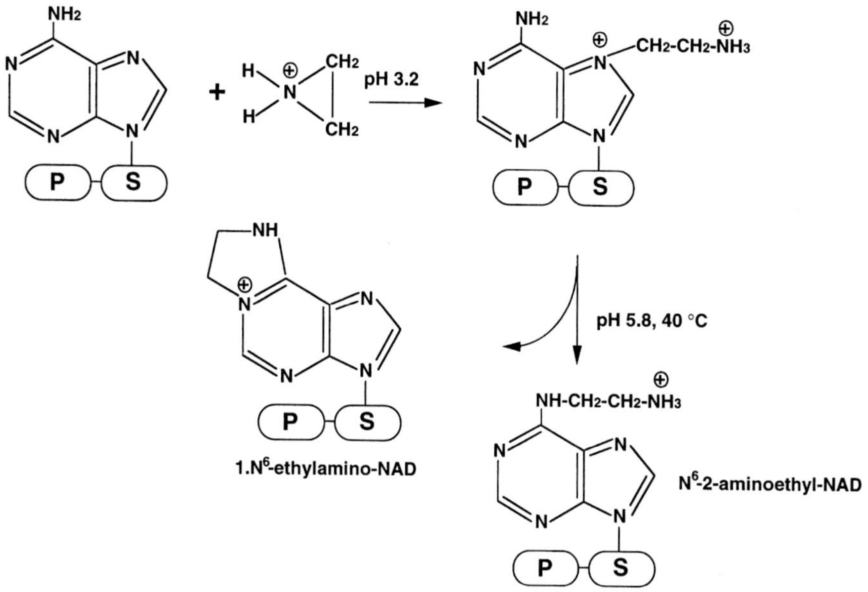

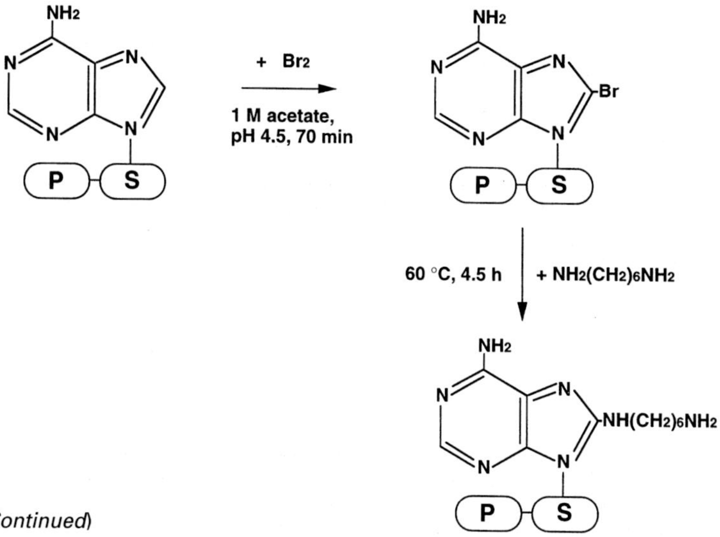

Figure 2. (*Continued*)

tives are available. A more general strategy of coupling amino group terminated spacers to nucleotides follows the typical procedural pattern (3, 9):

(a) Alkylation of the N^1-position of the adenine ring with ethylene imine.
(b) Dimroth rearrangement of N^1-functionalized nucleotide under harsh alkaline conditions (in the case of NAD modification, specific chemical reduction with sodium dithionite of the C^4-position of the nicotinamide ring is performed prior to the Dimroth rearrangement in order to achieve stability under alkaline conditions for the C–N bond between ribose and the nicotinamide).

A simpler procedure was proposed by Bückman and Wray (3) (*Protocol 1*, and *Figure 2c*). Using alkylating agents other than ethylene imine followed by the Dimroth rearrangement spacers can be synthesized which are terminated with different reactive groups. For example, alkylation of nucleotide with 3-propiolactone results in carboxy group terminated spacer (10).

Protocol 1. Preparation of N^6-2-aminoethyl–NAD (3)

Equipment and reagents
- Ethylene imine
- NAD
- 70% HClO$_4$

- Bio-Rex 70 cation exchange column (2.6 × 100 cm, 50–100 mesh) equilibrated with 1 mM HCl at 4°C

Method

1. Add 850 mmol ethylene imine slowly to a solution of 300 mmol NAD in distilled water maintaining pH at 3.2 by the simultaneous addition of HClO$_4$ (total volume 400 ml).

2. Stir gently the reaction mixture in the dark for 50 h maintaining pH at 3.2 by adding HClO$_4$.

3. Dilute reaction mixture to 1 litre with distilled water and precipitate the product with cold technical grade ethanol (4°C, 5 × 10 litres).

4. Recover the precipitate by subsequent centrifugation and dry in vacuum at 25°C. Store over NaOH in a desiccator at 4°C. The precipitate contained NAD (69 mmol), N^1-2-aminoethyl–NAD (195 mmol), and unidentified by-products.

5. Dissolve 20 g of the precipitate in 1 litre distilled water and adjust to pH 5.8 with 1 M LiOH.

6. Incubate with gentle stirring in the dark at 40°C for 80 h, keeping at pH 5.8 with 1 M LiOH. The reaction products consist of NAD (5.3 mmol), N^6-2-aminoethyl–NAD (10.3 mmol), 1.N^6-ethylamino–NAD (4.7 mmol), and by-products.

7. Reduce reaction volume to 50 ml by rotation evaporation.

6: Nucleotide- and dye-ligand chromatography

8. Adjust to pH 5.5 with 1 M HCl and apply to the Bio-Rex cation exchange column.
9. Collect four fractions by simple elution with 1 mM HCl:
    (a) Fraction 1 in 500 ml (44 ml after concentration) contains NAD (5.3 mmol) contaminated with considerable amount of N^6-2-aminoethyl–NAD, 1.N^6-ethylamino–NAD, and by-products.
    (b) Fraction 2 in 700 ml (62 ml after concentration) contains 1.N^6-ethylamino–NAD (3.54 mmol).
    (c) Fraction 3 in 300 ml (20 ml after concentration) contains overlapping 1.N^6-ethylamino–NAD and N^6-2-aminoethyl–NAD (total 1.05 mmol).
    (d) Fraction 4 in 100 ml (60 ml after concentration) contains N^6-2-aminoethyl–NAD (28.5 mmol).
10. Regenerate the column with 1 M HCl until the extinction at 254 nm is 0.
11. Pool concentrated fractions 1 and 3 and adjust to pH 5.5 with 1 M HCl.
12. Apply to the regenerated column and collect four fractions after elution with 1 M HCl at 4°C.
    (a) Fraction 1 in 300 ml (70 ml after concentration) contains NAD (5.3 mmol) contaminated with by-products.
    (b) Fraction 2 in 300 ml (20 ml after concentration) contains 1.N^6-ethylamino–NAD (1.15 mmol).
    (c) Fraction 3 in 150 ml (10 ml after concentration) contains overlapping 1.N^6-ethylamino–NAD and N^6-2-aminoethyl–NAD (total 0.56 mmol).
    (d) Fraction 4 in 800 ml (60 ml after concentration) contains N^6-2-aminoethyl–NAD (3.16 mmol).
13. Combine fraction 4 from step 9 and fraction 4 from step 12, and lyophilize obtaining 9.66 mmol N^6-2-aminoethyl–NAD (yield 3.2%).

The amino group terminated spacer has been introduced in the C^8-position via bromination followed by coupling with alkyldiamine (*Figure 2d, Protocol 2*).

Protocol 2. Preparation of C^8-6-aminohexyl–NADP$^+$ (11)

Equipment and reagents

- NADP$^+$
- Liquid bromine
- 1,6-diaminohexane
- Dowex 1-X8 column (180 ml bed) equilibrated with 1 M ammonium carbonate and extensively washed with water

Method

1. Dissolve 4 g NADP$^+$ in 15 ml 1 M acetate pH 4.5 and add 2.5 ml liquid bromine dropwise with vigorous stirring. Maintain pH between 3.9–4.5 by adding 1 M NaOH.

Protocol 2. *Continued*

2. After 70 min at room temperature extract unreacted bromine seven times with 30 ml carbon tetrachloride.
3. Add 350 ml cold acetone (–80°C) and store overnight at –80°C.
4. Wash the yellow precipitate with 60 ml cold acetone and redissolve in 20 ml water.
5. Add 12 g 1,6-diaminohexane in 5 ml water and heat at 60°C for 4.5 h. Monitor the reaction spectrophotometrically: A_{279}/A_{267} changes from 0.71 to 1.11.
6. Dilute the reaction mixture to 2 litres with HPLC grade water, load at a flow rate of 100 ml/h into a Dowex 1-X8 column, and elute by a linear (0–1.5 M) ammonium carbonate gradient.
7. Pool the fractions with absorbance maximum at 279 nm and lyophilize.

2.2 Nucleotides coupled via sugar

Coupling of nucleotides to agarose matrices was achieved via periodate oxidation followed by coupling to adipic acid dihydrazide–Sepharose 4B conjugate (*Protocol 3, Figure 3*) (12, 13).

Protocol 3. Preparation of sugar-coupled ATP–Sepharose 4B (12)

Equipment and reagents
- ATP
- CNBr-activated Sepharose 4B
- Diethyl adipate
- 98% hydrazine hydrate
- Metaperiodate

Method

1. Reflux 100 ml diethyl adipate, 200 ml hydrazine hydrate, and 200 ml ethanol for 3 h. Crystallize twice the resulting adipic acid dihydrazide precipitate from the ethanol/water mixture.
2. Suspend 100 ml of CNBr-activated Sepharose 4B in 100 ml of cold saturated solution of adipic acid dihydrazide (approx. 90 g/litre) in 0.1 M sodium carbonate buffer pH 9.5, and stir overnight at 4°C.
3. Wash the gel thoroughly with water and 0.2 M NaCl until washings show only slight reaction in the 2,4,6-trinitrosulfonate test.
4. Add metaperiodate solution to cold neutral solution of ATP to a final concentration of 0.01 M nucleotide and 0.01 M periodate and allow reaction to proceed for 1 h in the dark at 0°C.
5. Add 0.5 mole ATP to 100 ml adipic acid dihydrazide–Sepharose 4B

conjugate in 250 ml 0.1 M sodium acetate pH 5. Stir the suspension 3 h at 4°C.

6. Add 750 ml 2 M NaCl and continue stirring for another 30 min to remove unbound nucleotide from the matrix.
7. Wash the gel thoroughly with water.

2.3 Nucleotides coupled via phosphate

An established strategy for covalent binding of nucleotides via a terminal phosphate group is converting it into an ester or amide with a terminal amino

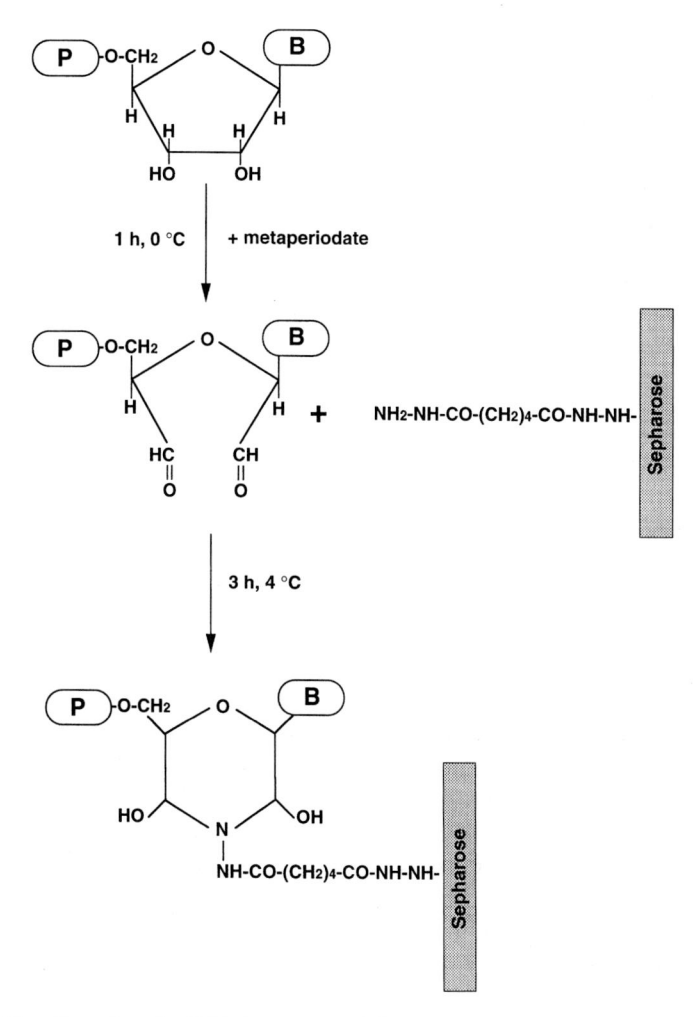

Figure 3. Coupling of nucleotide via sugar moiety.

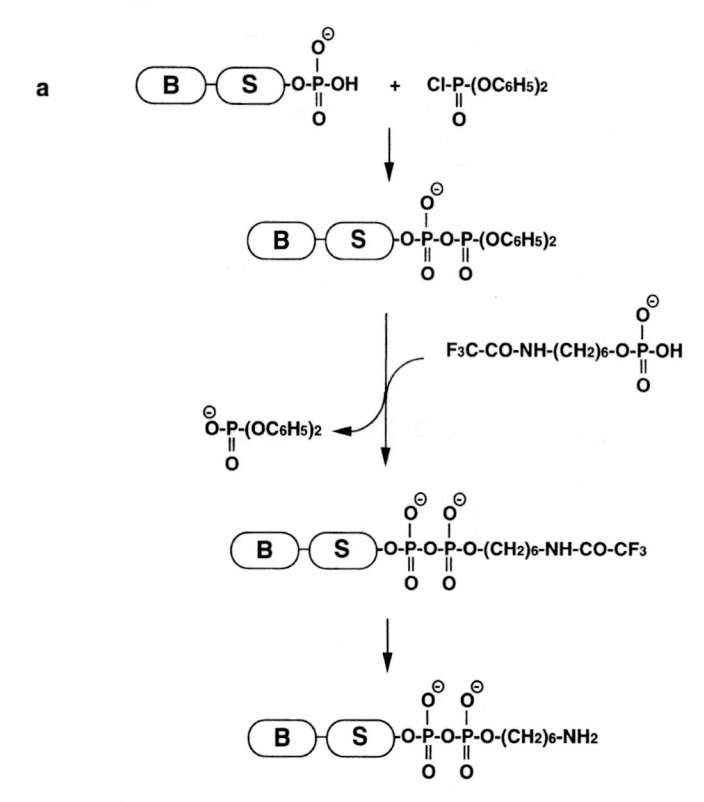

Figure 4. Modification of terminal phosphate group in nucleotides with amino group terminated spacer. (a) Anion displacement method, (b) imidazole method, and (c) carbodiimide method.

group. The latter is used further on for coupling of the nucleotide derivative to CNBr-, tresyl chloride-, or epoxy-activated agarose matrices or to carboxyl group-containing agarose matrix using the carbodiimide method. The reader is highly recommended the book of Hermanson *et al.* 1992 (14) for the protocols of ligand coupling to agarose matrices.

Two methods of synthesis appears to have general applicability for the synthesis of terminal phosphate esters (15). The anion displacement (*Figure 4a*) involves activation of the nucleotide phosphate group by conversion to the diphenylpyrophosphoryl derivative followed by the displacement of diphenylphosphate by *N*-trifluoroacetylhexanolamine phosphate and deprotection of the amino group.

The imidazole method (*Figure 4b*) provides a more convenient and generally applicable approach. This procedure allows the synthesis of *N*-trifluoroacetyl-6-amino-1-hexanol phosphate imidazolide, which is apparently quite stable if protected from moisture, and which reacts with phosphate or pyro-

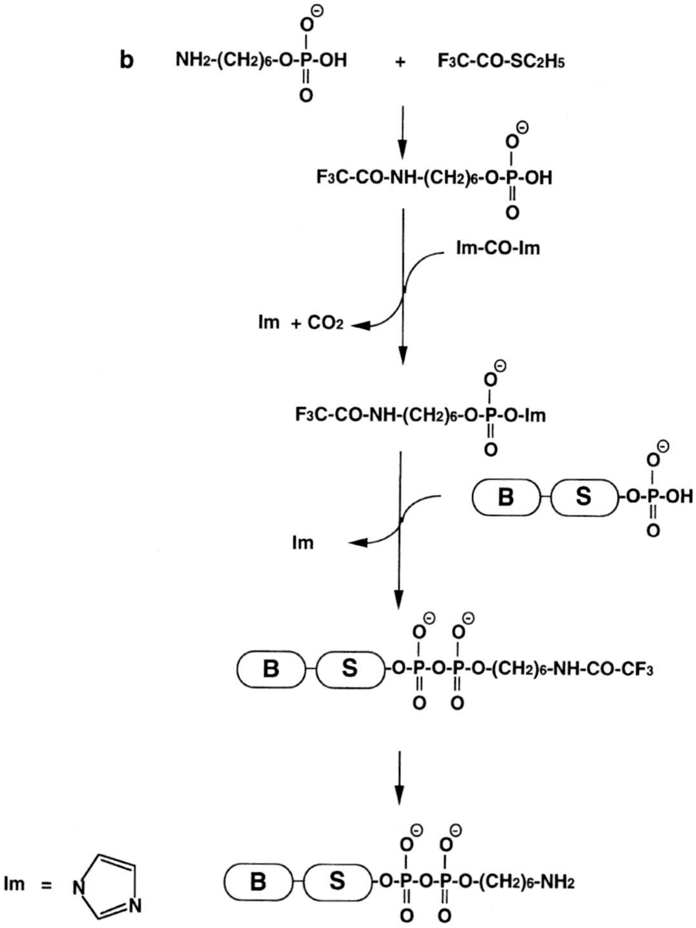

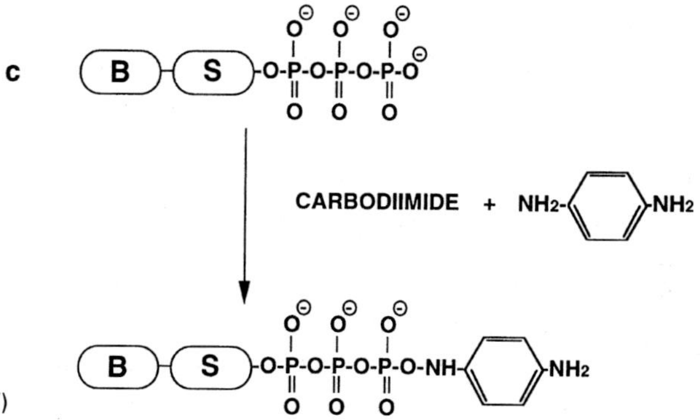

6: Nucleotide- and dye-ligand chromatography

Figure 4. (Continued)

phosphate esters to form the corresponding phosphate or pyrophosphate esters to form the corresponding di- and triphosphates in high yield (60–80%).

Terminal phosphate amides are easily obtained by coupling nucleotide triphosphate with an excess of *p*-phenylenediamine in the presence of carbodiimide (*Protocol 4* and *Figure 4c*). The 4-aminoanilido–GTP was obtained at greater than 80% yield using this method (16).

Protocol 4. Preparation of 4-aminoanilido–GTP and GTP–Sepharose (16)

Equipment and reagents
- Guanosine triphosphate tetrasodium salt
- *N*-ethyl-*N'*-(3-dimethylaminopropyl)carbodiimide hydrochloride
- *p*-Phenylenediamine
- Peroxide-free dioxane
- Poly(ethylene imine) cellulose for thin-layer chromatography (Merck, Darmstadt)
- [γ-^{32}P]GTP as a tracer
- Carboxypropylamino–Sepharose CL-4B

Method

1. Add guanosine triphosphate tetrasodium salt (33 μmol) and *N*-ethyl-*N'*-(3-dimethylaminopropyl)carbodiimide hydrochloride (20 μmol) dissolved in 1 ml distilled water to a solution of 60 μmol *p*-phenylenediamine in 1 ml dioxane and leave for 3 h at room temperature in the dark.

2. Monitor the reaction by thin-layer chromatography on poly(ethylene imine) cellulose in 0.75 M LiCl as solvent (R_f values: GTP, 0.1; GDP, 0.3; 4-aminoanilido–GTP, 0.32). When a trace of [γ-^{32}P]GTP is included in reaction mixture, over 90% of the radioactivity appears in the product. Owing to the aromatic NH_2 moiety the spot of 4-aminoanilido–GTP becomes coloured on exposure to air and light.

3. Pour the reddish brown reaction mixture into 50 ml ice-cold ethanol and collect the precipitate by centrifugation at 10 000 *g* at 4 °C.

4. Wash the precipitate with ice-cold ethanol until no colour remains in the supernatant.

5. Dissolve the product (obtained at greater than 80% final yield) in 4 ml water mixed with 4 ml packed carboxypropylamino–Sepharose CL-4B and adjust to pH 4.7.

6. Add 200 μmol *N*-ethyl-*N'*-(3-dimethylaminopropyl)carbodiimide hydrochloride and stir the mixture for 12 h at room temperature.

7. Filter the resin and wash with 2 litres each of 1 M NaCl and distilled water until the absorbtion at 252 nm of the wash fluid returns to zero. The GTP–Sepharose is stored in 50% ethanol at −20 °C.

6: *Nucleotide- and dye-ligand chromatography*

2.4 Protein purification strategy

Nucleotide–ligand chromatography is extensively used for purification of different enzymes belonging to all six known enzyme classes (*Table 2*). Most frequently it is used for purification of dehydrogenses, transferases, especially glycosyl transferases and phosphorus-containing group transferases, and hydrolases, especially acting on ester bonds.

The strategy of nucleotide–ligand chromatographic purification includes the application of the protein extract at pH close to the pH optimum of enzyme stability, in the presence of about 0.1–0.2 M salt to supress non-specific electrostatic interactions, and washing unbound protein with the equilibration buffer. The elution of bound target protein is achieved by the addition of millimolar concentrations of competing nucleotide (*Figure 5*). Nucleotides absorb strongly at 280 nm due to the aromatic moieties in their structure, thus rendering impossible the assay of eluted protein by measuring absorbance at this wavelength.

High purification folds with good recoveries are achieved with the proper choice of conditions. For instance, lactate dehydrogenase of *Fundulus heteroclitus* was purified 57-fold on a N^6-AMP–Sepharose 4B column with 98% recovery (17).

3. Dye–ligand chromatography

3.1 Dye coupling techniques

Triazine textile dyes are characterized by reactive triazine rings with one or two replaceable chlorine atoms capable of reacting with hydroxyl groups and hence binding strongly to the cellulose materials, cotton in particular, and conferring the specific colour to them. In general, the structure of the textile dye consists of an anthraquinone chromophore (Cibacron Blue F3GA) or azo chromophore (Procion Red HE-3B), diamino benzene ring and triazine ring (*Figure 1b* and *1c*, respectively). The chromophore endows each dye with its characteristic spectral properties which facilitates their identification. Different moieties in the structure of triazine dyes are capable of ionic, hydrophobic, or hydrogen bond interactions with proteins, and endow the dyes with the ability to bind these proteins. Up to 70–85% of the total protein present in a crude extract may bind to a column with covalently coupled dye–ligands (18).

Due to the active chlorine atom in the triazine ring, textile dyes can be conveniently coupled to any hydroxyl or amino group-containing matrix either directly or via a spacer. Bead agarose matrices are traditionally used in different modes of protein chromatography and it comes as no surprise that triazine dyes are mainly coupled to these matrices and used for protein purification. The chemistry of textile dye coupling to cellulose was studied in detail (19). Two general protocols emerged for dye coupling to bead agarose matrices,

Table 2. Subclasses of enzymes purified using nucleotide–ligand chromatography (adapted from ref. 4)

Enzyme subclasses	5′-AMP	2′,5′-ADP	2′,5′-ADP	5′-ADP	c-AMP	ATP	2′-AMP	GMP GDP GTP	NAD	NADP
1. Dehydrogenases										
1.1. Acting on the CH–OH groups of donors	+	+				+	+		+	+
1.2. Acting on the aldehyde or oxo groups of donors	+								+	+
1.4. Acting on the CH–NH$_2$ groups of donors	+	+					+	+		+
1.5. Acting on the CH–NH groups of donors		+								
1.6. Acting on NADH or NADPH	+	+							+	+
1.14. Acting on paired donors with incorporation of molecular oxygen		+								
1.18. Acting on reduced ferrodoxin as donor								+		
2. Transferases										
2.1. Transferring one-carbon groups	+			+						
2.4. Glycosyl transferases	+	+				+		+		
2.7. Transferring phosphorus-containing groups	+				+	+			+	
3. Hydrolases										
3.1. Acting on ester bonds	+			+	+					
3.6. Acting on anhydrides	+				+					
4. Lyases										
4.1. Carbon–carbon lyases	+		+							
4.2. Carbon–oxygen lyases			+							
5. Isomerases										
5.3. Intramolecular oxidoreductases						+				
6. Ligases (synthetases)										
6.1. Forming carbon–oxygen bonds	+									
6.2. Forming carbon–sulfur bonds										
6.3. Forming carbon–nitrogen bonds	+	+				+				
6.5. Forming phosphate ester bonds		+								

6: Nucleotide- and dye-ligand chromatography

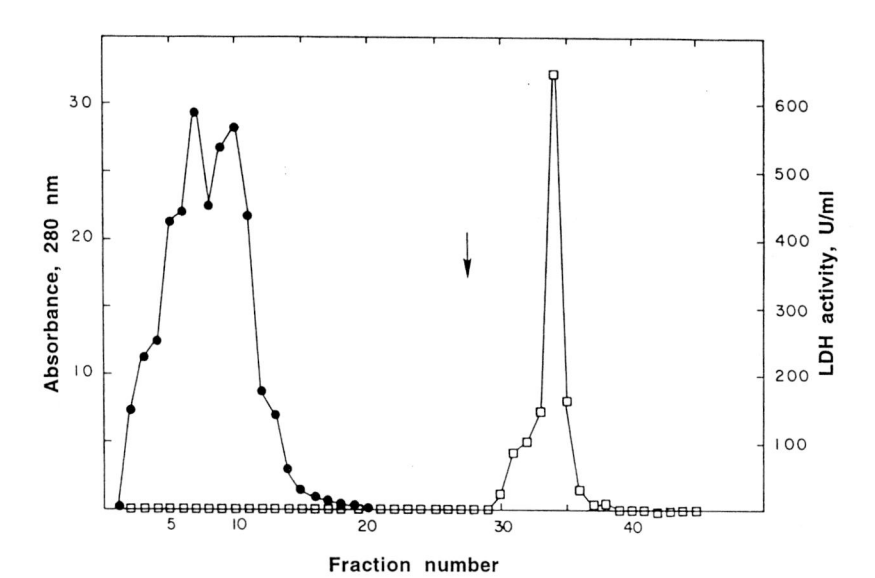

Figure 5. Nucleotide–ligand chromatography of the lactate dehydrogenase of *Fundulus heteroiclitus* on a N^6-AMP–Sepharose 4B column (17). The column (2.5 × 20 cm, 6 μmol of ligand/ml gel) was equilibrated with 20 mM sodium phosphate buffer pH 7 containing 1 mM 2-mercaptoethanol, 1 mM EDTA, and 0.2 M NaCl. 20 ml fractions were collected at a flow rate of 100 ml/h. The arrow indicates the addition of 0.25 mM NADH to the elution buffer. Open symbols for LDH activity and closed symbols for absorbance at 280 nm.

one for coupling the dye at room temperature, and the other for coupling at elevated temperature (*Protocol 5* and *Protocol 6*).

Cibacron Blue F3GA (or the closely related Procion Blue MX-3G and MX-R) is the most popular dye used in dye–ligand chromatography. Truly, Cibacron Blue, more so than any other immobilized ligand, has become almost a general purpose protein purification tool (14).

Protocol 5. Coupling of triazine dye to hydroxyl-containing matrix at elevated temperature (14)

Equipment and reagents
- Sepharose CL-6B or another cross-linked agrose matrix in bead form
- Solution of 1 g Cibacron Blue F3GA in 30 ml water

Method

1. Wash Sepharose CL-6B (100 ml settled gel) with 2 litres water, suction dry to a moist cake, and transfer to a 1 litre three-necked round-bottom flask provided with a paddle stirrer. The use of magnetic stirrer may result in crushing the beads.

Protocol 5. *Continued*

2. Suspend the gel in 100 ml water and, while stirring, heat to 60 °C.
3. Slowly add Cibacron Blue F3GA solution to the gel suspension and stir for 30 min at 60 °C.
4. Add 15 g NaCl to the reaction mixture and stir for 1 h at 60 °C.
5. Increase the temperature of the reaction mixture to 80 °C and add 1.5 g Na_2CO_3 to the gel suspension.
6. Continue the reaction for 2 h at 80 °C.
7. Cool the reaction mixture to room temperature and filter. Extensively wash the gel with water until washings are colourless. Warm water often speeds this washing process. Finally wash the gel with 2 litres each of 1 M NaCl and water. Immobilized Cibacron Blue F3GA can be stored in 0.02% sodium azide at 4 °C.

Protocol 6. Coupling of triazine dye to hydroxyl-containing matrix at room temperature (20)

Equipment and reagents

- Sepharose CL-6B or another cross-linked agrose matrix in bead form
- Solution of 1.2 g Cibacron Blue F3GA in 80 ml water

Method

1. Wash Sepharose CL-6B according to *Protocol 5*, step 1.
2. Suspend 80 g of the gel in 280 ml of water.
3. Add Cibacron Blue F3GA solution to the gel suspension.
4. Add 40 ml 4 M NaCl solution and 4 ml 1 M NaOH solution.
5. Stir at ambient temperature for 72 h.
6. Wash the gel according to *Protocol 5*, step 7.

   Immobilization of other triazine dyes is performed essentially by the same procedure.

Salt promotes the precipitation of the dye on the support while alkali conditions facilitate the formation of ether linkages between dye and support (21). Less time and alkali is required for the immobilization of dichloro-*s*-triazine dyes (Procion M series) than for monochloro-*s*-triazine dyes (Procion H, Procion P, Cibacron series) (19). The rate of dye fixation to agarose depends on the pH and temperature. Raising the pH value of the reaction medium results in acceleration in the fixation rate accompanied by acceler-

6: Nucleotide- and dye-ligand chromatography

ated hydrolysis of the dye. Empirically deduced conditions which favour adequate coupling of triazine dyes to agarose are 1% (w/v) sodium carbonate or 0.02–0.1 M NaOH (22).

Long treatment at low temperatures promotes cross-linking of the hydroxyl-containing polymer chains in the matrix by dichloro-s-triazine dyes, whereas dye fixation at high temperatures leads to hydrolysis of the final chlorine atom (19). Any remaining chloro groups on the immobilized dye should be replaced with amino groups by suspending the dye matrix in 2 M ammonium chloride at pH 8.5 for 4 h at ambient temperature followed by washing with water (20).

Cibacron Blue F3GA was also coupled to ion exchange agarose matrices (DEAE Affi-Gel blue and CM Affi-Gel blue produced by Bio-Rad). These bifunctional gels facilitate protein binding which occurs as a combination of dye–ligand and ion exchange interactions. According to the manufacturer, dye–ligands bind albumin, proteases, and other complement proteins, while DEAE groups bind remaining acidic proteins of ascites fluid allowing single step purification of IgG. CM Affi-Gel blue gel chromatography provides a convenient initial step in the purification of serum proteins allowing rapid removal of ~ 90% of the albumin and all the plasminogen in serum samples.

Coupling of triazine dyes to other than agarose chromatographic matrixes is achieved usually by introducing hydroxyl groups on the surface of the beads either by covering with polysaccharide, dextran (23), or by epoxy silylation of silica and conversion of epoxy silica to the corresponding diol-silica (24–26), followed by dye coupling via standard procedure. Triazine dyes were also immobilized on epoxy-activated silica via 1,6-diaminohexane spacer (26, 27).

The degree of substitution of the matrix with the dye may be regulated by altering the coupling temperature and time. The dye content in the agarose matrix is determined by hydrolysis of the matrix in strong acid following by spectrophotometric assay of the dye content (*Protocol 7*).

Protocol 7. Determination of dye content in the bead agarose matrix (28)

Equipment and reagents
- Dry agrose matrix with coupled Cibacron Blue F3GA in bead form
- Standard Cibacron Blue F3GA solution in water, 2 mg/ml

Method
1. Suspend 10 mg dry agrose matrix with coupled Cibacron Blue F3GA in 5 ml 6 M HCl.
2. Prepare standard solutions containing 0–200 µg Cibacron Blue F3GA by diluting appropriate aliquots of 2 mg/ml solution of Cibacron Blue F3GA to a final volume of 5 ml 6 M HCl.

Protocol 7. Continued

3. Incubate all solutions in a water-bath for 1 h at 40°C and mix every 15 min.
4. Cool the samples and measure their absorbance at 515 nm.
5. Prepare a calibration curve from reading of standards and calculate the dye content of the matrix.

Commercial dye–ligand matrices based on 4% or 6% cross-linked agarose contain 6.4–8.2 μmol Cibacron Blue 3GA per ml drained gel (Pharmacia Biotech) or 0.2–6 μmol Cibacron Blue 3GA per ml drained gel (Sigma). One should note that functional concentration of immobilized dye is less than 2% of the actual measured (29).

3.2 Purity, leaching, and toxicity of triazine dyes

All commercial dye preparations contain certain additives in order to standardize the final product. The most common additives are NaCl, which is added to dilute the dye content to a standard value, and a de-dusting agent such as dodecylbenzene. In addition, various side-products from the reaction stage as well as hydrolysed dye are present in the final preparation. In spite of this, commercial dye samples are commonly used directly in protein purification studies since any applications on a preparative scale would employ the dye as it is, and in most cases the additives and side-products would not immobilize to the support and are washed out after the immobilization procedure (30). In some analytical studies purified dye preparations have been used (31, 32).

Dyes are extremely resistant to biological and chemical attack (33). Leakage of dyes from the affinity matrix is mainly due to partial matrix degradation under extreme conditions of regeneration/cleaning-in-place procedures (34, 35). The dye–matrix covalent bond is quite stable but may be hydrolysed under certain conditions. The rate of hydrolysis is a function of the pH, ionic strength, and temperature (35).

Spectrophotometric methods which have a sensitivity limit of about 1 μg/ml for the commonly used dyes, are not sensitive enough for detecting leached dye from dye-linked adsorbents. Competitive ELISAs were developed to detect leaching of Cibacron Blue F3GA, Basilen Blue E3G, Reactive Yellow 13, and Procion Red HE3B from chromatography adsorbents (2, 36–38), and from a perfluoropolymer support (39). The ELISA had a sensitivity approximately three orders of magnitude higher than the spectrophotometric assay and was suitable for detection of the very low amounts of leached dye present when the matrices were subjected to extremes of pH. Leachable structures (from a Trisacryl matrix derivatized with either Cibacron Blue F3GA or Procion Red HE3B) as well as free dyes did not exhibit any cytotoxicity in *in vitro* studies (40, 41). However with agarose-based dye supports, the

6: Nucleotide- and dye-ligand chromatography

fragmented matrix present with the leached dyes was toxic towards human MRC-5 cells in *in vitro* studies. The carbohydrate component of the fragmented matrix is responsible for the cytotoxic response rather than the dye moiety itself (42). Dye leaching and toxicity is not a well studied area and this has often proved to be a bottle neck in the application of dye–affinity purification processes since the presence of leached dye–ligand in the final protein product is not acceptable.

3.3 Designer dyes

The ability of dyes to bind to an extensive range of proteins is a mixed blessing. On the one hand it allows wide applications of dye–affinity purifications, but on the other hand, the purification folds in the process can be rather low, due to co-purification of foreign proteins which are present in the crude extract. There has been considerable interest in understanding the interactions between dyes and proteins at a molecular level. X-ray crystallography, affinity labelling, kinetic studies, and difference spectroscopy have provided valuable insights into the structures of dye–protein complexes and the forces governing the binding. This information has been used in the *de novo* synthesis of dyes. Such 'designer dyes' can bind to the target protein with greater affinity and a higher specificity.

The first designer dyes were based on the structure of Cibacron Blue (*Figure 1b*) since the interactions of this dye with horse liver alcohol dehydrogenase were well studied. The dye binds to the enzyme at the NAD^+ binding site with the anthraquinone, diaminobenzene sulfonate, and triazine rings adopting positions similar to the adenine, adenyl ribose, and pyrophosphate groups of the cofactor, respectively (43). The major difference arises in the binding of the terminal aminobenzene sulfonate ring which binds in a cleft approximately 10 Å away from the nicotinamide ring binding site. The binding improves by altering the terminal ring so that it adopts a position similar to that of the nicotinamide base. X-ray crystallography studies showed that the terminal ring binding site is composed of residues which are located in the catalytic domain. These are mainly hydrophobic residues but there are also two arginine residues which should favour the binding of anionic groups (44). By preparation of the suitable terminal ring analogues of Cibacron Blue and measurement of their binding constants to alcohol dehydrogenase by difference spectroscopy, it was demonstrated that small anionic substituents at *ortho* or *meta* positions of the terminal ring gave a stronger binding to the enzyme in free solution (45). This is especially the case with carboxyl group-containing substituents which due to their small size are easily accommodated in the binding site. Substitution at the *para* position is not favoured especially if the substituent group is bulky, probably as a result of steric hindrance. An unsubstituted aniline ring at this position also facilitates tight binding of the dye to the enzyme (44). However, when these dye analogues are immobilized to agarose matrices via the triazine ring, the chromatography results are not greatly altered and the large changes in the binding constants that have been obtained in free solution are not reflected in the chromatographic results. Probably, the immobilization of the dyes to the matrix when carried out via the triazine ring renders the rest of the dye molecule rather inflexible resulting in a poorer fit between the immobilized ligand and the target protein. The nature of the terminal substituent in the triazine ring has also a minor effect on the dye–ligand purification of formate dehydrogenase from *Candida boidinii* (the purification fold using different biomimetic dyes varies from 2.6–3.4 for specific elution with NAD, or from 1.8–2.7 for non-specific elution with KCl) (46). The same effect is documented for bovine heart lactate dehydrogenase where purification fold varies from 6.4–10 (specific elution with NADH) or from 4.1–5.4 (non-specific elution with KCl) (47).

Immobilization via a spacer between the matrix and the triazine ring results in better agreement between the spectroscopy and the chromatography data (45). An alternative method is immobilization via a spacer linked to the 1-amino group of the anthraquinone ring (48). Not only does the binding affinity increase (in agreement with the difference spectroscopy data) but the specificity of the adsorbent for one of the enzymes is also improved.

Another approach towards the better fit of the dye–ligand is introducing a spacer between the bridging diaminobenzene sulfonate ring and the triazine ring. The affinity increases when a second aminobenzene sulfonate ring is introduced at this position. The diaminobenzene sulfonate ring is very rigid when compared with the phosphate ester linkage in the cofactor and with the introduction of a spacer at this point, greater flexibility is introduced in this region. When an ethyl group is introduced instead, the commercial sample of horse liver alcohol dehydrogenase is resolved into two active components (45).

These studies demonstrate how minor structural and positional changes on the dye can affect the strength and specificity of the interaction. Designer dyes have been used in the purification of calf intestinal alkaline phosphatase (49), pancreatic tissue trypsin (50), glucose oxidase from *A. niger* (51), chymosin from calf stomach rennet (51), and morphine dehydrogenase from *P. putida* (52). In these cases the product was obtained with a good recovery and high purity. Although the approach of designer dyes is extremely attractive, practical problems do exist, which have to be surmounted before the full potential of this technique is realized. X-ray crystallography data pertaining to the structure of dye–protein complexes is still very limited. In order to obtain information about the binding interactions, vast amounts of kinetic and spectroscopy data with different dye analogues have to be collected. Several dye analogues must be studied to find the optimal affinity ligand, since the designed dye with the strongest binding to the target protein may not be the best ligand for purification purposes. Preparation of each designer dye adsorbent requires complex chemistry. Since each dye adsorbent is optimized for a particular protein separation, the group specific nature of the adsorbent is questionable.

3.4 Polymer shielding

Proteins present in a crude extract are retained on the dye–matrix by a combination of specific and non-specific interactions, the sum of which may result in a very strong retentive force. Elution of proteins bound to an affinity column is brought about by changing the conditions in the mobile phase. This may be by the inclusion of a competitive substance, increase in pH or ionic strength, or addition of an organic solvent. The choice of the elution conditions would depend on which forces predominate in the binding of the target protein to the ligand. The net result would be the same, i.e. one component of the retentive force is decreased such that the residual forces are insufficient for the protein to remain bound to the column. If the retention force is not decreased sufficiently under the elution conditions, some protein molecules would remain bound on the column, resulting in lowered recoveries. Another outcome of such mixed interactions is that upon elution, as the protein molecules progress through the column, there could be some residual non-specific interactions leading to some retardation of the protein. This would result in the broad and tailed elution peak. It would be desirable to modulate the protein–ligand interaction by some means so that the non-specific interactions are reduced to an extent where they would be sufficient for providing the binding force for retention yet low enough to prevent tight binding and peak tailing.

The improvement of the protein elution can be achieved in two ways. First, by using as an eluent a compound which binds more efficiently to the ligands and hence displaces the bound protein. The second way is to reduce somewhat binding of the target protein to the matrix at the expense of non-specific interactions.

The first approach is realized using poly(ethylene imine) as a displacer of lactate dehydrogenase bound to Cibacron Blue F3GA–Sepharose or Procion Red HE3B–Sepharose (53). Poly(ethylene imine) forms very strong complexes with triazine dyes (54) and when applied to a dye–ligand column with bound protein, it displaces the latter. The yields during poly(ethylene imine) displacement are better comparing to the traditional elution with NADH plus oxamate or 1.5 M KCl, and the elution efficiency is significantly improved. Bound displacer was removed from the column by washing with high salt under basic conditions (Cibacron Blue F3GA–Sepharose) or by complexation with poly(acrylic acid) under basic conditions (Procion Red HE3B–Sepharose) (53).

The second approach is realized in the technique nick-named as 'polymer shielding' (54–61). Triazine dyes form complexes in solution with some non-ionic polymers, poly(vinyl pyrrolidone), poly(vinylcaprolactam), poly(vinyl alcohol) (54). When a solution of such a polymer is applied to a dye–matrix (*Protocol 8*), the polymer molecules attach to it by non-covalent multipoint interactions. With the introduction of a crude extract on such a polymer shielded matrix the target protein competes with the polymer present for binding to the ligand. It is unlikely that the bound polymer can mask the specific interactions of the target protein and ligand which are generally quite strong. A more plausible situation is where the weaker non-specific interactions of the target protein are masked by the bound polymer. Protein binding to the ligand causes only a local displacement of the bound polymer since the latter is attached to the matrix via multipoint interactions. Indications for this mechanism are based on the following observations. First, the inhibition constants of nucleotide-dependent enzymes in the presence of the dye are much smaller values (approx. 1 µM) as compared to the binding constants of non-nucleotide-dependent proteins such as BSA to the dyes (approx. 100 µM). This means that the specific interactions of an enzyme are much stronger than its non-specific ones. If the dissociation constants of the polymer–dye complexes are taken into account (10–20 µM), they fall in an intermediate range. This implies that the polymer can compete effectively with the enzyme's non-specific interactions but not with the specific ones (54).

Protocol 8. Polymer shielding of dye–ligand column

Equipment and reagents
- Cibacron Blue F3GA–Sepharose packed into a column
- 1% (v/w) solution of polymer, poly(vinyl pyrrolidone) in equilibration buffer

Method
1. Percolate column with 5 column volumes of polymer solution.
2. Wash the column extensively with 20 column volumes of 1.5 M KCl pH 3.5.
3. Re-equilibrate the column with 10 column volumes of equilibration buffer.

Polymer shielding therefore modulates the overall binding force. Retention of the target protein on the polymer shielded matrix will occur if the protein interactions in the presence of the polymer are sufficiently strong. Less efficiently binding proteins are not retained on the shielded column and are removed in the washing procedure. The masking of weak interactions by the polymer allows protein recovery in a sharp peak and smaller elution volume (55, 61).

A strategy for the optimal choice of polymer/dye combination to purify a given protein was developed. The application of the strategy to the purification of porcine muscle lactate dehydrogenase resulted in 30-fold enzyme purifica-

6: Nucleotide- and dye-ligand chromatography

tion with almost quantitative recovery (94%). The enzyme was recovered in an elution volume which was half that from the unshielded column (61).

Polymer shielded dye–affinity chromatography was applied for the purification of several enzymes. Phosphofructokinase from bakers yeast was eluted with a substrate (ATP) from a PVP shielded column with a 56% yield and a 27-fold purification. In contrast the yields from an unshielded column was 11% with a 24-fold purification. The elution volume from the polymer shielded column was less than one-third that of an unshielded column (56). Pyruvate kinase from porcine muscle was purified on a poly(vinyl alcohol) shielded Procion Red HE3B–Sepharose column. Yields with a specific and non-specific eluent were 84% and 95% respectively. The purification fold of four to six was an improvement on that obtained by chromatography on other dye–affinity adsorbents (62,63). In this system enzyme elution from an untreated column was not possible.

One can change from the polymer displacement mode to the polymer shielding mode provided the interactions of the polymer with the dye ligands are modulated in response to the changes in the environment. Poly(vinylcaprolactam) is a polymer with lower critical solution temperature (LCST) in aqueous solutions of about 32°C. PVCL adopts an extended hydrophilic conformation (coil) and is soluble in water at temperatures below LCST. At temperatures higher than LCST, polymer molecules undergo a reversible coil to globule transition due to the increased hydrophobic interactions. In the compact globule conformation poly(vinylcaprolactam) molecules aggregate and the polymer is insoluble in water. The decrease in temperature results in the globule to coil transition and poly(vinylcaprolactam) becomes soluble once again (64). When the Cibacron Blue F3GA–Sepharose column was percolated with poly(vinylcaprolactam) solution at room temperature (lower than LCST) the polymer binds strongly to the ligands and in the presence of 0.1 M KCl prevents the lactate dehydrogenase binding to the column. On increasing the temperature, the polymer molecules become more compact leaving some of the dye ligands free for the interaction with lactate dehydrogenase. At 40°C (higher than LCST) the enzyme binds to the column. Cooling the column back to the room temperature results in efficient elution of the bound lactate dehydrogenase due to the polymer displacement. Purification of the target enzyme is achieved from a crude extract with 90% recovery and 17-fold purification demonstrating the utility of this system. The elution occurs only as the result of the temperature shift without any changes in the buffer composition. Temperature shifts as a means for elution is an attractive alternative because it bypasses the need to remove eluent species from the enzyme preparation.

3.5 Routine purifications

In the first 10–15 years of dye–affinity separations, most dye–affinity purifications were centred around Cibacron Blue F3GA, the dye that had originally been coupled to dextran and used as a void volume marker. After some years, the red dye Procion Red HE3B was incorporated in purification procedures particularly since some workers claimed that it was selective towards those proteins which used NADPH/NADP$^+$ as cofactor (65). This observation was later revoked as a number of other proteins were seen to bind to this dye. In the late 1980s and early 1990s there has been a surge of publications in which new dyes (those which were common in the textile industry but had not been used for purification purposes) were evaluated as potential affinity ligands. The introduction of dye screening kits by various companies (which may contain up to 40 dye matrices) has also hastened the application of dye–affinity techniques in routine protein purifications.

Dye–ligand chromatography involves less specific interactions of target protein with the ligand as compared to nucleotide–ligand chromatography. Dye–ligand matrices are capable of binding proteins without defined nucleotide binding site. This fact implies a special strategy in the application of dye–ligand chromatography, the first step is usually screening for the dye–ligand capable of binding the target protein. When the appropriate dye–ligand is identified, the elution of the target protein is achieved either by addition of competing nucleotide (so-called specific elution) or by high salt (so-called non-specific elution). Due to the less specific nature of protein interactions with the dye–ligands, a lot of proteins other than target protein are capable of binding to the dye–ligand column. Elution with competing nucleotide results in elution of target protein probably bound to the dye–ligands via nucleotide binding sites while non-specifically bound proteins remain on the column. They can be eluted afterwards with high salt buffer (*Figure 6a*). Non-specific elution of target protein with high salt results first in elution of weakly bound impurities followed by elution of tightly bound target protein (*Figure 6b*).

Table 3 presents a collection of purification protocols at least one step of which was dye–affinity chromatography. Only the papers published in the 1990s are included. The choice of the papers was unbiased, all the papers on using dye–ligand chromatography we came across are included in the table. This table is intended to be representative, rather than exhaustive. We believe it reflects an actual situation of using dye–ligand chromatography in laboratory scale protein purifications and can be helpful for the choice of dye–ligand matrix and elution conditions when purifying proteins similar to that presented in *Table 3*.

Data from *Table 3* indicate that a dye–ligand chromatography has been incorporated in the protocols for purification of many proteins from different sources. These are the nucleotide-dependent enzymes as well the proteins with no specific nucleotide binding site. Mainly dye–ligand chromatography is used as a positive step, i.e. target protein binds to the matrix with contaminants passing through the column. However in many cases, it is a negative purification step, i.e. the target protein is in the breakthrough fraction while some contaminants bind to the matrix.

6: Nucleotide- and dye-ligand chromatography

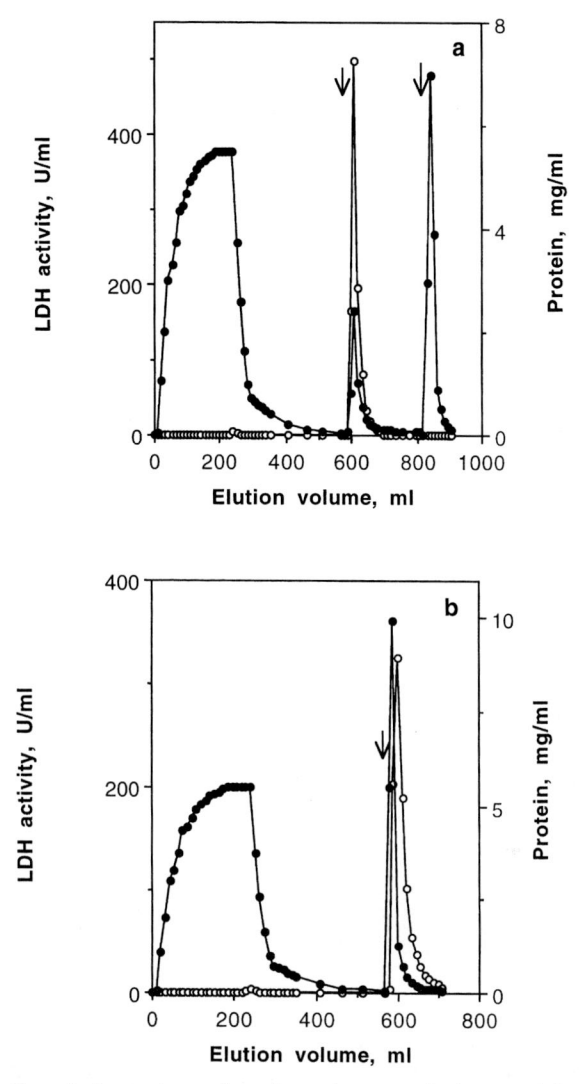

Figure 6. Dye–ligand chromatography of porcine muscle lactate dehydrogenase on a Cibacron Blue–Sepharose CL-4B column (66). The column (0.9 × 9.8 cm, 4.9 mmol of ligand/ml gel) was equilibrated with 20 mM Tris–HCl buffer pH 7.3. 11 ml fractions were collected at a flow rate of 33 ml/h. Arrows indicate (a) the addition of 0.1 mM NADH plus 10 mM oxamate, followed by elution with 1.5 M KCl and (b) the addition of 1.5 M KCl. Open symbols for LDH activity and closed symbols for protein.

The recoveries and purification folds achieved are extremely variable (*Figure 7*). Purification folds in the range of one to three are reported most frequently (*Figure 7a*). About half of all reported protocols have purification folds in the range of one to five, though the purification folds higher than 100

Table 3. Protein purification using dye-ligand chromatography

Protein/source	Dye	Matrix (company)	Yield (%)	Purification fold	Eluent	Ref.
Proteinase inhibitor from corn	P.Red H3B	Coupled directly to dextran coated Sperosil XO15 M (IBF Biotechnics)	89	6.4	6 M urea	67
Proteinase inhibitor	P.Red H3B	Coupled directly to Sephadex G50 F (Pharmacia)	89	4.1	6 M urea	67
Proteinase inhibitor	P.Red H3B	Coupled directly to Sepharose CL-4B (Pharmacia)	89	6.4	6 M urea	67
Homoserine dehydrogenase	C.Blue F3GA	Coupled directly to Sepharose 4B (Pharmacia)	38	41	2 mM NADP	68
Homoserine dehydrogenase	C.Blue F3GA	Coupled via 1,4-diaminobutane spacer to Sepharose 4B (Pharmacia)	62	95	2 mM NADP	68
Penicillin-binding protein	C.Navyblue 2GE	Coupled directly to TSK HW-65F Fractogel (Merck)	71	24	0.5 M NaCl	69
(+)-Pinene synthase	P.Red HE-3B	Matrex Gel Red A (Amicon)	> 100	> 5	10% (v/v) glycerol + 2 M KCl	70
(−)-Pinene synthase	P.Red HE-3B	Matrex Gel Red A (Amicon)	89	4.5	10% (v/v) glycerol + 2 M KCl	70
1,8-Cineole synthase	P.Red HE-3B	Matrex Gel Red A (Amicon)	88	4.5	10% (v/v) glycerol + 2 M KCl	70
(+)-Bornylpyrophos-phatase synthase	P.Red HE-3B	Matrex Gel Red A (Amicon)	> 100	> 5	10% (v/v) glycerol + 2 M KCl	70
(−)-Caryophyllene synthase	P.Red HE-3B	Matrex Gel Red A (Amicon)	83	4	10% (v/v) glycerol + 2 M KCl	70
Himulene synthase	P.Red HE-3B	Matrex Gel Red A (Amicon)	68	3.5	10% (v/v) glycerol + 2 M KCl	70
Geranyl pyrophos-phate synthase	P.Red HE-3B	Matrex Gel Red A (Amicon)	56	3	10% (v/v) glycerol + 2 M KCl	70
Farnesyl pyrophos-phate synthase	P.Red HE-3B	Matrex Gel Red A (Amicon)	86	4.5	10% (v/v) glycerol + 2 M KCl	70
Secondary alcohol dehydrogenase	P.Blue MX4GD	Coupled directly to Sepharose CL-4B (Pharmacia)	76	13	0.5 mM NADP$^+$	71

Enzyme	Dye	Matrix	Purification (fold)	Yield (%)	Elution	Ref.
Secondary alcohol dehydrogenase	P.Scarlet H2G	Coupled directly to Sepharose CL-4B (Pharmacia)	69	32	0.5 mM NADP$^+$	71
Transthyretin	C.Blue F3GA	Blue Sepharose CL-6B (Pharmacia)	95	1.2	Breakthrough	72
Ribonuclease C$_2$	C.Blue F3GA	Blue Sepharose CL-6B (Pharmacia)	76	900	0–0.5 M KCl gradient	73
Protein 12	C.Blue F3GA	Blue Sepharose CL-6B (Pharmacia)	92	1.1	Breakthrough	74
Calmodulin (lysine 115) N-methyltransferase	Re.Red 120	Reactive Red 120 agarose Type 3000-CL (Sigma)	55	173	0.5 M NaCl	75
D-*myo*-inositol (1,4,5)/(1,3,4,5)-polyphosphate 5-phosphatase	C.Blue F3GA	Blue Sepharose CL-6B (Pharmacia)	35	1.6	0.8 M NaCl	76
NMN adenylyltransferase	Re.Red A	Matrex Gel Red A (Amicon)	120	30	1–3 M NaCl linear gradient	77
Cholesteryl ester transfer protein	P.Red H-E3B	Reactive Red 120 agarose Type 3000-CL (Sigma)	203	62	0–1 M KSCN linear gradient	78
Cholesteryl ester transfer protein	P.Yellow M-8G	Reactive Yellow 86 agarose (Sigma)	110	37	30% (v/v) glycerol + 1 mM EDTA	78
Inositol 1,3,4,5,6-pentakisphosphate 2-kinase	C.Blue 3GA	Cibacron Blue 3GA agarose (Sigma)	93	8.3	0–1.5 M KCl linear gradient	79
Sialidase L	Re.Blue A	Matrex Gel Blue A (Amicon)	66	1.9	0.2 M NaCl	80
NAD$^+$-dependent acetaldehyde/alcohol dehydrogenase	C.Blue F3GA	Blue Sepharose CL-6B (Pharmacia)	2.8	71	10 mM NAD$^+$	81
Progesterone 5β-reductase	C.Blue F3GA	Blue Sepharose CL-6B (Pharmacia)	97	3.7	0–1 M NaCl linear gradient	82
Poly(hydroxy alkanoic acid) synthase	P.Blue H-ERD	Coupled directly to Sepharose CL-6B (Pharmacia)	61	1.3	0–2 M NaCl linear gradient	83
NADPH-dependent dehydroascorbate reductase	Re.Red 120	Reactive Red 120 agarose Type 3000-CL (Sigma)	87	1.2	1.5 M NaCl	84

Table 3. Continued

Protein/source	Dye	Matrix (company)	Yield (%)	Purification fold	Eluent	Ref.
Glutamyl-tRNA^{Glu} reductase	C.Blue F3GA	Blue Sepharose CL-6B (Pharmacia)	a	4	1 M NaCl	85
4-Chloro-benzoyl CoA dehalogenase	Re.Green 19	Reactive Green 19 agarose (Sigma)	105	2.0	Breakthrough	86
5'-Methylthio-adenosine phosphorylase	Re.Red A	Matrex Gel Red A (Amicon)	85	1.9	Breakthrough	87
NAD-dependent nucleotide diphospho-sugar epimerase	Re.Blue 2 CL-6B (Sigma)	Reactive Blue 2-Sepharose	62	1.9	5 mM NAD$^+$	88
Phosphofructokinase	Re.Blue 2	Blue 2-Sepharose CL-6B (Sigma)	69	34	1.5 mM MgATP	89
Alcohol dehydrogenase	Re.Red A	Matrex Gel Red A (Amicon)	47	1.9	0.5 mM NADP$^+$	90
Alcohol dehydrogenase	Blue MX-4GD	a	94	17	2 mM NAD$^+$	90
Secondary alcohol dehydrogenase	Re.Red A	Matrex Gel Red A (Amicon)	26	24	0.5 mM NADP$^+$	90
Acetaldehyde dehydrogenase	Re.Red A	Matrex Gel Red A (Amicon)	103	14.2	0.5 mM NADP$^+$	90
Recombinant ricin A	P.Red H3B P.Red HE3B P.Red HE7B P.Yellow HE4R	Coupled directly to Sepharose CL-4B (Pharmacia)	95	b	0.25–0.55 M stepwise gradient	91
Mn-dependent ADP-ribose pyrophosphatase	Re.Blue 2 CL-6B (Sigma)	Reactive Blue 2-Sepharose	72	77	Change pH from 7.5 to 9.7	92
ADP-ribose pyrophosphatase	Re.Green 19	Reactive Green 19 agarose (Sigma)	58	9.6	Change pH from 7.5 to 9.7	92
Ornithine acetyl-transferase	P.Scarlet H-E3G	Coupled directly to Sepharose CL-4B (Pharmacia)	52	8	Change pH from 5.5 to 6.5	93
1,2-Dioxygenase	C.Blue F3GA	DEAE Affi-Gel blue gel (Bio-Rad)	61	18	Breakthrough	94

Enzyme	Dye	Matrix			Elution	Ref
Luciferase	Re.Orange A	Dyematrex Orange A (Amicon)	140	4.3	Breakthrough	95
Recombinant complement C1s	Re.Orange A	Dyematrex Orange A (Amicon)	25	7.9	0–3 M NaCl linear gradient	96
Phosphoenol-pyruvate carboxykinase	C.Blue 3GA	Cibacron Blue 3GA agarose (Sigma)	41	6.3	0.05–0.65 M NaCl linear gradient	97
Glucose-6-phosphate dehydrogenase	Re.Red A	Matrex Gel Red A (Amicon)	71	3.3	0–1 M KCl linear gradient	98
Alkylation repair DNA glycosylase	C.Blue F3GA	Affi-Gel blue gel (Bio-Rad)	80	13	0.1–1 M KCl linear gradient	99
Tropine dehydrogenase	Mimetic Orange 1 A6XL	[a]	82	69	1 mM NADP$^+$	100
Tropine dehydrogenase	Re.Red 120	Reactive Red 120 agarose Type 3000-CL (Sigma)	64	3.4	5 mM NADP$^+$ + 0.1 M KCl	100
Citrate synthase isoenzyme I	Re.Red A	Matrex Gel Red A (Amicon)	c	c	0.2 mM oxaloacetate + 0.2 mM CoA	101
Citrate synthase isoenzyme II	Re.Red A	Matrex Gel Red A (Amicon)	c	c	0–3 M KCl linear gradient	101
Phosphatidic acid phosphohydrolase	C.Blue F3GA	Affi-Gel blue gel (Bio-Rad)	62	1.9	1.75 M NaCl	102
Lipoate protein ligase	Green HE4BD	[a]	61	95	0–0.2 M NaH$_2$PO$_4$ linear gradient	103
Lysophospholipase isoenzyme I	C.Blue F3GA	Blue Sepharose CL-6B (Pharmacia)	91	1.4	pH gradient from pH 6–8.5	104
α-Methylacyl-CoA racemase	Re.Blue 72	Reactive Blue 72-agarose (Sigma)	45	1.8	Change pH from 6.8 to 8.5	105
NMN adenylyltransferase	Re.Green A	Matrex Gel Green A (Amicon)	89	19	5–60 mM linear gradient of potassium phosphate pH 7.4	106
Malate dehydrogenase	Re.Red A	Matrex Gel Red A (Amicon)	76	8.2	5 mM NADH	107
Codeinone reductase (NADPH) from:	Re.Red A	Matrex Gel Red A (Amicon)			5 mM NADH	108
cultures			78	28.5		
plants			37	11.5		

Table 3. Continued

Protein/source	Dye	Matrix (company)	Yield (%)	Purification fold	Eluent	Ref.
CobU enzyme kinase activity guanyltransferase activity	C.Blue 3GA	Cibacron Blue 3GA agarose (Sigma) 51	37	1.1	Breakthrough	109
Toluene 2-mono-oxygenase (reductase component)	Re.Green 19 (Sigma)	Reactive Green 19 agarose	32	7.8	0.5 M NaCl	110
NAD(P)H: (quinone-acceptor) oxidoreductase	C.Blue F3GA	Blue Sepharose CL-6B (Pharmacia)	30	398	2–10 mM NADPH linear gradient	111
Mitochondrial NAD$^+$ glycohydrolase	Re.Green 19	Reactive Green 19 agarose (Sigma)	36	2.6	0.6 M NaCl	112
Trimeric aspartate transcarbamoylase	Re.Red A	Matrex Gel Red A (Amicon)	98	1.3	Breakthrough	113
Mg^{2+}-dependent, Ca^{2+}-inhibitable serine/threonine protein phosphatase	C.Blue F3GA	Affi-Gel blue gel (Bio-Rad)	39	8	Breakthrough	114
DNA polymerase β promoter initiator element-binding transcription factor	C.Blue F3GA	Affi-Gel blue gel (Bio-Rad)	26	2.7	0.1–1 M KCl linear gradient	115
Acetate forming enzyme, acetyl-CoA synthetase (ADP-forming)	Re.Red A	Matrex Gel Red A (Amicon)	44	11.5	0.2 M KCl	116
CI repressor protein	C.Blue F3GA	Affi-Gel blue gel (Bio-Rad)	32	2.6	0.1–0.5 M NaCl linear gradient	117
L(−)-Carnitine dehydrogenase	C.Blue F3GA	Blue Sepharose CL-4B (Pharmacia)	60	3.2	10–400 mM stepwise gradient of potassium phosphate pH 7.4	118

Fructokinase	P.Red HE-3B	Rainbow-Sorb High Dye number 25 (Center for Protein and Enzyme Technology)	96	1.9	Change from 20 mM K-Mes buffer pH 6 to 50 mM K-phosphate buffer pH 8	119
Recombinant human 'mini'-hexokinase	Re.Blue A	Matrex Gel Blue A (Amicon)	[a]	[a]	1.5 mM glucose 6-phosphate	120
Diacylglycerol pyrophosphate phosphatase	C.Blue F3GA	Affi-Gel blue gel (Bio-Rad)	98	5.8	0.3–0.9 M NaCl linear gradient	121
Diacylglycerol pyrophosphate phosphatase	Drimarene Rubine R/K 5BL (Sandoz)	Coupled directly to Sepharose CL-4B (Pharmacia)[d]	≈100	[e]	0.3–0.9 M NaCl linear gradient	122
Carbamoyl-phosphate synthetase from *Pyrococcus abyssi*	C.Blue F3GA	DEAE Affi-Gel blue gel (Bio-Rad)	99	4	0–0.3 M NaCl linear gradient	123
NADH oxidase	C.Blue F3GA	Affi-Gel blue gel (Bio-Rad)	30	5.1	0–0.25 M KCl exponential gradient	124
Theta glutathione S-transferase	Re.Orange A	Dyematrex Orange A (Amicon)	46	2	0.2–1 M NaCl linear gradient	125
Theta-class glutathione transferase (GSTT2–2)	Re.Orange A (Amicon)	Dyematrex Orange A	12	1.2	0–1 M KCl linear gradient	126
Dehydro-ascorbic acid-reductase (glutaroredoxin)	P.Red HE-3B	Red Sepharose CL-6B (Pharmacia)	77	3.6	0–2 M NaCl linear gradient	127
Xanthine dehydrogenase	C.Blue F3GA	Blue Sepharose CL-6B (Pharmacia)	72	7.1	0–0.6 M NaCl linear gradient	128
Lysophospho-lipase	C.Blue F3GA	Blue Sepharose CL-6B (Pharmacia)	120	3.4	Breakthrough	129
Cytosolic folypoly-γ-glutamate synthetase	C.Blue F3GA	Affi-Gel blue gel (Bio-Rad)	85	11.5	5–150 mM linear gradient of potassium phosphate pH 7.5	130

Table 3. Continued

Protein/source	Dye	Matrix (company)	Yield (%)	Purification fold	Eluent	Ref.
Phosphatidyl-inositol-3,4,5-triphosphate phosphatase	C.Blue F3GA	Hi-Trap (Pharmacia)	77	3.4	0–0.1 M NaCl linear gradient	131
Cytidine 5′-monophosphate N-acetyl-neuraminic acid synthetase	Re.Green 19	Reactive Green 19 agarose (Sigma)	53	2.1	0.5 M NaCl	132
Reversible artificial mediator accepting NADH oxidoreductase	C.Blue F3GA	Blue Sepharose CL-6B (Pharmacia)	44	3.3	0.2–0.5 M KCl linear gradient	133
ZAP-70 protein-tyrosine kinase	Re.Yellow 3	Reactive Yellow 3 agarose (Sigma)	83	10	0.15–1.5 M NaCl linear gradient	134

P.Blue/Yellow/Red/Navy = Procion Blue/Yellow/Red/Navy.
C.Blue/Navy = Cibacron Blue/Navy.
Re.Blue/Yellow/Red/Green = Reactive Blue/Yellow/Red/Green.
[a] Not presented.
[b] 94–98% pure.
[c] Was impossible to calculate because crude extract contained both isoenzymes I and II.
[d] Dye contains copper which was changed to zinc.
[e] Essentially pure.

6: Nucleotide- and dye-ligand chromatography

are not uncommon. Certainly, the purification fold depends greatly on the content of the target protein in the crude mixture applied to the column. Higher purification folds are obtained when dye–ligand chromatography is the first or one of the initial steps in the purification protocol, i.e. when more crude extracts with less content of the target protein are used. On many occasions, only one- to twofold purification is obtained. Incorporation of the

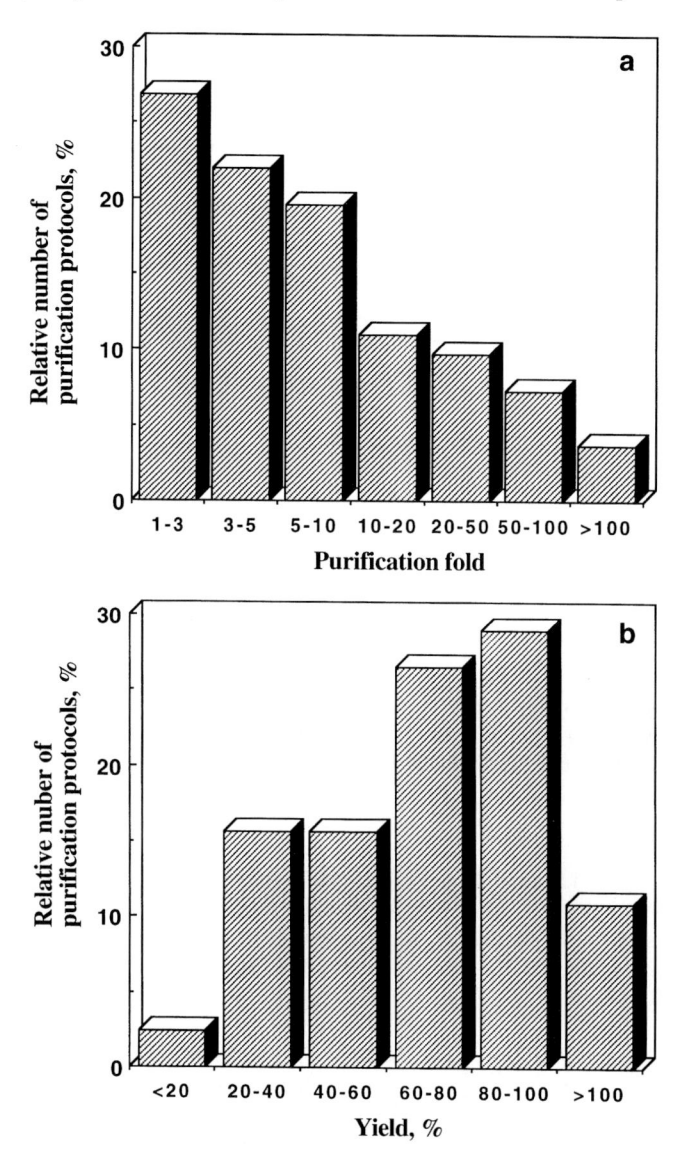

Figure 7. Relative number of purifications using immobilized dyes with given (a) purification fold and (b) yield. Data from *Table 3*.

dye–affinity step is still justified in these cases as it removes the components which may otherwise interfere with the subsequent purification.

Yields of the target protein are mainly in the range 60–100% (*Figure 7b*). About 10% of reported purification protocols have yields higher than 100%. Probably, the reason is removal of some inhibitors which decrease activity of the target protein in the crude extract. In most of the examples shown here the dye–affinity step has not been optimized, and it is likely that the results can be significantly improved.

From the above, it is obvious that both immobilized coenzyme analogues and textile dyes offer interesting binding properties to be used in affinity-mediated purification of proteins. As the trend today is towards use of affinity early in the purification, there is also a trend towards use of robust, stable, and preferably autoclavable ligands. This fact promotes the development in improving these ligands even further.

References

1. Walsh, G. and Headon, D. (1994). *Protein biotechnology*, p. 1. John Wiley & Sons, Chichester.
2. Jones, K. (1991). *Chromatographia*, **32**, 469.
3. Bückmann, A. F. and Wray, V. (1992). *Biotechnol. Appl. Biochem.*, **15**, 303.
4. Månsson, M.-O. and Mosbach, K. (1987). In *Pyridine nucleotide coenzymes. Chemical, biochemical, and medical aspects. Part B* (ed. D. D. O. Avramovic and R. Poulson), p. 217. John Wiley & Sons, New York.
5. Brodelius, P., Larsson, P.-O., and Mosbach, K. (1974). *Eur. J. Biochem.*, **47**, 81.
6. Craven, D. R., Harvey, M. J., Lowe, C. R., and Dean, P. D. G. (1974). *Eur. J. Biochem.*, **41**, 329.
7. Barry, S. and O'Carra, P. (1973). *FEBS Lett.*, **37**, 134.
8. Cautrecasas, P. (1970). *J. Biol. Chem.*, **245**, 3059.
9. Bückman, A. F. (1987). *Biocatalysis*, **1**, 173.
10. Okuda, K., Urabe, I., and Okada, H. (1985). *Eur. J. Biochem.*, **151**, 33.
11. Alhama, J., López-Barea, J., and Toribio, F. (1991). *J. Chromatogr.*, **586**, 51.
12. Lamed, R., Levin, Y., and Wilchek, M. (1973). *Biochim. Biophys. Acta*, **304**, 231.
13. Wilchek, M. and Lamed, R. (1974). In *Methods in enzymology* (ed. W. B. Jakoby and M. Wilchek), Vol. 34, p. 475. Academic Press, New York.
14. Hermanson, G. T., Mallia, A. K., and Smith, P. K. (1992). *Immobilized affinity ligand techniques*, p. 173. Academic Press, Inc., San Diego.
15. Baker, R., Olsen, K. W., Shaper, J. H., and Hill, R. L. (1972). *J. Biol. Chem.*, **247**, 7135.
16. Pfeuffer, E. and Pfeuffer, T. (1991). In *Methods in enzymology* (ed. R. A. Johnson and J. D. Corbin), Vol. 195, p. 171. Academic Press, Inc., San Diego.
17. Place, A. R. and Powers, D. A. (1984). *J. Biol. Chem.*, **259**, 1299.
18. Scopes, R. K. (1986). *J. Chromatogr.*, **376**, 131.
19. Stead, C. V. (1987). In *Reactive dyes in protein and enzyme technology* (ed. Y. D. Clonis, T. Atkinson, C. J. Bruton, and C. R. Lowe), p. 13. Stockton Press, NY.

6: Nucleotide- and dye-ligand chromatography

20. Stellwagen, E. (1990). In *Methods in enzymology* (ed. M. P. Deutscher), Vol. 182, p. 343. Academic Press, London.
21. Qadri, F. (1985). *Trends Biotechnol.*, **3**, 7.
22. Lowe, C. R. (1983). *Top. Enzyme Ferment. Biotechnol.*, **9**, 78.
23. Milovicova, D., Kovacova, M., Petro, M., and Bozek, P. (1995). *J. Liquid Chromatogr.*, **18**, 3061.
24. Clonis, Y. D. (1987). *J. Chromatogr.*, **407**, 179.
25. Clonis, Y. D., Jones, K., and Lowe, C. R. (1986). *J. Chromatogr.*, **363**, 31.
26. Small, D. A. P., Atkinson, A., and Lowe, C. R. (1981). *J. Chromatogr.*, **216**, 175.
27. Lowe, C. R., Small, D. A. P., and Atkinson, A. (1981). *J. Chromatogr.*, **215**, 303.
28. Chambers, G. K. (1977). *Anal. Biochem.*, **83**, 551.
29. Clonis, Y. D. (1987). In *Reactive dyes in protein and enzyme technology* (ed. Y. D. Clonis, T. Atkinson, C. J. Bruton, and C. R. Lowe), p. 33. Stockton Press, NY.
30. Stead, C. V. (1987). *J. Chem. Tech. Biotechnol.*, **37**, 55.
31. Weber, B. H., Willeford, K., Moe, J. G., and Piszkiewicz, D. (1979). *Biochem. Biophys. Res. Commun.*, **86**, 252.
32. Federici, M. M., Chock, P. B., and Stadtman, E. R. (1985). *Biochemistry*, **24**, 647.
33. Kelley, B. D. and Hatton, T. A. (1991). *Bioseparation*, **1**, 333.
34. Lowe, C. R. (1987). In *Reactive dyes in protein and enzyme technology* (ed. Y. D. Clonis, T. Atkinson, C. J. Bruton, and C. R. Lowe), p. 1. The Macmillan Press Ltd., London.
35. Boyer, P. M. and Hsu, J. T. (1993). In *Advances in biochemical engineering* (ed. A. Fiechter), Vol. 49, p. 1. Springer–Verlag, Berlin, Heidelberg.
36. Santambien, P., Hulak, I., Girot, P., and Boschetti, E. (1992). *Bioseparation*, **2**, 327.
37. Santambien, P., Girot, P., Hulak, I., and Boschetti, E. (1992). *J. Chromatogr.*, **597**, 315.
38. Santambien, P., Girot, P., Hulak, I., and Boschetti, E. (1992). *J. Biochem. Biophys. Methods*, **24**, 285.
39. Stewart, D. J., Purvis, D. R., Pitts, J. M., and Lowe, C. R. (1992). *J. Chromatogr.*, **623**, 1.
40. Bertrand, O., Boschetti, E., Cochet, S., Girot, P., Hebert, E., Monsigny, M., *et al.* (1994). *Bioseparation*, **4**, 299.
41. Santambien, P., Sdiqui, S., Hubert, E., Girot, P., Roche, A. C., Monsigny, M., *et al.* (1995). *J. Chromatogr.*, **664**, 241.
42. Hulak, I., Nguyen, C., Girot, P., and Boschetti, E. (1991). *J. Chromatogr.*, **539**, 355.
43. Biellmann, J.-F., Samama, J.-P., Bränden, C.-I., and Eklund, H. (1979). *Eur. J. Biochem.*, **102**, 107.
44. Burton, S. J., Stead, C. V., and Lowe, C. R. (1988). *J. Chromatogr.*, **455**, 201.
45. Lowe, C. R., Burton, S. J., Pearson, J. C., and Clonis, Y. D. (1986). *J. Chromatogr.*, **376**, 121.
46. Labrou, N. E., Karagouni, A., and Clonis, Y. D. (1995). *Biotechnol. Bioeng.*, **48**, 278.
47. Labrou, N. E. and Clonis, Y. D. (1995). *J. Chromatogr. A*, **718**, 35.
48. Burton, S. J., Stead, C. V., and Lowe, C. R. (1990). *J. Chromatogr.*, **508**, 109.
49. Lindner, N. M., Jeffcoat, R., and Lowe, C. R. (1989). *J. Chromatogr.*, **473**, 227.
50. Clonis, Y. D., Stead, C. V., and Lowe, C. R. (1987). *Biotechnol. Bioeng.*, **30**, 621.
51. Stead, C. V. (1991). *Bioseparation*, **2**, 129.

52. Bruce, N. C., Wilmot, C. J., Jordan, K. N., Stephens, L. D. G., and Lowe, C. R. (1991). *Biochem. J.*, **274**, 875.
53. Galaev, I. Y., Arvidsson, P., and Mattiasson, B. (1995). *J. Chromatogr.*, **710**, 259.
54. Galaev, I. Y. and Mattiasson, B. (1994). In *Separations for biotechnology* (ed. D. L. Pyle), Vol. 3, p. 179. SCI, Cambridge.
55. Galaev, I. Y. and Mattiasson, B. (1994). *J. Chromatogr. A*, **662**, 27.
56. Galaev, I. Y., Garg, N., and Mattiasson, B. (1994). *J. Chromatogr. A*, **684**, 45.
57. Galaev, I. Y. and Mattiasson, B. (1994). *Bio/Technology*, **12**, 1086.
58. Mattiasson, B., Galaev, I. Y., and Garg, N. (1996). *J. Mol. Recog.*, **9**, 259.
59. Garg, N., Galaev, I. Yu., and Mattiasson, B. (1994). *Biotech. Tech.*, **8**, 645.
60. Garg, N., Galaev, I. Y., and Mattiasson, B. (1996). *Bioseparation*, **6**, 193.
61. Garg, N., Galaev, I. Y., and Mattiasson, B. (1997). *Isolation Purification*, **2**, 301.
62. Makriyannis, T. and Clonis, Y. D. (1993). *Process Biochem.*, **28**, 179.
63. Tsamidis, G., Papageorgakopoulou, N., and Clonis, Y. D. (1992). *Bioprocess Eng.*, **7**, 213.
64. Galaev, I. Y. and Mattiasson, B. (1993). *Enzyme Microb. Technol.*, **15**, 354.
65. Watson, D. H., Harvey, M. J., and Dean, P. D. G. (1978). *Biochem. J.*, **173**, 591.
66. Galaev, I. Y. and Mattiasson, B. (1993). *J. Chromatogr.*, **648**, 367.
67. Algiman, E., Kroviarski, Y., Cochet, S., and Sing, Y. L. K. (1990). *J. Chromatogr.*, **510**, 165.
68. Costa-Ferreira, M. and Duarte, J. C. (1991). *J. Chromatogr.*, **539**, 507.
69. Mottl, H. and Keck, W. (1992). *Protein Expression Purification*, **3**, 403.
70. Lanznaster, N. and Croteau, R. (1991). *Protein Expression Purification*, **2**, 69.
71. Nagata, Y., Maeda, K., and Scopes, R. K. (1992). *Bioseparation*, **2**, 353.
72. Regnault, V., Rivat, C., Vallar, L., and Stoltz, J.-F. (1992). *J. Chromatogr. Biomed. Appl.*, **576**, 87.
73. Berkovsky, A. L. and Potapov, P. P. (1994). *J. Chromatogr.*, **656**, 432.
74. Mikhailov, A. M., Serebrennikov, V. M., Kobzeva, N. Y., Gorbacheva, I. V., and Bezborodova, S. I. (1993). *Appl. Biochem. Microbiol.*, **29**, 149.
75. Pech, L. L. and Nelson, D. L. (1994). *Biochim. Biophys. Acta*, **1199**, 183.
76. Hansbro, P. M., Foster, P. S., Hogan, S. P., Ozaki, S., and Dedborough, M. (1994). *Arch. Biochem. Biophys.*, **311**, 47.
77. Raffaelli, N., Amici, A., Emanuelli, M., Ruggieri, S., and Magni, G. (1994). *FEBS Lett.*, **355**, 233.
78. Rehberg, E. F., Greenlee, K. A., Melchior, G. W., and Marotti, K. R. (1994). *Protein Expression Purification*, **5**, 285.
79. Phillippy, B. Q., Ullah, A. H. J., and Ehrlich, K. C. (1994). *J. Biol. Chem.*, **269**, 28393.
80. Chou, M.-Y., Li, S.-C., Kiso, M., Hasegawas, A., and Li, Y.-T. (1994). *J. Biol. Chem.*, **269**, 18821.
81. Bruchhaus, I. and Tannich, E. (1994). *Biochem. J.*, **303**, 743.
82. Gärtner, D. E., Kielholz, W., and Seitz, H. U. (1994). *Eur. J. Biochem.*, **225**, 1125.
83. Liebergesell, M., Sonomoto, K., Madkour, M., Mayer, F., and Steinbuchel, A. (1994). *Eur. J. Biochem.*, **226**, 71.
84. Bello, B. D., Maellaro, E., Sugherini, L., Santucci, A., Comporti, M., and Cassini, A. F. (1994). *Biochem. J.*, **304**, 385.
85. Pontoppidan, B. and Kannangara, C. G. (1994). *Eur. J. Biochem.*, **225**, 529.
86. Crooks, G. P. and Copley, S. D. (1994). *Biochemistry*, **33**, 11645.

87. Cacciapuoti, G., Porcelli, M., Bertoldo, C., Rosa, M. D., and Zappia, V. (1994). *J. Biol. Chem.*, **269**, 24762.
88. Ding, L., Seto, B. L., Ahmed, S. A., and Coleman, J. (1994). *J. Biol. Chem.*, **269**, 24384.
89. Martinez-Costa, O. H., Estevez, A. M., Sanchez, V., and Aragon, J. R. (1994). *Eur. J. Biochem.*, **226**, 1007.
90. Burdette, D. and Zeikus, J. G. (1994). *Biochem. J.*, **302**, 163.
91. Alderton, W. K., Lowe, C. R., and Thatcher, D. R. (1994). *J. Chromatogr.*, **677**, 289.
92. Canales, J., Pinto, R. M., Costas, M. J., Hernandez, M. T., Miró, A., Bernet, D., *et al.* (1995). *Biochim. Biophys. Acta*, **1246**, 167.
93. Liu, Y., Heeswijck, R. V., Hoj, P., and Hoogenraad, N. (1995). *Eur. J. Biochem.*, **228**, 291.
94. Schmidt, S. R., Muller, C. R., and Kress, W. (1995). *Eur. J. Biochem.*, **228**, 425.
95. Belinga, H. F., Steghens, J. P., and Collombel, C. (1995). *J. Chromatogr.*, **695**, 33.
96. Toyoguchi, T., Yamaguchi, K., Imajoh-Ohmi, S., Kato, N., Kusunoki, M., Kageyama, H., *et al.* (1995). *Biochim. Biophys. Acta*, **1250**, 90.
97. Hunt, M. and Köhler, P. (1995). *Biochim. Biophys. Acta*, **1249**, 15.
98. Anderson, B. M. and Anderson, C. D. (1995). *Arch. Biochem. Biophys.*, **321**, 94.
99. Bjorås, N., Klungland, A., Johansen, R. F., and Seeberg, E. (1995). *Biochemistry*, **34**, 4577.
100. Bartholomew, B. A., Smith, M. J., Long, M. T., Darcy, P. J., Trudgill, P. W., and Hopper, D. J. (1995). *Biochem. J.*, **307**, 603.
101. Mitchell, C. G., Anderson, S. C. K., and El-Mansi, E. M. T. (1995). *Biochem. J.*, **309**, 507.
102. Fleming, I. N. and Yeaman, S. J. (1995). *Biochem. J.*, **308**, 983.
103. Green, D. E., Morris, T. W., Green, Jr. J. E. C., and Guest, J. R. (1995). *Biochem. J.*, **309**, 853.
104. Sunaga, H., Sugimoto, H., Nagamachi, Y., and Yamashita, S. (1995). *Biochem. J.*, **308**, 551.
105. Schmitz, W., Albers, C., Fingerhut, R., and Conzelmann, E. (1995). *Eur. J. Biochem.*, **231**, 815.
106. Balducci, E., Orsomando, G., Polzonetti, V., Vita, A., Emanuelli, M., Raffaelli, N., *et al.* (1995). *Biochem. J.*, **310**, 395.
107. Mahmoud, Y. A.-G., Souod, S. M. A. E., and Niehaus, W. (1995). *Arch. Biochem. Biophys.*, **322**, 69.
108. Lenz, R. and Zenk, M. H. (1995). *Eur. J. Biochem.*, **233**, 132.
109. O'Toole, G. A. and Escalante-Semerena, J. (1995). *J. Biol. Chem.*, **270**, 23560.
110. Newman, L. M. and Wackett, L. P. (1995). *Biochemistry*, **34**, 14066.
111. Trost, P., Bonora, P., Scagliarini, S., and Pupillo, P. (1995). *Eur. J. Biochem.*, **234**, 452.
112. Zhang, J., Ziegler, M., Schneider, R., Klocker, H., Auer, B., and Schweiger, M. (1995). *FEBS Lett.*, **377**, 530.
113. Baker, D. P., Aucoin, J. M., Williams, M. K., DeMello, L. A., and Kantrowitz, E. R. (1995). *Protein Expression Purification*, **6**, 679.
114. Wang, Y., Santini, F., Qin, K., and Huang, C. Y. (1995). *J. Biol. Chem.*, **34**, 25607.
115. He, F., Narayan, S., and Wilson, S. H. (1996). *Biochemistry*, **35**, 1775.
116. Sanchez, L. B. and Müller, M. (1996). *FEBS Lett.*, **378**, 240.

117. Shearwin, K. E. and Egan, J. B. (1996). *J. Biol. Chem.*, **271**, 11525.
118. Hanschmann, H., Ehricht, R., and Kleber, H.-P. (1996). *Biochim. Biophys. Acta*, **1290**, 177.
119. King, K., Phan, P., Rellos, P., and Scopes, R. K. (1996). *Protein Expression Purification*, **7**, 373.
120. Bianchi, M., Serafini, G., Corsi, D., and Magnani, M. (1996). *Protein Expression Purification*, **7**, 58.
121. Wu, W.-I., Liu, Y., Riedel, B., Wissing, J. B., Fischl, A. S., and Carman, G. M. (1996). *J. Biol. Chem.*, **271**, 1868.
122. Cochet, S., Pesliakas, H., Kroviarski, Y., Carton, J. P., and Bertrand, O. (1996). *J. Chromatogr. A*, **725**, 237.
123. Purcarea, C., Simon, V., Prieur, D., and Herve, G. (1996). *Eur. J. Biochem*, **236**, 189.
124. Masullo, M., Raimo, G., Ruso, A. D., Bocchini, V., and Bannister, J. V. (1996). *Biotechnol. Appl. Biochem.*, **23**, 47.
125. Mainwarning, G. W., Nash, J., Davidson, M., and Green, T. (1996). *Biochem. J.*, **314**, 445.
126. Tan, K.-L. and Board, P. G. (1996). *Biochem. J.*, **315**, 727.
127. Park, J. B. and Levine, M. (1996). *Biochem. J.*, **315**, 931.
128. Xiang, Q. and Edmondson, D. E. (1996). *Biochemistry*, **35**, 5441.
129. Sugimoto, H., Hayashi, H., and Yamashita, S. (1996). *J. Biol. Chem.*, **271**, 7705.
130. Chen, L., Qi, H., Korenberg, J., Garrow, T. A., Choi, Y.-J., and Shane, B. (1996). *J. Biol. Chem.*, **271**, 13077.
131. Kabuyama, Y., Nakatsu, N., Homma, Y., and Fukui, Y. (1996). *Eur. J. Biochem.*, **238**, 350.
132. Tullius, M. V., Munson, R. S., Wang, J., and Gibson, B. W. (1996). *J. Biol. Chem.*, **271**, 15373.
133. Bayer, M., Walter, K., and Simon, H. (1996). *Eur. J. Biochem.*, **239**, 686.
134. Isakov, N., Wange, R. L., Watts, J. D., Aebersold, R., and Samuelson, L. E. (1996). *J. Biol. Chem.*, **271**, 15753.

7

Synthetic peptides as affinity ligands

GARY J. REYNOLDS and PAUL A. MILLNER

1. Introduction

There are many reasons for wanting to construct an immobilized peptide matrix. Purification of a specific antibody is one typical application. In this approach the antibody to be purified would either be raised against peptide conjugated to a carrier of some sort and the same peptide used to construct the matrix for purification. Alternatively, the antibody may have been raised against a whole protein and the peptide–matrix used to select antibodies binding to a specific epitope. The latter is the basis for many of today's highly specific and powerful protein purification procedures using peptide affinity chromatography.

1.1 Peptides (production, purification, and handling)

Synthetic peptides are usually prepared by a solid phase synthesis (1). The level of sophistication of modern synthesis facilities means that peptides up to and over 100 residues can be made (2) with reasonable care, in quantities up to hundreds of milligrams. In practice most peptide synthesis involves peptides of 30 amino acids or less due mainly to the propensity of larger peptides to form secondary structures and the expense. Additional advantage are that shorter peptides tend not to exhibit preferred conformations and possess substantial segmental flexibility (3).

Since the solvents for synthesis are organic, peptides containing mostly hydrophobic residues cause few problems during synthesis. However, the aqueous solvents used for affinity chromatography can cause solubility problems for such peptides when used as ligands. Techniques dealing with coupling hydrophobic peptides will be dealt with within each activation/coupling chemistry as appropriate, but it is worth mentioning that some peptides are just too hydrophobic to be practically useful. When coupling of peptides to activated matrices does take place in an organic environment, however, coupling efficiency is nearly always improved since hydrolysis of the activated matrix cannot occur (although the expense of coupling under organic conditions, the solvent waste generated, and the associated solvation problems with gel hydration/dehydration often acts as a strong disincentive). When a peptide exhibits solubility problems, this does not preclude its use for affinity chromatography as solubility can often be improved by dissolution in up to 10% (v/v) acetic acid for basic peptides or in dilute NH_4OH or basic buffers (e.g. $NaHCO_3$) for acidic peptides. Harsher conditions to aid dissolution include solutions such of urea and guanidium hydrochloride.

A major consideration when preparing synthetic peptides for coupling to matrices is the level of purity required. In terms of quality control, the minimum assessment of the crude peptide should be mass determination and desalting. Probably the most common contaminants encountered are peptides containing additions or deletions; it is important to consider whether such contaminant peptides are important to the chromatography to be performed since they will also couple to activated matrices. Homogeneity and quantitative purification can usually be achieved in one step by reverse-phase (RP) HPLC and milligram quantities are achievable using analytical scale HPLC equipment. When the availability of peptide to be coupled is limited, the use of salting-out conditions can still enable high coupling yield and minimal losses. To achieve this, saturated solution of non-chaotropic salts, e.g. NaCl, should be add to the coupling solution until the solution just starts to become cloudy (ammonium salts should be avoided as primary amines interfere with many of the coupling procedures described within this chapter). The peptide will then partially precipitate on the matrix surface raising the local concentration and driving the coupling reaction.

Special attention should always be paid to the quality of buffers and reagents used. Peptides should be dissolved in deionized water that has been degassed or N_2 gassed in order to avoid oxidation problems and AnalaR grade reagents should be used since metal ions can catalyse tryptophan degradation and peptide bond cleavage. Whenever possible, growth of micro-organisms should be retarded by the inclusion of 0.05% (w/v) NaN_3. Neutral or slightly basic buffers are also recommended. This is particularly so for peptides containing Asp–peptide bonds which are acid labile, with Asp–Pro showing the greatest sensitivity. Finally, peptides should be stored only for short periods in solution and for longer periods should be stored dehydrated at –20°C. Care should be taken to allow peptide stocks to come to ambient temperature before manipulation, weighing, etc. as most peptides are fairly hygroscopic.

1.2 Immobilization chemistries

The most useful immobilization chemistries are those that lead to a single defined peptide to matrix link. However, random coupling can be of use under certain circumstances and some chemistries are described in Section 4. Synthetic peptides can in many cases be designed to contain a single coupling group along their length although this may be impractical for longer peptides where residue duplication is inevitable and multiple point coupling would ensue.

7: Synthetic peptides as affinity ligands

Groups that are available for directed coupling within peptides are:

(a) α-Amine of the N-terminal residue and ε-amine of lysine.
(b) α-Carboxyl groups of the C-terminal residue and β- and γ-carboxyls of aspartic acid or glutamic acid.
(c) Sulfydryl group of cysteine.

Other reactive groups can be used for immobilization but tend to be avoided because of the difficult chemical steps or undesirable side-reactions. These include:

(a) Thioether group of methionine.
(b) Secondary amine of proline (pyrrolidine ring).
(c) Active hydrogen of tyrosine.
(d) Aromatic ring of phenylalanine.
(e) Indolyl ring of tryptophan.
(f) The polar side chains of asparagine, glutamine, serine, and threonine.

Because of the propensity of peptides to contain more than one amine or carboxyl group, directed coupling to either of these groups can result in coupling of the peptide in more than one orientation. However, this is still preferable to random coupling.

In some cases, the complete peptide sequence is required for functionality of the affinity matrix. In this case, spacer arms (sometimes known as 'leashes') can be incorporated between the matrix and the peptide. Alternatively, during peptide synthesis extra amino acids can be incorporated at the C- or N-terminus to allow greater mobility of the peptide and reduce steric hindrance. In many cases, both of these strategies lead to an increase in the effectiveness of the affinity interaction. The extra (spacer) residues are selected for lack of side chain reactivity; glycine and alanine respectively are particularly suitable. However, if spacer residues are incorporated to the peptide it is still worth terminating the peptide with an appropriate reactive residue, e.g. N-terminal lysine increases the chances of N-terminal coupling since both ε-NH$_2$ and α-NH$_2$ groups can couple. Finally, some workers have reported that over-long spacer arms can promote non-specific interactions between ligand and target and reduce the selectivity of the separation (4). It should also be mentioned that deleterious steric effects can also be invoked by 'over-coupling' the peptide to the matrix; this is often performed to compensate for ligand leaching from the affinity matrix. However, there are two options to counteract this:

(a) Lowering the number of activated groups on the matrix.
(b) Decreasing the amount of peptide to be coupled to the activated groups, and in general too high coupling levels are best avoided since most of the currently used chemistries result in a highly stable matrix.

Table 1. Commercially available derivatized and activated matrices for affinity chromatography

Supplier	Gel name	Reactive side chain	Activation chemistry	Spacer arm length	Coupling agent
Pharmacia	CNBr Sepharose 6MB (HB agarose)	NH$_2$	CNBr	None	None
Pharmacia	NHS Superose HR and HiTrap NHS (cross-linked agarose)	NH$_2$	NHS	12 atom	None
Pharmacia	CH Sepharose 4B (HB agarose)	NH$_2$	NHS	6 carbon	None
Pharmacia	EAH Sepharose 4B (HB agarose)	NH$_2$		10 atom	Carbodiimide
Pharmacia	ECH Sepharose 4B (HB agarose)	COOH		9 atom	Carbodiimide
Pharmacia	Epoxy activated Sepharose 4B (HB agarose)	OH, NH$_2$, SH	Epoxy	12 atom	None
Pharmacia	Thiol Sepharose 4B (HB agarose)	SH		4 atom	Various[a]
Pharmacia	Thiopropyl Sepharose 6B (HB agarose)	SH		7 atom	Various[a]
Pierce	AminoLink (cross-linked (CL) agarose)	NH$_2$	Glutaraldehyde	None	None[b]
Pierce	UltraLink iodoacetyl (bisacrylamide/azalactone copolymer)	SH	Iodoacetyl	15 atom	None
Pierce	Reacti-Gel 6× (CL agarose)	NH$_2$	CDI	2 atom	None
Pierce	SulfoLink (CL agarose)	SH	Iodoacetyl	12 atom	None
Pierce	Immobilized TNB-thiol	SH	SH	12 carbon	Various[a]
Pierce	UltraLink carboxy (bisacrylamide/azalactone copolymer)	NH$_2$		2 carbon	Carbodiimide
Pierce	Immobilized diaminodipropylamine (cross-linked agarose)	COOH		9 atom	Carbodiimide
Pierce	ImmunoPure epoxy activated agarose (CL agarose)	OH, NH$_2$, SH	Epoxy	13 atom	None
Bio-Rad	Affi-Gel 10 (CL agarose)	NH$_2$ (pI 6.5–11)	NHS	10 atom (uncharged)	None
Bio-Rad	Affi-Gel 15 (CL agarose)	NH$_2$ (below pI 6.5)	NHS	15 atom (positive)	None
Bio-Rad	CM BioGel A (CL agarose)	NH$_2$		2 carbon	Carbodiimide
Bio-Rad	Affi-Gel 102 (CL agarose)	COOH		6 atom	Carbodiimide
Bio-Rad	Affi-Gel 501 (CL agarose)	SH	HgCl	8 carbon	None
PerSeptive Biosystems	POROS EP (cross-linked polystyrene/divinylbenzene)	OH, NH$_2$, SH	Epoxy	3 atom	None

Supplier	Product	Functional groups	Cross-linker	Spacer	Alternative cross-linker
PerSeptive Biosystems	POROS AL (cross-linked polystyrene/divinylbenzene)	NH$_2$	Glutaraldehyde	None	None[b]
PerSeptive	POROS NH (cross-linked polystyrene/divinylbenzene)	COOH		2 atom	Carbodiimide
Biosystems Presearch	BIOSPHER GM 1000 Epoxy[5] (glycidylmethacrylate)	OH, NH$_2$, SH	Epoxy	3 atom	None
Rohm Pharma[c]	Eupergit C (methacrylate)	OH, NH$_2$, SH	Epoxy	None	None
Rohm Pharma[c]	Eupergit C 250 L (methacrylate)	OH, NH$_2$, SH	Epoxy	None	None
Alltech Associates	HEMA-AFC BIO 1000 EL and EH (methacrylate)	OH, NH$_2$, SH	Epoxy	10 atoms	None
Alltech Associates	HEMA-AFC BIO 1000[8] VS (1%) and VSL (0.1%)[10] (methacrylate)	NH$_2$, OH	Vinyl sulfone	10 atoms	None
Waters	Protein-Pak affinity epoxy activated[d] (silica)	OH, NH$_2$, SH	Epoxy	7 atom	None
Merck	Fractogel epoxy 650(S)	OH, NH$_2$, SH	Epoxy	?	None
Toso Haas	Toyopearl AF-Tresyl-650M (Toyopearl HW-65 F - methacrylate)	NH$_2$	Tresyl chloride	?	None
Toso Haas	Toyopearl AF-Epoxy-650M (Toyopearl HW-65 F - methacrylate)	OH, NH$_2$	Epoxy	?	None
Sigma	Tosyl chloride activated agarose (CL agarose)	NH$_2$	Tosyl chloride	None	None
Sigma	Tresyl chloride activated agarose (CL agarose)	NH$_2$	Tresyl chloride	None	None
Sigma	N,N'-carbonyldiimidazole activated agarose (CL agarose)	NH$_2$	CDI	1 atom	None
Sigma	Epichlorohydrin activated agarose (CL 6% and 4% beaded agarose)	OH, NH$_2$, SH	Epichlorohydrin	3 atom	None
Sigma	Vinyl sulfone activated agarose (CL agarose)	NH$_2$, OH	Vinyl sulfone	5 atom	None

[a] SH containing peptides will form reversible -SS- bonds by oxidation, but homobifunctional -SH-directed cross-linkers must be used to produce an irreversible link.
[b] Needs reducing with sodium borohydride to form an irreversible link.
[c] Rohm Pharma (Darmstadt, Germany) distribute Eupergit via several suppliers, e.g. Fluka.
[d] Basically an HPLC packing material available loose or in pre-packed microcolumns.

1.3 Choice of matrix

The range of commercially available activated matrices is now very comprehensive, as the sophistication of chromatographic methods has increased, many researchers have sought matrices that are useful in terms of physical properties of the actual gel. For example, gels based on cross-linked agarose, although still popular, cannot be incorporated into HPLC methods that offer advantages in speed, efficiency, and reproducibility. In addition, cross-linking agarose has the effect of 'using up' hydroxyls that would be used for activation purposes. Ideally, then a matrix should possess the following attributes:

(a) Physical stability, in order to avoid matrix compression.

(b) Hydrophilic character, to avoid non-specific hydrophobic interactions.

(c) Chemically inert and capable of resisting extreme pH changes, denaturants, etc.

(d) Macroporous, to provide a high gel surface area.

(e) Highly activated or capable of being highly activated.

(f) Inexpensive to purchase and use.

With these attributes in mind, some companies (e.g. PerSeptive Biosystems and Presearch) have developed matrices (POROS and BIOSPHER respectively) which offer high flow rates and are available in both underivatized and activated forms (*Table 1*). However, a general feature of such modern matrix materials is a lower activation level (typically between 20–40%) than that found with agarose. At the analytical level this maybe acceptable, but may mean that scaling up becomes financially prohibitive.

2. Coupling peptides via amine groups directly to the matrix

There are several useful activation procedures that permit direct coupling of amines to the matrix without spacers although coupling of small peptides (< ten residues) is not recommended using such protocols unless something is known of the nature of binding concerned. Some of the protocols are useful for peptides that are not normally soluble in aqueous media. Only a selection of the most useful will be dealt with here.

2.1 Cyanogen bromide (CNBr) activation

The cyanogen bromide method (5) was the most widespread technique for attaching peptides to matrices but it is being slowly replaced by modern activation techniques.

However, it is still popular due to ease of use, good immobilization yield,

7: Synthetic peptides as affinity ligands

and almost any synthetic polymer containing hydroxyl groups can be activated with CNBr. Moreover, the covalent coupling is quite mild and the size of peptide does not affect coupling efficiency. Many companies supply pre-activated matrices for even greater convenience. On the down side, there is always some leaching of peptide from the support at basic and acidic elution conditions. If the peptide contains multiple amines then multiple cross-linking cannot be avoided, but this has the advantage of reducing ligand leaching and cross-linking the matrix leading to enhanced chemical stability (6).

Traditional activation of gels involved the addition of CNBr to a stirred aqueous slurry of agarose beads (7, 8) requiring manual titration to maintain a basic pH. Activation can be achieved as effectively using the method of March et al. (9) (see Protocol 1, steps 1–4) which has the advantage of dispensing with the need for pH monitoring of a reaction that lacks a real end-point and is thus difficult to reproduce. As a convenience pre-activated gels are now available from a number of sources (Table 1) and are usually formulated as a lyophilized powder. Such gels require swelling, which can be accomplished by incubation for 20 min in 1 mM HCl in order to preserve the active groups that will hydrolyse under coupling (basic) conditions. The gel should then be washed quickly with the coupling buffer (approx. 5 ml buffer/3.5 ml swollen gel) prior to coupling. Both activated gels and commercially available pre-activated gels can then coupled to peptide as detailed in Protocol 1, steps 5–8. An alternative activation scheme using CNBr is described by Cutler (Chapter 9, Section 5.1). Finally, in many of the protocols the recommended wash steps are performed on a sintered glass funnel. However, for smaller volumes of gel, a disposable plastic or glass column (e.g. Pierce disposable polypropylene column, Bio-Rad Econo-column) is often a better alternative and has the advantage that the peptide affinity matrix can be left within the column for subsequent chromatography.

Protocol 1. CNBr activation of hydroxyl containing gels and subsequent peptide conjugation

Caution: CNBr is extremely toxic! This procedure must always be performed in a suitable fume-hood. All glassware and solutions should be treated with 10 M NaOH to destroy CNBr prior to disposal.

Equipment and reagents

- Agarose or Sepharose CL-4B
- 2 M Na_2CO_3
- 0.1 M Na_2CO_3 pH 9.5
- CNBr solution: 2 g/ml in AnalaR acetonitrile
- Coupling buffer: 50 mM MOPS pH 7.5, 0.1 M NaCl
- Peptide in coupling buffer, 2–10 mg/ml
- 1 M ethanolamine pH 8
- 0.1 M sodium acetate buffer pH 4, 0.5 M NaCl
- Sintered glass filter funnel

Protocol 1. *Continued*

Method

1. Wash 1 vol. of an agarose slurry (settled gel volume) with an excess of deionized water. Estimate settled gel volume quickly by centrifugation at 200 g for 2 min.
2. Add 1 vol. 2 M Na_2CO_3 and swirl continuously to mix and keep the slurry in suspension. Do not stir or the gel will be damaged.
3. Add 0.05 vol. 20% (w/v) CNBr in acetonitrile and swirl vigorously for 2 min.
4. Immediately recover the gel by filtration under vacuum on a sintered glass funnel followed immediately by 5–10 vol. each of 0.1 M sodium bicarbonate pH 9.5, deionized water, and then the buffer[a] which is to be used in the subsequent coupling reaction. Proceed to coupling steps (5–8) immediately.
5. Transfer the coupling buffer washed gel to coupling buffer containing peptide at a gel to buffer ratio 1:2. The peptide should at a concentration of about 2–10 mg/ml gel.
6. Mix the gel/peptide slurry gently on a rocking table for 2 h at room temperature or overnight at 4°C.
7. Block the remaining reactive groups by adding an excess of 1 M ethanolamine pH 8 for 2 h at room temperature.
8. Wash the gel in an excess of coupling buffer and then in 0.1 M acetate buffer pH 4, 0.5 M NaCl followed by a further wash in coupling buffer.

[a] Do not use buffers containing amino groups such as Tris or glycine since these will couple to the activated gel.

Controlling or lowering the efficiency of coupling of peptide to the matrix can be achieved in several ways. Coupling at a lower pH is less efficient and lowers the concentration of ligand coupled whilst coupling at lower temperatures has the same effect. Alternatively, a lower concentration of CNBr will produce fewer active groups for coupling.

2.2 Cyanuric chloride (2,4,6-trichloro-1,3,5-triazine)

Finlay *et al.* (10) developed cyanuric chloride chemistry to activate chromatographic supports for coupling via amine groups. The activation chemistry is of comparable efficiency to CNBr but has the advantage of being neither volatile or lacrimatous in use (although still toxic). The activation process requires organic solvents since cyanuric chloride is insoluble in aqueous buffers, but subsequent peptide coupling can be carried out under organic or aqueous conditions and is very useful for coupling hydrophobic peptides. As with

7: Synthetic peptides as affinity ligands

CNBr, almost any polysaccharide (-OH containing) matrix can be activated and the resulting covalent link is very stable even in alkaline conditions giving minimal leaching of ligand. The triazine groups on the gel can, however, lead to ionic and aromatic non-specific adsorption whilst using the gel. Typically, the use of recrystallized cyanuric chloride is recommended and whilst the reagent can be purchased 99%+ pure and will work well without further purification, highest reactivity is always achieved by firstly removing the hydrolysis products and then dissolving the purified cyanuric chloride in ice-cold acetone just prior to gel activation (11).

2.2.1 Activation of hydroxyl containing gels with cyanuric chloride: high chloride substitution

Protocol 2 is after Makriyannis and Clonis (11), who used the cyanuric chloride method to determine performance of the immobilization reaction of tripeptide ligands to agarose gel for the purpose of studying affinity matrix production. In this activation procedure two chlorides are left on the triazine ring to react, thus leading to high yields of immobilized peptides. The first chloride reacts with an amine within three minutes (12) but leaving the coupling reaction to proceed for 48 hours allows the second chloride to react.

Protocol 2. Activation of hydroxyl containing gels with cyanuric chloride: high chloride substitution

Equipment and reagents

- AnalaR cyanuric chloride
- AnalaR chloroform
- Agarose 6% (e.g. Ulrogel A6R, Sepharose CL-6B)
- 0.1 M NaOH
- AnalaR acetone (ice-cold)
- 50% (v/v) acetone in water

Method

1. Recrystallize cyanuric chloride by dissolving 1 g/10 ml chloroform, decanting the soluble chloride, and recovering it by removing the chloroform by rotary evaporation under reduced pressure.[a]
2. Dissolve cyanuric chloride in ice-cold acetone to a final concentration of 20 μM.
3. Agarose 6% (Ultrogel A6R) is prepared by washing with an excess of 0.1 M NaOH and dried briefly on a sintered glass funnel under suction.
4. Add 1.5 ml cyanuric chloride solution per 1 ml wet gel[b] whilst swirling vigorously, then immediately add 3 ml ice-cold deionized water.[c]
5. After 30 min at room temperature the activated matrix is washed with an excess of ice-cold acetone (20 ml/ml wet gel).
6. Further wash the matrix with ice-cold 50% (v/v) acetone/water (20 ml/ml gel).[d]

Protocol 2. *Continued*

7. Dry the product under vacuum and store (up to several months) desiccated or else proceed immediately to the coupling step (*Protocol 4*).

[a] Storing the hygroscopic product desiccated will keep the product stable for several months.
[b] The activated gel will remain effective even when substituted at low cyanuric chloride concentrations (e.g. 3.5 µmol/g).
[c] Hydrolysis of cyanuric chloride generates enough acid to prevent hydrolysis of the second triazinyl chloride.
[d] At this stage the activated gel can be stored at 4^°C for several weeks without an appreciable loss of its peptide coupling ability.

2.2.2 Activation of hydroxyl containing gels with cyanuric chloride: low chloride substitution

Protocol 3 is after Finlay *et al.* (10) who described a method of utilizing cyanuric chloride activation that avoided hydrolysis and permitted controlled activation of the support. Briefly, aniline (under anhydrous conditions) displaces only one of the two remaining chlorides on the triazine ring leaving the final chlorine of the mono-chloro-*s*-triazine–Sepharose available for replacement by a ligand containing a suitable nucleophilic group.

Protocol 3. Activation of hydroxyl containing gels with cyanuric chloride: low chloride substitution

Equipment and reagents

- Sepharose CL-6B
- AnalaR dioxane
- Dioxane/water (70:30)
- Dioxane/water (30:70)
- 2 M *N,N'*-diisopropylethylamine in dioxane
- 1 M cyanuric chloride
- 2 M aniline in dioxane
- Coarse sintered glass funnel
- Three-necked round-bottomed flask equipped with a water-jacketed condenser and stirrer
- Oil-bath

Method

1. Wash 100 ml Sepharose CL-6B with 200 ml deionized water/dioxane (30:70) without suction on a coarse sintered glass funnel.

2. Continue to wash gel with 200 ml deionized water/dioxane (70:30) followed by 1 litre dioxane.[a]

3. Transfer the gel to 60 ml dioxane in a 500 ml three-necked round-bottomed flask equipped with a water-jacketed condenser and stirrer.

4. Transfer the flask to an oil-bath maintained at 50 °C and stir gently (≈ 100 r.p.m.).

5. Add 20 ml 2 M *N,N'*-diisopropylethylamine in dioxane and continue to stir for 30 min.

7: *Synthetic peptides as affinity ligands*

6. Add 20 ml 1 M cyanuric chloride[b] in dioxane and continue to stir for 60 min.
7. Wash the activated gel on a coarse sintered glass funnel without suction with 1 litre dioxane.
8. Stir the activated gel with 200 ml 2 M aniline in dioxane at room temperature for 30 min.
9. Wash extensively in dioxane and either react the gel with a dioxane soluble peptide or transfer the gel gradually back to the aqueous phase (steps 2 and 1) and react with a water soluble peptide (*Protocol 4*).
10. Alternatively, the dioxane can be removed under vacuum and the dried activated support stored at 4°C for several months.

[a] The gradual transfer of gel into anhydrous conditions is required so that the structural integrity of the matrix is not lost, although CL-Sepharose is markedly more stable than uncross-linked agarose towards organic solvents.
[b] Cyanuric chloride should be recrystallized as described in *Protocol 2*.

2.2.3 Coupling peptides to cyanuric chloride activated gels

Peptides can be coupled to cyanuric chloride activated gels via their N-terminal or side chain amines in organic or aqueous conditions depending on the solubility of the peptide used. However, dehydration of the gel under organic conditions may lead to decreased porosity and poor flow characteristics during use. The activated gels are extremely reactive towards primary amines so care should be taken to exclude amine containing buffers. The level of peptide coupled can be controlled at this stage since the coupling reaction is temperature-, pH-, and peptide concentration-dependent although at lower temperature, pH, or peptide concentrations care should be taken to ensure that the reaction is given enough time to complete. Notwithstanding this, the coupling procedure is extremely robust and gives acceptable coupling ($>90\%$) across a wide range of conditions.

Protocol 4. Coupling peptides to cyanuric chloride activated gels

Equipment and reagents
- Activated agarose gel
- 0.5 M KH_2PO_4 buffer pH 7.9
- 1 M NH_4Cl buffer pH 8.5
- Sintered glass filter funnel

Method

1. Activated agarose gel (if dry) should be rehydrated in 50 vol. coupling buffer at 4°C for 10 min with gentle swirling.
2. 1 ml activated gel should be suspended in 2 ml peptide in coupling buffer (0.5 M KH_2PO_4 pH 7.9).[a]

Protocol 4. Continued

3. Swirl the gel gently for 48 h at 25°C then wash extensively with deionized water to remove uncoupled peptide.
4. Unreacted solid phase triazinyl chlorines are eliminated by treating the gel with 1 M NH$_4$Cl buffer pH 8.5 for 3 h at room temperature.
5. Wash extensively with water prior to equilibration and use.

[a] Between 2–20 mg peptide/ml gel should be conjugated.

2.3 Glutaraldehyde

Glutaraldehyde was popularly used as a homobifunctional cross-linking agent for the conjugation of enzymes to antibodies. Whilst it has largely been superseded by heterobifunctional cross-linkers, it is still used successfully to conjugate peptides via their primary amines to gels having amine or amide groups. Cambiaso et al. (13) described an efficient procedure for the derivatization of aminohexyl–Sepharose 4B that yielded a 13 atom spacer arm; the method can be applied to any gel containing a free amine. Amine containing gels themselves can be quickly and easily created by derivatizing epoxy activated gels (detailed in Section 4) by suspending the gel in 1.5 vol. concentrated ('880') ammonia solution at 40°C for 1.5 h with gentle shaking, after which time the gel is simply washed copiously in water. Ternynck and Avrameas (14) however, used glutaraldehyde to activate the amide groups of various polyacrylamide beads and whilst the method (*Protocol 5*) is time-consuming and was originally optimized for BioGel P300 (no longer commercially available) the method is efficient, reproducible, and reliable. Apart from the activation of amine/amide containing gels, glutaraldehyde can also be used to activate hydrazide containing gels to produce formyl groups that will readily react with peptide amine groups. Whilst the protocol specifies recovery of the gel beads after washing, by centrifugation, it also works well using filtration on a sintered glass funnel to wash the beads.

Protocol 5. Activation of amide containing gels with glutaraldehyde and conjugation of peptide

Equipment and reagents

- Acrylamide gel beads
- 6% (v/v) glutaraldehyde in 0.1 M KH$_2$PO$_4$ buffer pH 7
- 0.1 M lysine pH 7.4
- 0.2 M KH$_2$PO$_4$ pH 7.4
- 0.2 M glycine–HCl pH 2.8
- Phosphate-buffered saline (PBS)
- Low speed centrifuge

Method

1. Hydrate gel filtration media in deionized water for 24 h.
2. Wash extensively in several volumes of deionized water.

7: Synthetic peptides as affinity ligands

3. To 100 ml hydrated gel add 500 ml 6% (v/v) solution of glutaraldehyde[a] in 0.1 M KH_2PO_4 pH 7.4.
4. Incubate overnight at 37°C.
5. Wash the gel extensively (10–20 times) with 500 ml volumes of deionized water, and then centrifuge (3000 g for 10 min at 4°C) to sediment the gel.
6. Store for no more than one week[b] in deionized water at 4°C or proceed to coupling steps (steps 7–13).
7. To 10 ml 0.1 M KH_2PO_4 buffer pH 7.4 containing between 15–30 mg peptide add 10 ml activated gel.
8. Agitate gently at room temperature overnight.
9. Centrifuge the suspension at 3000 g for 10 min at 4°C and remove the supernatant.
10. Wash the gel pellet gently with an excess of PBS and repeat steps 9 and 10 until the absorbance of the supernatant is less than 0.05 at 280 nm.[c]
11. Block remaining free active aldehyde groups on the gel by suspending it in an equal volume of 0.1 M lysine pH 7.4, and gently agitating at room temperature for 18 h.
12. Wash extensively in PBS followed by 2 × 50 ml ice-cold 0.2 M glycine–HCl pH 2.8, then 20 ml 0.2 M KH_2PO_4 pH 7.4.
13. Wash with PBS until the optical density at 280 nm is 0.05 or less.

[a] Grade 1 glutaraldehyde (specially purified and stored at −20°C) is routinely used to avoid excessive polymer accumulation that would cause a loss of controlled conjugation.
[b] Ternynck and Avrameas (14) detected no appreciable loss of the gel's capacity to bind proteins after this time period.
[c] Only phenylalanine, tyrosine, and tryptophan have side chains which permit direct estimation in solution. If the peptide to be coupled does not contain aromatic amino acids the supernatant should be assayed in some other way, e.g. protein assay.

2.4 N,N'-carbonyldiimidazole (CDI)

CDI can be used to activate both carboxylic acids and hydroxyl groups on gels for the purpose of spontaneously conjugating amine containing peptides to the gel but by two subtly different chemical routes. CDI will act as a zero length cross-linker if the activated group on the gel is a carboxylic acid but when gels containing hydroxyls are activated, this generates a one atom spacer. Commercially available pre-activated matrices (such as N,N'-carbonyldiimidazole activated agarose from Sigma) are usually the latter formulation. The link created by activation of hydroxyl groups is not stable in alkaline medium (pH > 10) and also undergoes slow hydrolysis in any aqueous environment. *Protocol 6* (15) describes the production of N,N'-carbonyldiimidazole activated agarose. Because of the reproducibility of the activation reaction it is easy to control the amount of bound peptide to matrix. Depending on the amount of reagent used, 0.1–3 mmol of active groups can be introduced per dry gram of matrix with 45% yields readily obtainable for amines coupled throughout this concentration range. Amines are efficiently coupled to CDI activated gels in alkaline buffers within the pH range 9–11 but alkaline-sensitive peptides can be coupled at pH 8.5 in 0.1 M borate buffer. Higher coupling yields are obtainable by coupling in an organic environment. *Protocol 6*, steps 5–11 shows a general coupling method.

Protocol 6. Activation of hydroxyl containing gels with *N,N'*-carbonyldiimidazole and conjugation to peptide

Equipment and reagents
- Sepharose CL-6B gel
- AnalaR dioxane
- Dioxane/water (30:70)
- Dioxane/water (70:30)
- *N,N'*-carbonyldiimidazole (CDI)
- 0.1 M Na_2CO_3 pH 9 coupling buffer
- 0.1 M Na_2CO_3 pH 9, 1 M NaCl
- 0.1 M ethanolamine pH 8
- Sintered glass funnel

Method
1. Wash 3 ml moist Sepharose CL-6B gel that has been hydrated in deionized water sequentially with 20 ml each of water, dioxane/water (30:70), dioxane/water (70:30), and dioxane.
2. Suspend gel in 5 ml dioxane.[a]
3. Add 0.12 g CDI[b] and swirl gently at room temperature for 15 min.
4. Wash the gel extensively, with about 100 ml dioxane, and store in dioxane or couple peptide to the gel (steps 5–11).
5. After washing the activated gel with dioxane, remove residual dioxane by filtration on a sintered funnel, taking care not to completely dry the gel.
6. Dissolve the peptide in an equal volume of 0.1 M Na_2CO_3 pH 9 coupling buffer. Approximately double the molar ratio of peptide to active links per ml of gel should be used.
7. Swirl gently overnight at 4°C.
8. Wash the gel extensively with coupling buffer containing 1 M NaCl.
9. Wash the gel extensively with coupling buffer.
10. Resuspend the gel in an equal volume of 0.1 M ethanolamine pH 8 for 3–5 h at 4–25°C.
11. Wash extensively in running buffer.

[a] Most organic solvents contain enough water for the CDI to hydrolyse to CO_2 and imidazole and good quality anhydrous organic solvents (dioxane, acetone, DMF, DMSO, etc.) are advocated. However, once activated, hydrolysis of CDI activated gels is very slow and occurs in hours rather than minutes.
[b] This amount of CDI will give an activated gel containing 46.4 μmol active groups/ml moist gel. Commercially available gels (such as those from Pierce, i.e. CDI activated 6% cross-linked beaded agarose, Trisacryl GF-2000, and TSK HW-65F) have comparable activation levels of around 50 μmol/ml gel.

2.5 *N*-hydroxysuccinimide (NHS) esters

N-hydroxysuccinimide (NHS) esters are very popular as homobifunctional cross-linkers for amines and as a means to immobilize ligands onto solid supports for affinity chromatography (16, 17). They have largely superseded the imidoesters (imidates) that, although soluble in aqueous solutions, do not react as well at physiological pHs. Many different NHS esters are available with subtle modifications for specific purposes. These include:

(a) *N*-hydroxysulfosuccinimide (sulfo-NHS); designed to increase aqueous solubility.
(b) Disuccinimidyl suberate; contains an eight atom spacer that is non-cleavable.
(c) Dithiobis(succinimidylpropionate); contains an eight atom spacer that is cleavable under reducing conditions.

A wide range of pre-activated matrices, which contain an eclectic collection of spacers, are commercially available (*Table 1*; the structure of the linker for Affi-Gel 10 is shown in Chapter 5, *Figure 6*). However, hydrolysis of many commercial gels over time can lead to the formation of carboxylate groups that can cause ionic interactions to occur during chromatography. Also, because of this, and other problems, the NHS ester, N,N,N',N'-tetramethyl-(succinimido)uronium tetrafluroborate (TSTU) is advocated to activate hydroxyl containing gels for coupling amine containing peptides. *Protocol 7* details the preparation of an *N*-hydroxysuccimimide activated support that, when hydrolysed prior to conjugation, reverts back to the native hydroxyls. An alternative method for *N*-hydroxysuccimimide activation is given in Chapter 5. TSTU is insoluble in water so activation takes place in DMF. The TSTU activation procedure yields high levels of esters for coupling and the coupling process is very efficient (typically 1–5 mg peptide/ml gel with coupling of 90%). The TSTU procedure (both activation and coupling) can also be completed in mixed aqueous/organic solvent (18). This is a further advantage since expensive anhydrous solvents are not required. Both α-amines at the N-terminal and the ε-amines of lysine will be coupled.

Protocol 7. Activation of hydroxyl containing gels with TSTU and conjugation to peptide

Equipment and reagents

- Sepharose CL-6B
- AnalaR dioxane
- Dioxane/water (30:70)
- Dioxane/water (50:50)
- Dioxane/water (70:30)
- AnalaR dimethylformamide (DMF)
- AnalaR methanol
- 0.1 M N,N,N',N'-tetramethyl-(succinimido) uronium tetrafluoroborate (TSTU) in DMF
- 0.1 M 4-(dimethylamino)-pyridine (DMAP) in DMF
- AnalaR isopropanol
- 0.1 M K_2HPO_4 buffer pH 7
- Sintered glass filter funnel

Protocol 7. *Continued*

Method

1. Hydrate 1 g Sepharose CL-6B in deionized water.
2. Sequentially dehydrate the gel with 20 ml washes of 30:70 (v/v) dioxane/water, then 50:50 (v/v) dioxane/water, followed by 70:30 (v/v) dioxane/water, and finally dry dioxane.
3. Resuspend 1 ml gel in 0.5 ml 0.1 M TSTU in DMF with continuous swirling.
4. Gently swirl the gel continuously and add 0.5 ml 0.1 M 4-(dimethyl-amino)-pyridine (DMAP)[a] in DMF dropwise.[b]
5. Add a further 2 mL DMF and swirl for 1 h at room temperature.
6. Filter the gel and wash it successively with 20 ml DMF, methanol, and isopropanol.
7. Resuspend the slurry in an equal volume of isopropanol.
8. The active matrix can be stored at 4°C for several months or couple to peptide (steps 9–11).
9. Wash the isopropanol from 1 ml gel with excess ice-cold deionized water.[c]
10. Immediately add 1–10 mg peptide dissolved in 0.1 M K_2HPO_4 buffer pH 7 and swirl overnight at 4°C.
11. Wash extensively with 0.1 M M K_2HPO_4 buffer pH 7.

[a] DMAP is recommended as base since diisopropylethylamine or triethylamine causes extensive discoloration of carriers such as Sepharose.
[b] This quantity of DMAP should yield between 49–55 μmol/g active ester.
[c] NHS activated matrices generally have short half-lives in aqueous solutions; the use of ice-cold water at this point reduces unwanted hydrolysis.
[d] The peptide coupling steps (steps 9–11) take place in phosphate buffer. However, if the peptide is too hydrophobic to be soluble in this buffer it can be coupled in DMF to gel that has been washed in methanol then DMF to remove isopropanol. Rehydrating after coupling is simply a matter of reversing the dehydration procedure as described in step 2.

3. Coupling peptides via sulfydryl groups

Fewer procedures exist to couple peptides to matrices via sulfydryl groups as compared to amines. However, many synthetic peptides are prepared with either a C- or N-terminal cysteine to allow directional coupling of the peptide though its sulfydryl group. Partly, this reflects the lower abundance of Cys residues compared with Lys residues, but also the sulfydryl side chain of single Cys residue represents a unique linkage point whereas in Lys containing peptide both α- and ε-amino groups are reactive. Since Cys containing peptides tend to undergo some oxidation in storage, producing -S-S- linked dimers

7: Synthetic peptides as affinity ligands

prior to coupling free sulfydryls need to be regenerated. This is accomplished by reducing any disulfide bonds with an immobilized reducing agent, such as dihydrolipoamide (19) or a suitable soluble reducing agent such as DTT (20, 30), DTE, mercaptoethylamine, and other alkanethiols. The former offers the advantage of easy separation of reductant and peptide. Also, dihydrolipoyls are much more reactive due to the proximity of the SH groups which favours closure of a stable dithiolane ring, and which accomplishes quickly the quantitative reduction of disulfide bonds in peptides. *Protocol 8* details preparation of dihydrolipoamide gel whilst *Protocol 9* describes the use of DTT as a soluble reductant. The peptide -SH groups created should then be estimated as described in *Protocol 10*.

If the peptide of interest lacks appropriate cysteine residues there are also a number of procedures for introducing thiols into peptides. However, the simplest method is with 2-iminothiolane–HCl ('Traut's reagent') (21) that reacts with primary amines and leaves an exposed sulfydryl group available for coupling. For further information on introducing sulfydryls into peptides the reader is referred to refs 22–25.

Finally, the reduced peptide should be immediately coupled to a sulfydryl reactive matrix. Commercially available matrices are available (*Table 1*), whilst *Protocol 11* and *Protocol 12* detail preparation of two such matrices. In addition to covalent coupling, cysteinyl peptides can also be reversibly coupled to thiol containing matrices such as Thiol Sepharose 4B (Pharmacia). Protocols will not be issued here but the reader will find such media detailed in *Table 1* and should refer to manufacturer's instructions regarding suitable cross-linking agents to couple peptide and Thiol Sepharose.

3.1 Reduction of disulfide bonds

Protocol 8. Preparation and use of immobilized (solid phase) reductant (dihydrolipoamide)

Equipment and reagents
- EAH Sepharose (Pharmacia), or equivalent amine containing matrix
- 0.1 M Na_2HPO_4 pH 7.3
- DL-6,8-thioctic acid
- 0.1 M Na_2HPO_4 pH 10
- 1 M NaCl
- 1-ethyl-3-(3-dimethylaminopropyl)carbodiimide (EDC)
- 0.05 M Na_2HPO_4 pH 8, 1 mM EDTA, 10 mM DTT
- 0.05 M Na_2HPO_4 pH 8, 1 mM EDTA
- Sintered glass filter funnel

Method

1. Wash 100 ml hydrated EAH Sepharose (Pharmacia) or other amine containing gel[a] with 1 litre ice-cold deionized water.

2. Wash further with 200 ml ice-cold 0.1 M Na_2HPO_4 pH 7.3 and leave the gel cake moist.

Protocol 8.   Continued

3. Dissolve 1 g DL-6,8-thioctic acid in 100 ml 0.1 M Na_2HPO_4 pH 10.
4. Lower the pH of the DL-6,8-thioctic acid solution to pH 7.3 with HCl and add to the moist gel cake swirling continuously.[b]
5. Add 3 g EDC and continue to swirl for 20 h at room temperature.
6. Wash the gel successively with 1 litre of each of ice-cold 0.1 M sodium phosphate pH 7.3, 1 M NaCl, then deionized water.
7. Store as a slurry at 4°C in 0.02% (w/v) NaN_3 pH 6.5[c] until use (steps 8–11).
8. Dissolve 3 μmol peptide in 0.5 ml 0.05 M Na_2HPO_4 pH 8.[d]
9. Wash an amount of gel equivalent to 30 μmol of SH groups in an excess of 0.5 M Na_2HPO_4 pH 8 containing 1 mM EDTA and 10 mM DTT.[e]
10. Wash free DTT from the gel with an excess of 0.5 M phosphate buffer pH 8 containing 1 mM EDTA, and add the gel to dissolved peptide under nitrogen and continuous swirling.
11. After 30 min at room temperature, filter the solution to recover reduced peptide free from reductant.
12. Determine the concentration of peptide -SH using *Protocol 10*.

[a] Any commercially available gel whose spacer terminates with an amine is acceptable.
[b] The pH is raised then lowered since DL-6,8-thioctic acid is only partially soluble at pH 7.3.
[c] At greater than pH 8 the reductant is readily oxidized; keep the gel at low pH until *just* prior to use.
[d] The reductant is effective between pH 6–9 but reaction times should be increased at lower pHs.
[e] DTT is required to convert the immobilized reductant into a fully reduced form. To regenerate the gel prior to reuse repeat step 9.

Protocol 9.   Reduction of peptide sulfydryls using DTT

Equipment and reagents

- 50 mM dithiothreitol (DTT) in phosphate-buffered saline (PBS) pH 8
- 1 mM EDTA in PBS pH 7
- Sephadex G10 column: 30–50 cm × 1 cm

Method

1. Dissolve 5 mg peptide in 1 ml PBS[a] containing 20 μl 50 mM DTT.[b]
2. Stand for 30 min at room temperature.
3. Separate the peptide from excess DTT by chomatography on a Sephadex G10[c] column (see Chapter 1) equilibrated with PBS containing 1 mM EDTA.

7: Synthetic peptides as affinity ligands

4. Check free sulfydryls if required using *Protocol 10*.

[a] To minimize oxidation bubble nitrogen through buffers before use.
[b] For a peptide of mass 2000 Da, this is a 20-fold molar excess of DTT.
[c] The exclusion limit of Sephadex G10 is ~ 700 Da which ensures that the peptide is not retarded by the gel. The more commonly used Sephadex G25 is more suitable for desalting proteins.

3.2 Determination of free sulfydryl groups using DTNB

The most reliable way of determining free sulfydryl content is using Ellman's protocol (26) (*Protocol 10*) that relies on the stoichiometric release of the chomophore 5-mercapto-2-nitrobenzoic acid when free thiols react with the substrate 5,5'-dithiobis(2-nitrobenzoic acid) (DTNB).

Protocol 10. Determination of free sulfydryl groups using (DTNB)

Equipment and reagents

- 5,5'-dithiobis(2-nitrobenzoic acid)
- AnalaR methanol
- 0.1 M Na_2HPO_4 pH 8
- 2-mercaptoethanol, or 50 mM cysteine
- UV visible spectrophotometer

Method

1. Dissolve 4 mg DTNB in 1 ml methanol.
2. Dissolve 500 μg reduced peptide[a] in 0.1 ml PBS[b] immediately prior to the assay.
3. Prepare 0.1 ml standards ranging from 25–400 μM of 2-mercaptoethanol[c] in 0.1 M Na_2HPO_4 buffer pH 8.
4. Mix gently together[b] the 0.1 ml 2-mercaptoethanol standard or peptide, with 1 ml 0.1 M Na_2HPO_4 buffer pH 8, and 0.1 ml DTNB solution.
5. Incubate for 15 min at room temperature then determine the absorbance at 420 nm.

[a] Or use 20–100 μl of desalted reduced peptide (see *Protocols 8* and *9*).
[b] Use degassed buffers and do not agitate the assay vessel so as to avoid oxidation.
[c] Cysteine can be substituted here.

3.3 Activation

Although there are a range of activated matrices that will couple to sulfydryl groups, generally one of two chemistries are employed when linking cysteinyl peptides to gels. These involve iodoacetyl or maleimide active groups. Both form permanent thioether bonds with peptides containing free sulfydryls and form excellent affinity chromatography media. *Protocol 11* and *Protocol 12* provide details of gel activation to produce iodoacetyl and maleimide functional groups respectively. Commercially, iodoacetyl derivatized matrices are more commonly available (*Table 1*) than are resins with maleimide groups. However, a wide range of maleimide homobifunctional and heterobifunctional cross-linking reagents are available that permit derivatization of a range of gels to yield a sulfydryl reactive leaving group. Hermanson (27) provides an excellent review of the field of cross-linking reagents.

3.3.1 Activation of amine containing gels with iodoacetic acid

In *Protocol 11* the activation method relies on the conjugation of the acid to an amine containing matrix using EDC. Accordingly, any matrix which has free amine groups can be used.

Protocol 11. Activation of amine containing gels with iodoacetic acid

Equipment and reagents
- EAH Sepharose (Pharmacia), or equivalent amine containing matrix
- Iodoacetic acid, or bromoacetic acid
- 1 M NaOH
- 1 M HCl
- 1-ethyl-3-(3-dimethylaminopropyl)carbodiimide (EDC)
- Sintered glass filter funnel

Method

1. Wash 100 ml settled bed volume of an amine containing gel with 1 litre deionized water.

2. Dissolve 12 g iodoacetic acid[a] in 100 ml deionized water and adjust to pH 4.5 with 1 M NaOH.

3. Resuspend the gel in the iodoacetic acid solution with constant swirling.

4. Add 10 g EDC and maintain the solution at pH 4.5 for 2 h under constant swirling with additions of either 1 M HCl or 1 M NaOH.

5. Wash extensively in deionized water and proceed to couple peptide or store at pH 7 in darkness.[b]

[a] Iodoacetic acid can be replaced by bromoacetic acid to give gels that should be chemically identical with regards to coupling characteristics. Bromoacetyl agaroses are also commercially available.
[b] Haloacetyl derivatives are extremely light-sensitive.

3.3.2 Activation of amine containing gels with sulfo-MBS

Protocol 12 is one that is regularly used for conjugation of sulfydryl containing peptides to carrier proteins prior to the production of polyclonal antibodies. It works equally as well in our hands for the immobilization of sulfydryl

7: Synthetic peptides as affinity ligands

containing peptides to amine containing matrices. Sulfo-MBS is a water soluble heterobifunctional reagent with a NHS ester reactive towards amines at one end, and a maleimide reactive towards free sulfydryls at the other.

Protocol 12. Activation of amine containing gels with sulfo-MBS

Equipment and reagents
- EAH Sepharose (Pharmacia), or equivalent amine containing matrix
- 0.1 M NaH$_2$PO$_4$ pH 7.2
- *m*-Maleimidobenzoyl-*N*-hydroxysulfosuccinimide ester (sulfo-MBS)

Method
1. Wash 10 ml settled bed volume of amine containing gel with ice-cold deionized water.
2. Resuspend the gel in an equal volume of ice-cold 0.1 M NaH$_2$PO$_4$ pH 7.2.
3. With constant swirling at 4°C add 208 mg (50 μmol/ml gel) sulfo-MBS.
4. After 1 h rapidly wash the gel with ice-cold 0.1 M NaH$_2$PO$_4$ pH 7.2 to remove unreacted cross-linker. Proceed immediately to coupling (*Protocol 13*).

3.4 Coupling cysteinyl peptides to iodoacetyl or maleimide activated gels

Both iodoacetyl and maleimide activated gels react spontaneously with free sulfydryls in reduced peptides (*Protocol 13*). The only significant precaution that should be taken is that free reactive groups are blocked after coupling so that non-specific binding does not take place during chromatography.

Protocol 13. Coupling cysteinyl peptides to iodoacetyl or maleimide activated gels

Equipment and reagents
- Iodoacetyl activated (*Protocol 11*) or sulfo-MBS activated (*Protocol 12*) gel
- Coupling buffer: 50 mM Tris pH 8.5, 5 mM EDTA
- 50 mM 2-mercaptoethanol in coupling buffer
- 0.05% (w/v) NaN$_3$ in coupling buffer
- 1 M NaCl
- 10 ml disposable column
- Blood wheel
- UV visible spectrophotometer

Method
1. Pack 5 ml iodoacetyl (or sulfo-MBS) activated gel into a disposable 10 ml column.

Protocol 13. *Continued*

2. Wash with > 6 vol. coupling buffer.
3. Dissolve 5–10 mg peptide in 5 ml coupling buffer.
4. Take 20 µl peptide solution to 1 ml coupling buffer and measure A_{280}.
5. Drain gel and add peptide solution. Mix on blood wheel[a] for 15 min at room temperature (~ 20°C) then stand for a further 30 min at room temperature.
6. Drain the gel and retain the supernatant.
7. Wash gel with 3 vol. coupling buffer and retain the wash liquid.
8. Add 5 ml 50 mM 2-mercaptoethanol[b] in coupling buffer to block unreacted sites, mix on blood wheel for 30 min, then stand at room temperature for 30 min.
9. Wash column with 20 ml 1 M NaCl, then with 100 ml buffer to be used for chromatography prior to use. If not immediately using column, wash with 0.05% NaN_3 in coupling buffer and store at 4°C.
10. Combine supernatants from step 6 and 7 and measure volume and A_{280}. Calculate proportion of peptide coupled by comparison with A_{280} measurement in step 4.

[a] This provides gentle, end-over-end mixing.
[b] 50 mM cysteine in coupling buffer is an alternative blocking agent.

4. Coupling peptides and introducing spacer arms using carbodiimides

The carbodiimides are a versatile group of cross-linkers that are used extensively for the immobilization of hormones and peptides and immobilization of peptides onto ELISA plates (28). They comprise a group of compounds whose general formula is R–N=C=N–R' where R and R' are aliphatic or aromatic groups. Specifically, they mediate the formation of amide linkages between a carboxylate and an amine (29) and because of this can be used for:

(a) Coupling C-terminal or side chain (Asp, Glu) carboxyls to amine containing gels.
(b) Coupling N-terminal or side chain amines to carboxyl containing gels.
(c) Introducing or extending spacer arms such as 6-aminocaproic acid or 1,6-diaminohexane.

However, because of the abundance of cross-reacting groups in peptides, carbodiimides find their principle use in the introduction of spacers arms and care should be taken if choosing this chemistry for coupling peptides.

7: Synthetic peptides as affinity ligands

5. Chromatography on peptide affinity gels

Separation of macromolecules on an affinity gel which has a peptide as ligand is usually performed on a relatively small (few millilitres) scale, owing to the cost of synthetic peptides. Accordingly, a disposable polypropylene column may be appropriate if elution is to take place in batch mode. However, if a continuous gradient is to be run, then a glass column with flow adapter (see Chapter 1) is worth using to maximize resolution.

5.1 Designing the peptide ligand

The design of the particular peptide to be immobilized must be obviously tailored to the separation intended. However, there are several factors that can be usefully be taken into account. The first of these is that ideally, if the peptide represents a region of protein sequence then it should correspond to a known region of protein–protein interaction in order to maximize its selectivity in the separation. An example of this is shown in *Figure 1* where a peptide, (C)DETRRRNLLEAGLL, corresponding to the C-terminus of the plant G-protein α-subunit was used as a ligand to isolate a candidate receptor protein

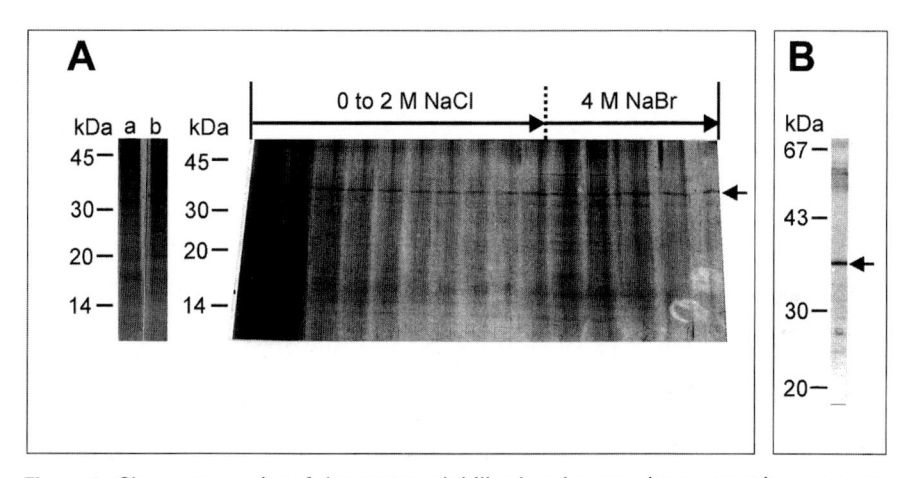

Figure 1. Chromatography of detergent solubilized maize membrane proteins on a peptide affinity matrix. Approx. 25 mg solubilized proteins in 20 ml buffer comprising 25 mM Tris pH 8, 1 mM EDTA, 10 mM $MgCl_2$, 1 mM DTT, 1% (w/v) sodium cholate, 30 mM MEGA 9, 0.1 mM PMSF, 1 μM leupeptin, and 1 μM pepstatin (TCM buffer), were passed through a 1 ml SulfoLink resin column to which the peptide (C)DETRRRNLLEAGLL had been coupled. Proteins were then eluted using a 20 ml linear gradient of 0–2 M NaCl in TCM (but with 0.33% cholate and 10 mM MEGA 9) followed by 5 ml 4 M NaBr in the same buffer. The proteins were then subjected to SDS–polyacrylamide gel electrophoresis and the gel silver stained. The arrow indicates the putative 37 kDa receptor. (A) Lane a, 1 μg membrane proteins; lane b, 1 μg solubilized proteins. Subsequent lanes contain protein eluting with 0–2 M NaCl and 4 M NaBr. (B) Proteins eluted by 1–2 M NaCl were combined and concentrated by acetone precipitation before electrophoresis.

(31). This region of Gα-subunits is known to be the main contiguous stretch of sequence responsible for Gα–receptor interaction. Note that an additional Cys residue was added at the N-terminus of the peptide to enable coupling of the peptide to SulfoLink resin (*Table 1*) in an appropriate orientation. The protein isolated emerged over a range of fractions (*Figure 1A*) but could be seen to be > 95% pure when concentrated (*Figure 1B*). However, notice should be taken of the chemical attributes of the peptide ligand. A key consideration is the hydrophobic and ionic character of the peptide. Very hydrophobic peptides will show a tendency for non-specific absorption whilst strongly acidic or basic peptides will act as cation or anion exchangers respectively.

5.2 Choosing the elution conditions

Ideally, the most specific eluent for a macromolecule bound to a peptide affinity matrix is a solution of free peptide. However, except for very small scale columns this will usually prove prohibitively expensive and a more usual approach is to utilize a linear gradient of NaCl (see *Figure 1*) or similar salt. For method development it is usually sufficient to use a 0–2 M NaCl gradient in loading buffer, over 10–20 column volumes. In situations where a high concentration of NaCl is required to release the target molecule, detergents can also be incorporated into the eluent to decrease the NaCl concentration required for elution. Sodium cholate is a useful and inexpensive cationic detergent which is fairly gentle whilst zwitterionic and neutral detergents are represented by CHAPS and Triton X-100 respectively.

In many cases retardation of proteins eluting from peptide affinity columns (due to the non-specific interactions indicated above) is unavoidable. Although we have not tried polymer screening (Chapter 6) to suppress low affinity interactions, it seems likely that this technique might result in elution of target proteins from peptide columns in a reduced volume of eluent. Once a chromatographic run has been completed, it is also preferable to elute the peptide column using 4 M NaBr (or 4 M NaI) which act as chaotropes and remove any strongly, and possibly non-specifically bound protein. The column should then be stored in a neutral pH buffer containing 0.05% (w/v) NaN_3 to prevent microbial growth. Finally, some care should be taken to prevent the action of proteases on the peptide ligand. This is particularly so when chromatographing crude extracts. Although there are many protease inhibitor cocktails that might be used, we have found the following combination effective in most situations: 1 mM EDTA, 0.1 mM phenylmethylsulfonyl chloride (PMSF), 1 µM pepstatin, and 1 µM leupeptin. Except for EDTA, all of the protease inhibitors should be added just before use.

References

1. Atherton, E. and Sheppard, R. C. (1989). In *Solid phase peptide synthesis: a practical approach* (ed. E. Atherton and R. C. Sheppard). IRL Press, Oxford.

7: Synthetic peptides as affinity ligands

2. Nutt, R. F., Brady, S. F., Darke, P. L., Ciccarone, T. M., Colton, C. D., Nutt, E. M., et al. (1988). *Proc. Natl. Acad. Sci. USA*, **85**, 7129.
3. Grant, G. A. (1992). In *Synthetic peptides: a users guide* (ed. G. A. Grant), p. 185. Freeman, New York.
4. O'Carra, P., Barry, S., and Griffin, T. (1973). *Biochem. Soc. Trans.*, **1**, 289.
5. Axen, P., Porath, J., and Ernback, S. (1967). *Nature*, **214**, 1302.
6. Van Eijk, H. G. and van Noort, W. L. (1976). *J. Clin. Chem. Biochem.*, **14**, 475.
7. Porath, J., Axen, R., and Ernback, S. (1967). *Nature*, **215**, 1491.
8. Cuatrecasas, P. (1970). *J. Biol. Chem.*, **245**, 3059.
9. March, S. C., Paritch, I., and Cuatrecasas, P. (1974). *Anal. Biochem.*, **60**, 149.
10. Finlay, T. H., Troll, V., Levy, M., Johnson, A. J., and Hodgins, L. T. (1978). *Anal. Biochem.*, **87**, 77.
11. Makriyannis, T. and Clonis, Y. D. (1997). *Biotechnol. Bioeng.*, **53**, 49.
12. Kay, G. and Crook, E. M. (1967). *Nature*, **216**, 514.
13. Cambiaso, C. L., Goffinet, A., Vaerman, J.-P., and Heremans, J. F. (1975). *Immunochemistry*, **12**, 273.
14. Ternynck, T. and Avrameas, S. (1972). *FEBS Lett.*, **23**, 24.
15. Bethell, G. S., Ayers, J. S., Hancock, W. S., and Hearn, M. T. W. (1979). *J. Biol. Chem.*, **254**, 1683.
16. Cuatrecasas, P. and Parikh, I. (1972). *Biochemistry*, **11**, 2291.
17. Wilchek, M., Knudsen, K. L., and Miron, T. (1994). *Bioconjugate Chem.*, **5**, 491.
18. Bannwarth, W. and Knorr, R. (1991). *Tetrahedron Lett.*, **32**, 1157.
19. Gorecki, M. and Patchornik, A. (1973). *Biochim. Biophys. Acta*, **303**, 36.
20. Sears, D. W., Mohrer, J., and Beychok, S. (1977). *Biochemistry*, **16**, 2031.
21. Traut, R. R., Bollen, A., Sun, T.-T., Hershey, J. W. B., Sundberg, J., and Pierce, L. R. (1973). *Biochemistry*, **12**, 3266.
22. Carlsson, J., Drevin, H., and Axen, R. (1978). *Biochem. J.*, **173**, 723.
23. Imagawa, M., Joshitake, S., Hamaguchi, Y., Ishigawa, E., Nutsu, Y., Urusizaki, I., et al. (1982). *J. Appl. Biochem.*, **4**, 41.
24. Duncan, R. J. S., Weston, P. D., and Wrigglesworth, R. (1983). *Anal. Biochem.*, **132**, 68.
25. Kitagawa, T. and Aikawa, T. (1976). *J. Biochem. (Tokyo)*, **79**, 233.
26. Ellman, G. L. (1959). *Arch. Biochem. Biophys.*, **82**, 70.
27. Hermanson, G. T. (1996). *Bioconjugate techniques*. Academic Press, London.
28. Dagenais, P., Desprez, B., Albert, J., and Escher. E. (1994). *Anal. Biochem.*, **222**, 149.
29. Hoare, D. G. and Koshland, D. E. (1966). *J. Am. Chem. Soc.*, **88**, 2057.
30. Ellman, G. L. (1959). *Arch. Biochem. Biophys.*, **82**, 70.
31. Wise, A., Thomas, P. G., White, I. R, and Millner, P. A. (1994). *FEBS Lett.*, **356**, 233.

8

Drugs and inhibitors as affinity ligands

NIGEL M. HOOPER

1. Introduction

With the pharmaceutical industry's search to identify ever more therapeutic agents, numerous drugs and inhibitors are being synthesized. Several of these compounds can be used as affinity ligands for the purification of a particular target protein. However, some of the earliest affinity ligands were those based on the dyes used in the textile industry, such as the synthetic polycyclic dyes Cibacron Blue and Procion Red which bind to a wide range of proteins including dehydrogenases, kinases, and other enzymes requiring adenylyl containing groups (1); the use of such dyes is described in Chapter 6. However, other proteins, e.g. albumin and interferon, bind in a less specific manner due to electrostatic and/or hydrophobic interactions with the compounds. More recently numerous drugs and enzyme inhibitors have been used as selective affinity ligands. In some cases the interaction between the ligand and the protein has been mapped in detail, often through co-crystallization of the compound with the protein and subsequent X-ray crystallographic analysis. However, details of how a compound interacts with a particular protein need not be known for its use as an affinity ligand.

In this chapter I provide general considerations to be taken into account when using drugs or inhibitors as affinity ligands, including the choice of ligand, matrix, spacer arm, coupling method, and elution method. In addition, I detail protocols from our work on the use of inhibitors as affinity ligands to purify zinc metalloproteases. Further information can be found in the companion volumes in this series (2, 3).

2. Choice of ligand

To a large extent the choice of ligand will be dependent on the protein that is to be purified. However, some general considerations need to be borne in mind.

2.1 Specificity

The ligand should ideally be specific for the protein of interest. Obviously if only one protein is going to interact with the ligand the need for preliminary or subsequent chromatographic steps is circumvented. If the ligand is known (or is found) to bind to several proteins (which may be structurally and/or functionally related) then other chromatographic steps may be required to achieve complete purification.

2.2 Maintenance of binding affinity when coupled

In order for the ligand to be used successfully in an affinity chromatography application it has to be linked to an insoluble matrix without destroying the interaction with the protein. As long as the ligand contains a reactive group (e.g. amino, carboxyl, hydroxyl, or thiol) (see *Table 1*) which is not essential for its interaction with the protein then it should be possible to attach it to an insoluble matrix. For example, the angiotensin converting inhibitor enalaprilat (*N*-[(*S*)-1-carboxy-3-phenylpropyl]-L-alanyl-L-proline) (*Figure 1a*) (4) contains two reactive carboxyl groups that could be used to couple the compound to a matrix. However, one of these carboxyl groups is essential for the interaction of the inhibitor with the active site of this zinc metallopeptidase where it co-ordinates to the zinc ion. Thus, if coupled through this group the inhibitor would fail to interact with the protein. This problem is overcome with the structurally related compound lisinopril (*N*-[(*S*)-1-carboxy-3-phenylpropyl]-L-

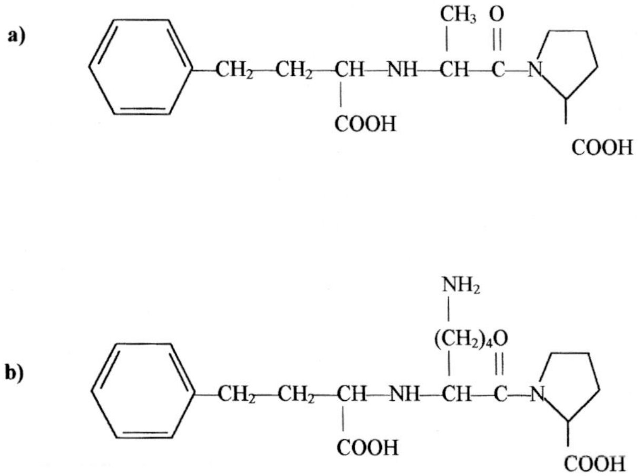

Figure 1. Structures of the angiotensin converting enzyme inhibitors (a) enalaprilat and (b) lisinopril. Enalaprilat (MK422; *N*-[(*S*)-1-carboxy-3-phenylpropyl]-L-alanyl-L-proline). Lisinopril (MK521; *N*-[(*S*)-1-carboxy-3-phenylpropyl]-L-lysyl-L-proline) (4). Lisinopril, but not enalaprilat, can be coupled through its ε-amino group to an insoluble matrix (see *Figure 2*).

8: Drugs and inhibitors as affinity ligands

lysyl-L-proline) (*Figure 1b*) (4) in which the primary amino group on the lysyl side chain is not involved in its interaction with the enzyme, and through which the inhibitor can be linked to a matrix and subsequently used to purify the enzyme.

2.3 Affinity and reversibility of binding

The ligand should have a high enough affinity for the protein of interest. Ideally the dissociation constant (K_d) for the ligand–protein complex should be in the range 10^{-4} to 10^{-8} M in free solution. If the dissociation constant is greater than 10^{-4} M the interaction may be too weak for successful affinity chromatography. If, on the other hand, the dissociation constant is lower than 10^{-8} M it may be difficult to disrupt the ligand–potein interaction. If no information is available on the strength of the ligand–protein interaction, then a trial and error approach should be used. It should also be borne in mind that once bound to the matrix the affinity of the immobilized ligand for the protein may be less than that for the free ligand. In the case of angiotensin converting enzyme, the affinity of the enzyme for the immobilized lisinopril ligand (10^{-5} M) is weak compared to its affinity for free lisinopril (10^{-10} M), but still sufficient to effect purification (5). Ideally the ligand–protein interaction should be reversible so that the protein can be eluted from the affinity matrix without requiring denaturing conditions (see Section 8). However, the interaction should be strong enough that non-specifically bound protein can be removed from the affinity matrix without disrupting the specific ligand–protein interaction.

2.4 Stability

The ligand should be stable under the conditions employed in the chromatographic procedure. As the aim of most purifications is to isolate the protein in a biologically active form, the chromatography will probably be done at or near neutral pH and at a low temperature (4°C), conditions under which most ligands are likely to be stable. Certain ligands, however, may be susceptible to metabolism by enzymes in the starting mixture, thus destroying or altering their specificity. For example, peptide-based ligands may be susceptible to hydrolysis by peptidases/proteases. Performing the chromatography at 4°C will minimize such unwanted reactions, and protease inhibitors can be included in the buffers if required (6, 7).

3. Choice of matrix

There are numerous matrices available from several different suppliers (e.g. Sigma, Bio-Rad, Pierce, Pharmacia) on which the ligand can be immobilized. A comparative study of matrices has indicated that agarose is one of the

Table 1. Details of some of the affinity matrix and spacer arm length combinations commercially available

Ligand reactive group	Matrix	Spacer arm length[a]	Coupling agent	Supplier
-NH$_2$	CNBr–Sepharose	1	CNBr	Pharmacia, Sigma
	Affi-Gel 10 or 15 gel	1	N-hydroxysuccinimide	Bio-Rad
	AminoLink Plus coupling gel	1	Aldehyde/borohydride	Pierce
	Reacti-Gel	1	Imidazolyl-carbamate	Pierce
	CM BioGel A gel	1	Carbodiimide	Pierce
	Activated CH Sepharose 4B	6	Carbodiimide	Pharmacia
	Aldehyde agarose	6	Aldehyde/borohydrde	Sigma
	ECH Sepharose 4B	10	Carbodiimide	Pharmacia, Sigma
	NHS activated Superose	10	N-hydroxysuccinimide	Pharmacia
	HiTrap NHS activated gel	10	N-hydroxysuccinimide	Pharmacia
	Epoxy activated Sepharose 6B	12	Epoxy	Pharmacia
-COOH	Affi-Gel 102 gel	6	Carbodiimide	Bio-Rad
	Diaminodipropylamine gel	9	Carbodiimide	Pierce
	EAH–Sepharose	10		Pharmacia
	Adipic acid dihydrazide–agarose	11	Hydrazide	Sigma
-SH	Thiopropyl Sepharose 6B	7	Thiol (reversible)	Pharmacia
	Epoxy activated Sepharose 6B	12	Epoxy	Pharmacia
	SulfoLink gel	2	Iodoacetyl	Pierce
-OH	Epoxy activated Sepharose 6B	12	Epoxy	Pharmacia
-CHO	Agarose adipic acid hydrazide	9	Hydrazide	Pharmacia
	Adipic acid dihydrazide–agarose	11	Hydrazide	Sigma
-active H	PharmaLink	–	Mannich reaction	Pierce

[a] Number of carbon atom equivalents.

better matrix materials (8). The choice of matrix may be determined by the coupling procedure that is to be used to immobilize the ligand and the length of the spacer arm between the ligand and the matrix (see Sections 4 and 5). Pre-activated matrices are commercially avaliable with varying length spacer arms and activated groups (see *Table 1* for details).

4. Choice of spacer arm

A spacer arm is used to position the ligand away from the surface of the matrix so as to allow the protein relatively unhindered access. Most spacer arms have six to eight methylene groups and are hydrophilic so as to minimize non-specific protein interactions. A longer spacer arm may increase the binding

8: Drugs and inhibitors as affinity ligands

interaction, especially if the binding site is within a deep cleft or pocket in the protein. However, there is the increasing chance of unwanted hydrophobic interactions between the arm and proteins in the sample.

With angiotensin converting enzyme, it was observed that increasing the length of the spacer arm on which the ligand was attached increased the binding capacity of the affinity matrix (9). Decreasing the length of the spacer arm from 2.8 nm to 2.2 nm or less decreased the binding affinity by some 350-fold or more. Using the selective peptide-based inhibitor lisinopril (*Figure 1b*) we constructed two affinity resins for the purification of angiotensin converting enzyme: one with a 1.4 nm spacer arm (5) and one with a 2.8 nm spacer arm (*Figure 2*) (10).

As can be seen in *Table 2*, the longer spacer arm affinity matrix yielded a larger quantity of purified enzyme, an increase of 28-fold as compared to the shorter spacer arm affinity matrix. This increased yield appeared to be primarily due to the increased binding capacity of the 2.8 nm spacer arm affinity matrix with nearly 100% of the applied activity binding to the column, with a significantly larger amount of activity subsequently eluted. These observations suggest that the active site of angiotensin converting enzyme is recessed deep within the interior of the protein rather than at its surface, such that

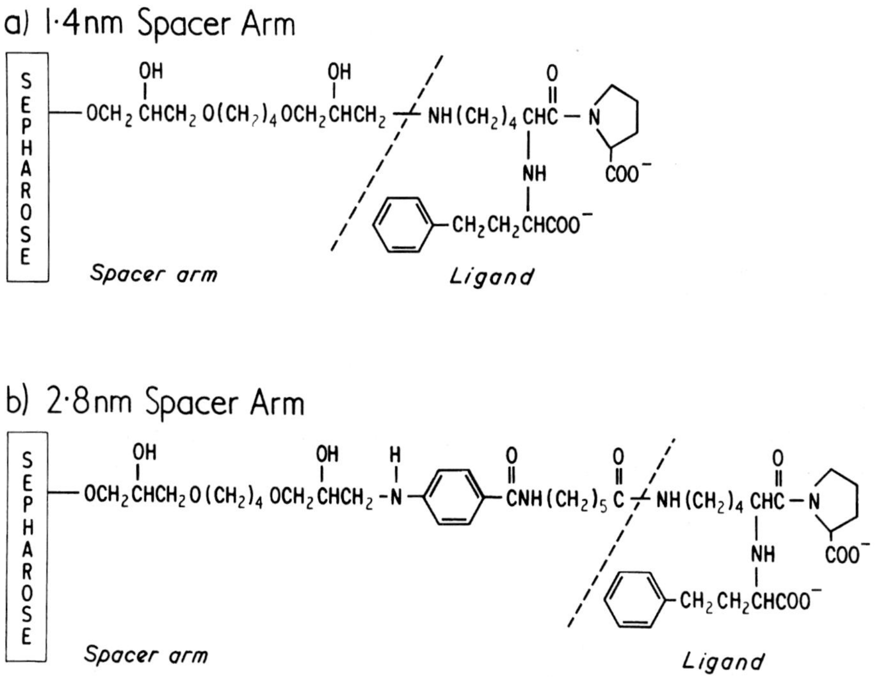

a) 1·4 nm Spacer Arm

SEPHAROSE —OCH$_2$CHCH$_2$O(CH$_2$)$_4$OCH$_2$CHCH$_2$ — NH(CH$_2$)$_4$CHC — N ...
(with OH groups on the CHCH$_2$ carbons; ligand contains NH, COO$^-$, and phenyl-CH$_2$CH$_2$CHCOO$^-$ group)

Spacer arm   Ligand

b) 2·8 nm Spacer Arm

SEPHAROSE —OCH$_2$CHCH$_2$O(CH$_2$)$_4$OCH$_2$CHCH$_2$—N—⌬—CNH(CH$_2$)$_5$C—NH(CH$_2$)$_4$CHC — N ...
(with OH, OH, H, O, O, and O substituents as shown; ligand contains NH, COO$^-$, and phenyl-CH$_2$CH$_2$CHCOO$^-$ group)

Spacer arm   Ligand

Figure 2. Structures of the spacer arms in (a) lisinopril–1.4 nm–Sepharose and (b) lisinopril–2.8 nm–Sepharose.

Table 2. Comparison of the purification efficiency of the 1.4 nm and 2.8 nm spacer arms used in lisinopril–Sepharose affinity chromatography[a]

Spacer arm length (nm)	Concentration of lisinopril ligand (μmol/ml of gel)	Activity bound (% of applied activity)	Activity eluted (% of applied activity)	Yield of purified enzyme (mg)
1.4	1 ± 0.1[b]	36.9	5.4	0.36
2.8	1–2	99	43	9.97

[a] Results are the mean of eight purifications for the 1.4 nm spacer arm and three purifications for the 2.8 nm spacer arm, and refer to the purification of angiotensin converting enzyme from porcine kidney cortex. Data from ref. 10.
[b] Data from ref. 5.

shorter length spacer arms do not project the ligand far enough off the surface of the matrix to allow the protein to bind optimally. More recent studies with biotinylated ligands of varying spacer arm length have confirmed that the active site of angiotensin converting enzyme must lie at least 1.1 nm below the outer surface of the molecule (11). In contrast with another zinc metallopeptidase, membrane dipeptidase, increasing the length of the spacer arm that was used to couple the inhibitor cilastatin to the affinity matrix resulted in a decrease in the binding affinity (12). The lisinopril–Sepharose affinity matrices allowed angiotensin converting enzyme to be purified to apparent homogeneity in a single step with a 53 500-fold enrichment from porcine brain striatum (10) and a 110 000-fold enrichment from human plasma (5), indicating that lisinopril is one of the most selective and effective ligands for affinity chromatography described to date.

5. Coupling method

The coupling method used will depend on the reactive group on the ligand through which it is to be attached to the matrix. The earliest coupling procedure involved the linking of compounds containing a primary amino group to polysaccharide matrices activated by cyanogen bromide (13). This coupling method is still widely used today, especially as CNBr–Sepharose is available commercially overcoming the hazards involved in activating the matrix with CNBr. *Protocol 1* describes the basic procedure for coupling up a ligand with a primary amino group to CNBr–Sepharose. This method is applicable to both small ligands and larger proteins. We routinely use it for the coupling of IgG for immunoaffinity chromatography applications (14) and in immobilizing the small inhibitor cilastatin for the affinity purification of membrane dipeptidase (EC 3.4.13.19) (12, 15).

Nowadays numerous matrices are available commercially that have differing length spacer arms and activated groups, such that almost any combination of reactive group on a ligand can be combined with a particular length

8: Drugs and inhibitors as affinity ligands

spacer arm (see *Table 1*). For details of the application and coupling procedure for a particular matrix the reader should refer to the manufacturer's literature. *Protocols 2* and *3* detail procedures for introducing different length (12 atom, 1.4 nm and 24 atom, 2.8 nm) spacer arms between the ligand lisinopril and the matrix Sepharose for the affinity chromatographic purification of angiotensin converting enzyme (10). These two procedures which utilize epoxy activated and *N*-hydroxysuccinimide activated coupling procedures, respectively, are applicable for the coupling of any ligand that has a primary amino group, with the longer spacer arm being particularly useful as commercially available matrices do not have such long spacer arms.

Protocol 1. Coupling of ligand to CNBr–Sepharose

Equipment and reagents
- Flask shaker
- Sintered glass funnel (G2)
- CNBr–Sepharose (Pharmacia, Sigma)
- 1 mM HCl
- Coupling buffer: 0.1 M NaHCO$_3$, 0.5 M NaCl pH 8.3

Method
1. Add 1 g CNBr–Sepharose to 20 ml 1 mM HCl and allow suspension to swell for 5 min.
2. Filter Sepharose on a sintered glass funnel attached to a water vacuum pump[a] and wash four times with 20 ml aliquots of 1 mM HCl.
3. Add Sepharose suspension to 10 ml coupling buffer containing 2–5 mg ligand, and mix for 6 h at room temperature on a flask shaker.[b]
4. Filter ligand–Sepharose on a sintered glass funnel[a] and wash with 50 ml coupling buffer.
5. Finally resuspend ligand–Sepharose in coupling buffer and store at 4°C.[c]

[a] Alternatively, allow the Sepharose suspension to settle and then decant off the liquid.
[b] Do not use a magnetic stirrer with a flea as this will fragment the Sepharose.
[c] Do not freeze the Sepharose suspension. For prolonged periods of storage 0.2% sodium azide can be added to prevent bacterial growth.

Protocol 2. Preparation of lisinopril–1.4 nm–Sepharose (5)

Equipment and reagents
- Flask shaker
- Sintered glass funnel (G2)
- Sepharose CL-4B
- 1,4-butanediol diglycidyl ether (70% nominal solution)
- 0.6 M NaOH
- Sodium borohydride
- 0.3 M K$_2$CO$_3$ pH 11
- 1 M glycine pH 10
- 1 M NaCl

Protocol 2. *Continued*

Method

1. Mix 20 ml Sepharose CL-4B, 20 ml 1,4-butanediol diglycidyl ether, and 20 ml 0.6 M NaOH containing 2 mg/ml sodium borohydride on a flask shaker[a] for 8 h at room temperature in a fume-hood.
2. Wash the resulting epoxy activated Sepharose[b] with 1 litre H_2O on a sintered glass funnel attached to a water vacuum pump.
3. Add the epoxy activated Sepharose to 20 ml 0.3 M K_2CO_3 pH 11 containing 2.2 mM lisinopril, and mix in a flask shaker for three days at room temperature.
4. Remove the lisinopril coupling solution by filtration on a sintered glass funnel and keep for determination of coupling efficiency (see Section 6). Resuspend the Sepharose in 20 ml 1 M glycine pH 10 and mix overnight at room temperature on a flask shaker to block unreacted groups.
5. Wash the resulting lisinopril–1.4 nm–Sepharose extensively with 1 M NaCl and H_2O, resuspend in H_2O, and store at 4°C.[c]

[a] Do not use a magnetic stirrer with a flea as this will fragment the Sepharose.
[b] Alternatively, epoxy activated Sepharose can be obtained commercially (see *Table 1*).
[c] Do not freeze the Sepharose suspension. For prolonged periods of storage 0.2% sodium azide can be added to prevent bacterial growth.

Protocol 3. Preparation of lisinopril–2.8 nm–Sepharose (10)

Equipment and reagents

- Flask shaker
- Sintered glass funnel (G2)
- 90 mM *N*-(4-aminobenzoyl)-6-aminocaproic acid in 0.1 M NaOH
- 1 M glycine pH 10
- 0.1 M sodium acetate pH 3.8
- 0.1 M sodium borate pH 8.7
- 0.5 M NaCl
- Dioxane
- *N*-hydroxysuccinimide
- *N,N'*-dicyclohexylcarbodiimide
- 0.3 M K_2CO_3 pH 11

Method

1. Epoxy activate 20 ml Sepharose CL-4B as detailed in *Protocol 2*, steps 1–2 or alternatively, epoxy activated Sepharose can be obtained commercially (see *Table 1*).
2. Add the epoxy activated Sepharose to 40 ml 90 mM *N*-(4-aminobenzoyl)-6-aminocaproic acid in 0.1 M NaOH for 16 h at room temperature on a flask shaker.[a]
3. Add 20 ml 1 M glycine pH 10 to the Sepharose suspension and mix for a further 4 h at room temperature to block the unreacted groups.

8: Drugs and inhibitors as affinity ligands

4. Wash the modified Sepharose on a sintered glass funnel attached to a water vacuum pump with 100 ml of each of the following:
   (a) 0.1 M sodium acetate pH 3.8.
   (b) 0.5 M NaCl.
   (c) 0.1 M sodium borate pH 8.7.
   (d) 0.5 M NaCl.
5. Wash the modified Sepharose with 100 ml anhydrous dioxane on a sintered glass funnel in a fume-hood, and then resuspend it in 60 ml dioxane.
6. To the modified Sepharose in a flask add 0.691 g N-hydroxysuccinimide (final concentration 0.1 M) stirring continuously.
7. Immediately add 1.24 g N,N'-dicyclohexylcarbodiimide (final concentration 0.1 M) stirring continuously, and then mix the suspension on a flask shaker for 70 min at room temperature.
8. Wash the resulting N-hydroxysuccinimide ester–Sepharose on a sintered glass funnel with the following:
   (a) 150 ml dioxane.
   (b) 60 ml methanol.
   (c) 60 ml dioxane.
   (d) 60 ml water.
9. Resuspend the activated Sepharose in 50 ml 0.3 M K_2CO_3 pH 11 containing 2.2 mM lisinopril and mix on a flask shaker for two days at 4°C.
10. Remove the lisinopril coupling solution by filtration on a sintered glass funnel and keep for determination of coupling efficiency (see Section 6). Resuspend the Sepharose in 50 ml 1 M glycine pH 10 and mix for 2 h at room temperature on a flask shaker to block unreacted groups.
11. Wash the resulting lisinopril–2.8 nm–Sepharose extensively with H_2O and 0.5 M NaCl, and store at 4°C.[b]

[a] Do not use a magnetic stirrer with a flea as this will fragment the Sepharose.
[b] Do not freeze the Sepharose suspension. For prolonged periods of storage 0.2% sodium azide can be added to prevent bacterial growth.

6. Determination of ligand coupling efficiency

It is often advisable to assess how much of the ligand has been coupled up to the matrix before attempting the chromatography, since lack of efficient protein binding may be due to inefficient ligand coupling. An aliquot of the ligand solution before coupling should be retained as well as an aliquot following coupling. The amount of ligand in the two samples should then be compared in order to determine the amount of ligand coupled to the matrix. The method used to determine the amount of ligand present will depend on the individual ligand, but may involve spectrophotometry, activity assay, immunoassay, etc. If a radiolabelled form of the ligand is available then spiking the coupling solution with a small amount of this would allow determination of the coupling efficiency by monitoring the incorporation of radioactivity into the matrix. For lisinopril and other peptide-based ligands we monitor coupling efficiency by comparing the absorbance at 214 nm of the lisinopril peak in the ligand solutions using reverse-phase HPLC (10).

7. Affinity chromatography

The actual procedure of affinity chromatography will depend to a large extent on the nature of the ligand and its interaction with the protein to be purified. Although *Protocol 4* details the procedure for the affinity purification of angiotensin converting enzyme from mammalian tissue or fluid samples (10), some general principles apply to any affinity chromatographic method. First, as long as it is not detrimental to the ligand–protein interaction, the presence of salt in the running buffer will decrease the likelihood of non-specific electrostatic interactions between the matrix and other proteins in the sample. Secondly, once all the sample has been applied it is essential to wash the affinity matrix extensively with buffer prior to elution so as to remove non-specifically bound protein. This can be monitored by measuring the absorbance of the wash fractions at 280 nm, and elution started only when the absorbance has fallen to near zero. In practice, we routinely include a pre-column of unmodified Sepharose linked in series between the pump and the affinity column so as to reduce non-specific binding to the affinity column. In addition, the use of a pre-column prolongs the life of the affinity column and prevents it clogging up, and hence reducing the flow rate, during the wash and elution steps of the chromatography.

Protocol 4. Affinity purification of angiotensin converting enzyme on lisinopril–2.8 nm–Sepharose (10)

Equipment and reagents

- Peristaltic pump
- Glass or plastic columns
- Narrow bore (1–2 mm) tubing to connect the pump to the column
- Fraction collector
- Running buffer: 10 mM Hepes, 0.3 M KCl, 0.1 mM ZnCl$_2$ pH 7.5[a,b]
- Lisinopril-2.8 nm–Sepharose (see *Protocol 3*)
- Unmodified Sepharose CL-4B

8: Drugs and inhibitors as affinity ligands

Method

1. Pre-equilibrate a 1 × 6 cm (approx. 5–10 ml) column of lisinopril–2.8 nm–Sepharose and a 1.5 × 10 cm (approx. 10–20 ml) pre-column of unmodified Sepharose CL-4B with 200 ml running buffer at a flow rate of 10–20 ml/h.
2. Apply protein sample at a flow rate of 10–20 ml/h and collect run-through for determination of unbound protein.
3. Once all the sample has been applied, remove the pre-column and wash the affinity column with 400 ml running buffer.
4. Elute the bound protein from the column by including 10 μM lisinopril in the running buffer or by using 50 mM sodium borate buffer pH 9.5. Collect 2 ml fractions and monitor the absorbance at 280 nm.
5. Pool those fractions with high absorbance at 280 nm[c] and dialyse extensively against 5 mM Tris–HCl pH 8 with dialysis tubing of molecular weight cut-off > 10 000.
6. Concentrate protein using centrifugal concentrators, e.g. Vivascience Vivaspin or Amicon Centricon.[d] Alternatively apply the protein to a small (0.5 ml) column of DEAE–cellulose pre-equilibrated with 5 mM Tris–HCl pH 8 and elute into 0.5–1 ml with the same buffer containing 0.7 M NaCl.

[a] 0.1% Triton X-100 can be included in the running buffer when isolating integral membrane proteins.
[b] ZnCl$_2$ is included here to maximize the binding of angiotensin converting enzyme to the ligand.
[c] As only the protein of interest should have bound and then been eluted, measurement of absorbance at 280 nm is the simplest method of determining which fractions contain the protein. If the buffers contain Triton X-100, alternative detection methods will be required as this detergent absorbs strongly at 280 nm.
[d] For integral membrane proteins isolated in the presence of detergent, we have found that the protein can bind irreversibly to the membrane in the centrifugal concentrators and thus recommend concentration on DEAE columns.

8. Method of elution

In order for affinity chromatography to be successful, not only must the protein of interest bind the ligand and be washed free of other proteins, but the bound protein must be eluted off the immobilized ligand, preferably in its native, biologically active state. Elution of a protein from an affinity resin can quite often be achieved by including free ligand in the column buffer (see *Protocol 4*). The free ligand then competes with the immobilized ligand for the protein and, if at high enough concentration, will displace it. The protein then elutes from the column bound to the free ligand. As most drugs and inhibitors used in affinity chromatography are relatively small molecules (M_r

< 5000) the active protein can be recovered following dialysis (see *Protocol 4*) or centrifugation using sample concentrators (e.g. Vivascience Vivaspin or Amicon Centricon).

Alternatively the protein may be eluted from the affinity resin by altering the ionic strength or the pH of the buffer if the ligand–protein interaction is sensitive to changes in the salt concentration or pH (see *Protocol 4*). If a very high or low pH is used for elution, the protein can be collected directly into tubes containing a strong buffer (e.g. 1 M Tris–HCl pH 7) in order to minimize irreversible denaturation. If the protein cannot be recovered from the affinity column by one of these procedures, then elution with a strong chaotropic agent (e.g. urea) could be used. A more drastic approach is to boil the sample in the presence of SDS gel dissociation buffer, then pellet the Sepharose by centrifugation in a microcentrifuge and remove the supernatant containing the protein. This method is especially useful if the sample is to be loaded directly onto an SDS–polyacrylamide gel for determination of its molecular weight, or prior to transfer to a membrane for blotting or N-terminal sequencing.

Once eluted from the affinity column the protein is often in a relatively large volume and needs to be concentrated prior to further analysis or storage. The most universal procedure that we use for the concentration of a wide variety of proteins is to adsorb it onto a small column of DEAE–cellulose and then to elute off into a small volume by including 0.5–0.7 M NaCl in the buffer. Alternatively, the protein can be concentrated using a centrifugal sample concentrator (see *Protocol 4*).

9. Affinity chromatography with biotinylated ligands

Affinity chromatography with biotinylated ligands exploits the reversible interaction between biotin and (strept)avidin (16). The vitamin biotin has an extremely strong affinity ($K_a = 10^{15}$ M^{-1}) for the egg white protein avidin (or streptavidin, its bacterial relative from *Streptomyces avidinii*). The rationale in using biotinylated ligands is that when biotin is modified, it still binds with high affinity to avidin. Thus the ligand is covalently attached to the biotin molecule, and either the biotinylated ligand is immobilized on an avidin column through which the sample containing the protein to be purified is passed or, especially if the ligand binds slowly to the protein, the biotinylated ligand is mixed first with the sample and then passed through an avidin column to which the biotinylated ligand–protein complex binds (*Figure 3*). Non-specific protein can then be washed through before elution, either by adding free biotin to displace the biotinylated ligand–protein complex from the avidin, or by altering the pH or the ionic strength of the buffer in order to dissociate the ligand–protein interaction.

8: Drugs and inhibitors as affinity ligands

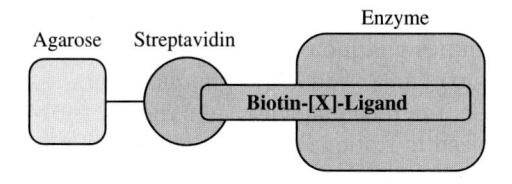

Figure 3. Principle of the use of biotinylated ligands for affinity chromatography on streptavidin–Sepharose.

Figure 4. Structures of the biotinylated peptide aldehyde used in the purification of apopain (17). The square brackets enclose the 0.9 nm spacer present when using biotinamidocaproic acid as opposed to biotin. The structure of the peptide is: Asp–Glu–Val–Asp–CHO.

This procedure was recently used to isolate apopain, the protease responsible for the cleavage of poly(ADP-ribose) polymerase and necessary for apoptosis (17). A potent peptide aldehyde inhibitor was biotinylated either without or with a 0.9 nm spacer arm (*Figure 4*). The protein was then purified as described in *Protocol 5*. In this case the inclusion of the longer spacer arm had no significant effect on the purification of apopain. Details for the coupling of biotin to the angiotensin converting enzyme inhibitor lisinopril with different length spacer arms can be found in ref. 11. Pierce have recently introduced biotinylation reagents incorporating a 3.1 nm spacer arm (EZ-Link™ Sulfo-NHS-LC-LC-Biotin and EZ-Link™ NHS-LC-LC-Biotin).

Protocol 5. Purification of apopain using a biotinylated ligand (17)

Equipment and reagents
- Biotin or biotinamidocaproic acid (Sigma)
- Streptavidin–Sepharose (Sigma)
- Peristaltic pump
- Glass or plastic columns
- Narrow bore (1–2 mm) tubing to connect the pump to the column

Method

1. Biotinylate the ligand on the free amino group with either biotin or biotinamidocaproic acid following the manufacturer's instructions.

Protocol 5. *Continued*

2. Add 20 nmol biotinylated ligand to the protein sample (2.5 ml, 3 mg protein) in 50 mM Tris–HCl, 0.15 M NaCl pH 7.4.
3. Pass the biotinylated ligand–protein sample through a column of streptavidin–Sepharose (1 ml bed volume, binding capacity approx. 60 nmol biotin/ml).
4. Wash the streptavidin–Sepharose extensively with 50 mM Tris–HCl, 0.15 M NaCl pH 7.4.
5. Elute bound protein by adding 2 mM D-biotin in 50 mM Tris–HCl, 0.15 M NaCl pH 7.4 to the streptavidin–Sepharose and incubating overnight at 4°C.

10. Regeneration of the affinity matrix

Once the protein of interest has been recovered from the affinity column, it is important to extensively wash the column to remove non-specifically bound and irreversibly bound protein before reuse. The procedure is detailed in *Protocol 6* on a variety of inhibitor-based affinity columns. Results in the columns have a prolonged lifetime and repeatedly give a high yield of purified protein. Between chromatographic runs it is a good idea to repack the affinity matrix in the column so as to ensure optimal flow rates. When storing affinity matrices for long periods we remove them from the column and store them at 4°C in the presence of sodium azide to prevent bacterial growth (see *Protocol 6*).

Protocol 6. Regeration of the affinity matrix

Equipment and reagents

- Peristaltic pump
- Glass or plastic columns
- Narrow bore (1–2 mm) tubing to connect the pump to the column
- Wash buffer A: 5 mM Hepes–NaOH, 0.5 M NaCl, 0.5 mM EDTA, 2% Triton X-100 pH 7.5
- Wash buffer B: 10 mM glycine, 0.3 M KCl, 1 M sodium thiocyanate pH 9.5
- Wash buffer C: 5 mM Hepes–NaOH, 0.15 M KCl, 0.1 mM EDTA, 0.1% Triton X-100, 0.1% sodium azide pH 7.5

Method

1. Wash the affinity matrix with 50–100 ml of each of wash buffers A, B, and C.
2. Store the affinity matrix in wash buffer C at 4°C.

Acknowledgements

I gratefully acknowledge the help of S. Parvathy and W. M. Moore for preparation of the figures.

References

1. Turner, A.J. (1981). *Trends Biochem. Sci.*, **6**, 171.
2. Dean, P.D.G., Johnson, W.S., and Middle, F.A. (ed.) (1985). *Affinity chromatography: a practical approach*, p. 215. IRL Press, Oxford.
3. Harris, E.L.V. and Angal, S. (ed.) (1989). *Protein purification methods: a practical approach*, p. 317. IRL Press, Oxford.
4. Bull, M.G., Thornberry, N.A., Cordes, M.H.J., Patchett, A.A., and Cordes, E.H. (1985). *J. Biol. Chem.*, **260**, 2952.
5. Bull, H.G., Thornberry, N.A., and Cordes, E.H. (1985). *J. Biol. Chem.*, **260**, 2963.
6. North, M.J. (1989). In *Proteolytic enzymes: a practical approach* (ed. R.J. Beynon and J.S. Bond), p. 105. IRL Press, Oxford.
7. Hooper, N.M. (1997). In *Neuropeptide protocols* (ed. B. Irvine and C.H. Williams), p. 369. Humana Press, Totowa.
8. Angal, S. and Dean, P.D.G. (1977). *Biochem. J.*, **167**, 301.
9. Pantoliano, M.W., Holmquist, B., and Riordan, J.F. (1984). *Biochemistry*, **23**, 1037.
10. Hooper, N.M. and Turner, A.J. (1987). *Biochem. J.*, **241**, 625.
11. Bernstein, K.E., Welsh, S.L., and Inman, J.K. (1990). *Biochem. Biophys. Res. Commun.*, **167**, 310.
12. Littlewood, G.M., Hooper, N.M., and Turner, A.J. (1989). *Biochem. J.*, **257**, 361.
13. Axen, R., Porath, J., and Ernback, S. (1967). *Nature*, **214**, 1302.
14. Lloyd, G.S., Hryszko, J., Hooper, N.M., and Turner, A.J. (1996). *Biochem. Pharmacol.*, **52**, 229.
15. Campbell, B.J., Forrester, L.J., Zahler, W.L., and Burks, M. (1984). *J. Biol. Chem.*, **259**, 14586.
16. Wilchek, M. and Bayer, E.A. (ed.) (1990). *Methods in enzymology*, Vol. 184, p. 746. Academic Press, San Diego.
17. Nicholson, D.W., Ali, A., Thornberry, N.A., Vaillancourt, J.P., Ding, C.K., Gallant, M., *et al.* (1995). *Nature*, **376**, 37.

9

Immunoaffinity chromatography

PAUL CUTLER

1. Introduction

The principle of immunoaffinity or immunoadsorption chromatography is based on the highly specific interaction of an antigen with its antibody (1). Immunoaffinity chromatography is a specialized form of affinity chromatography and as such utilizes a ligand (either antigen or antibody) immobilized onto a solid support matrix in a manner which retains its binding capacity. Although the technique includes the separation of antibodies using immobilized antigens (2–4), it is more commonly performed for the identification, quantification, or purification of antigens (*Figure 1*). The crude extract is pumped through the column and the unbound material washed clear prior to elution of the retained antigen by alterations to the mobile phase conditions which weaken the antibody–antigen interaction.

As with other affinity chromatography methods, the power of immunoaffinity chromatography lies in the ability to separate a target biomolecule from a crude mixture such as a cell extract or culture medium with a high degree of specificity and in high yield. The target can be concentrated from a dilute solution to a high degree of purity, typically 10^3- to 10^4-fold purification in a single chromatographic step. The most common format for both analytical and preparative separations is column chromatography (*Table 1*) (5).

Whilst affinity chromatography exploits known physiological interactions such as receptors binding hormones, enzymes binding inhibitors, antibodies are now available for immunoaffinity purification of target proteins, e.g. structural proteins, for which biological ligands were previously not available. The technique can be applied to any molecule capable of eliciting an antibody response. Whilst it was a long stated axiom that any molecule of a molecular weight greater than 5 kDa is theoretically capable of such a response, synthetic peptides as short as hexamers have been used to raise antibodies. In certain cases the technique has been applied to small molecules by raising antibodies to epitopes linked to larger molecules. This has been exploited in the generation of immunoaffinity methodology for highly sensitive drug residue analysis (6–8). The specificity of the matrices also means that columns

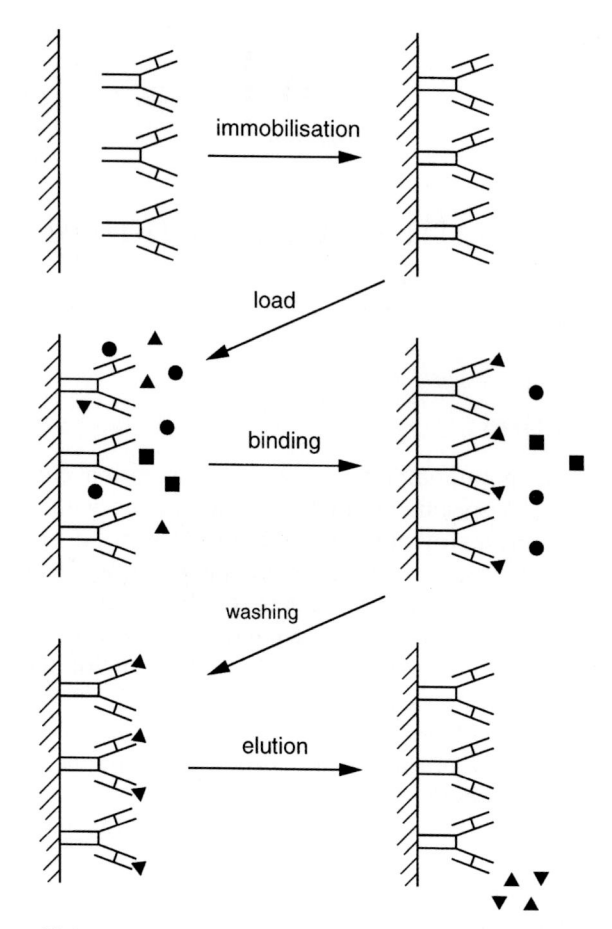

Figure 1. Immunoaffinity chromatography. A diagrammatic representation of immunoafffinity chromatography. The matrix is generated by chemically immobilizing an antibody to an activated support matrix. A crude mixture containing the target (antigen) is loaded and the impurities are washed away by the mobile phase leaving the target attached to the matrix. The mobile phase conditions are altered to elute the bound material in a discrete band.

of differing selectivities can be used in series to analyse or prepare a number of different solutes in one experiment (9).

The exquisite specificity of antibodies has led to the technique being used to separate native proteins and autoantigens produced by the mutation of endogenous proteins responsible for autoimmune diseases. Similar molecules such as antibodies of differing subclasses can even be resolved (10). Anti-idiotypic antibodies raised against specific antibodies can be also used to mimic the antigen (11).

Initial separations were based on low pressure purification systems, how-

Table 1. Applications of immunoaffinity techniques

Application	Format	Issues/comment
Lab scale purification	Production of < 100 mg of product from crude extract	High purity with good yield and good capacity. Activated agarose acrylamide matrices often used.
Bio-assay	Rapid detection and quantitation of known analytes	High throughput often required demanding stability. Matrix must withstand HPLC operating pressures. Silica often used.
Process scale	Isolation of multigram batches of product perhaps to therapeutic levels of purity	High purity low ligand leakage for therapeutics. Large particle/pore size. Stability for reuse and sanitation an important factor.
Subtractive specific impurities	Removal of small levels of leakage.	High affinity antibodies required with low ligand chromatography.
Bio-panning/ immunoprecipitation	Searching for known compounds in low titre from complex mixtures	High affinity antibodies required. Harsh conditions often used for elution.

ever the method has been successfully employed in high performance liquid chromatography (HPLC). This has led to the term high performance immunoaffinity chromatography (HPIAC) being introduced (1). By virtue of the innate specificity of the immunological reaction between antibody and antigen, resolution is high in both low and high pressure formats. The major advantage of HPIAC is speed, with separations attainable in minutes. The quantitative analytical techniques have been applied to sensitive assays for compounds in complex mixtures (12), and rapid separation of proteins susceptible to proteolytic cleavage (13).

2. Antibody selection

Antibodies are produced by B lymphocytes primarily as part of the adaptive immune system. They are produced in response to specific agents (antigens) and as such display selective affinity in the antigen binding (Fab) region (*Figure 2*). The constant region (Fc) may mobilize the immune system by

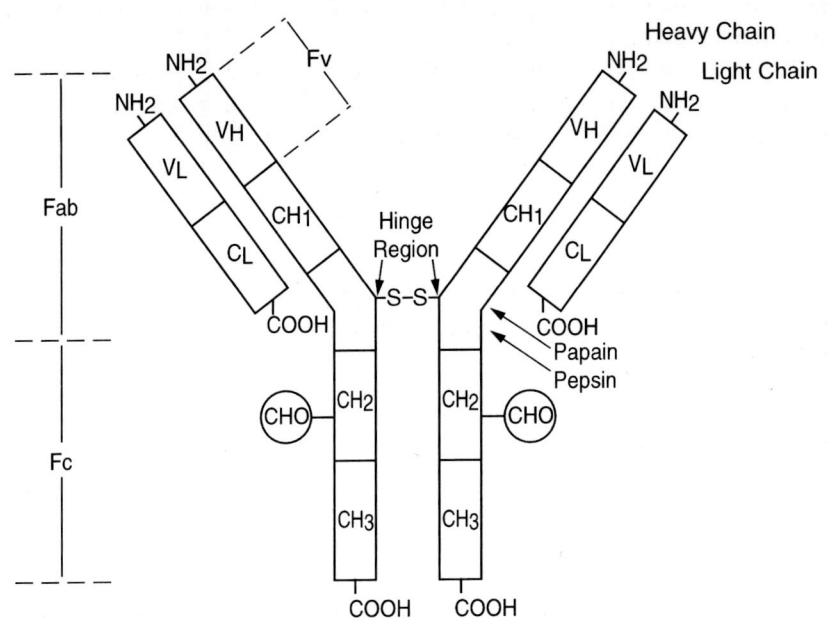

Figure 2. Immunoglobulin G (IgG) structure. IgG molecules are tetramers of heavy and light chains. The Fab (antigen binding fragment) region mediates selective target isolation. The Fc (crystallizable fragment) region controls receptor binding and complement activation. The Fc region also plays a role in binding to protein A and protein G. N-linked oligosaccharide moieties are found on Asn 297 of the heavy chain in the CH2 region but may be found at additional sites. Antibodies can be digested into smaller fragments by pepsin below the disulfide in the hinge region to yield Fc and (Fab)′$_2$ fragments or by papain above the hinge disulfide to yield Fc and Fab fragments.

9: Immunoaffinity chromatography

activation of complement and binding to the Fc receptors on host tissue cells such as phagocytes. Antibodies are raised to specific regions (epitopes) on antigens and hence antibodies can be induced in the laboratory by the introduction of a specific antigen (immunogen) into an animal giving rise to polyclonal antibodies of differing affinities and selectivities, i.e. to different epitopes (14). An individual clone from a lymphocyte can be made immortal by fusion with an appropriate myeloma cell line to produce a hybridoma and hence, via cell culture a stable supply of monoclonal antibodies (mAbs) (15). Once the genetic code is known for the specific mAb or an active fragment of antibody (e.g. a Fab region) the molecule can be generated by recombinant technology.

The selection of the antibody for use in immunoaffinity purification is critical. Initial immunopurification studies were performed using polyclonal antisera. This makes selection of affinities suitable for binding and elution difficult. With the advent of mAbs these problems were to a large extent eliminated. Despite the initial work-up of a hybridoma cell line for the production of mAbs the increased control and supply of material from mAbs has resulted in a rapid growth in the use of immunopurification reagents. The methodology is now being applied to process scale isolations (8, 16) and to the isolation of therapeutic grade proteins (9).

The antibody must be capable of stable binding at neutral pH and yet be readily dissociated at either mild acid or alkali conditions. At equilibrium the reaction between antibody and antigen can be expressed as:

$$Ab + Ag \underset{k_d}{\overset{k_a}{\longleftrightarrow}} Ab.Ag \qquad [1]$$

where Ab is free antibody, Ag is free antigen, and Ab.Ag represents the complex. k_a and k_d represent the association and dissociation constant. Under the law of mass action:

$$k_a [Ab][Ag] = k_d [Ab.Ag]. \qquad [2]$$

From which an affinity constant (K) can be defined:

$$K = \frac{[Ab.Ag]}{[Ab][Ag]} \qquad [3]$$

The value of K can vary within a range of 10^{-3} M^{-1} to 10^{-14} M^{-1}. The ideal range for an immunoaffinity reagent must be determined empirically but if $K > 10^{-4}$ M^{-1} then the antigen may not be retained sufficiently to be practicable. When $K < 10^{-8}$ M^{-1} the affinity may be too strong and harsh elution conditions may be necessary to elicit dissociation. A value in the range 10^{-7} to 10^{-8} M^{-1} is often considered optimal. This criteria has meant that IgMs are seldom used as high affinity antibodies; although in immuno co-precipitation high biological affinity (avidity) can be advantageous. Various methods exist for determination of a monoclonal antibody's avidity including the BIAcore™

affinity electrophoresis, and enzyme-linked immunosorbent assay (ELISA) (17–19). In simple format the antibody–antigen is allowed to complex to an equilibrium state before the unreacted antibody is measured by classical ELISA techniques (19). Such methods avoid the need for equilibrium dialysis but do not consider the effects of immobilization on binding kinetics. Comparison of affinity constants (K) derived by ELISA methods indicate that K under non-dissociating conditions may be a poor reflection of performance (18). These findings should not be taken as a call to disregard available affinity data but to emphasize the need, where possible, to test the immunoaffinity matrix at a small scale prior to final production.

An alternative to both polyclonal antibodies and monoclonal antibodies are recombinantly expressed antibodies and antibody fragments. One example is coliphage cloning in which Fv fragments expressed in *Escherichia coli* as gene fusions carry additional sequences specific for secretion (leader sequences), purification (e.g. hexahistidine sequences for immobilized metal affinity chromatography), and/or functional groups for immobilization (20, 21). These molecules can be produced in a tailored format and the use of bacterial expression indicates the potential for abundant and relatively inexpensive supplies of reagents. Additionally their reduced size aids penetration into matrices with a low pore size. Early expectations were that such matrices could be more stable as the antibody is a tetrameric molecule held together by intrachain and interchain disulfides whereas Fv regions are less prone to denaturation. However studies have indicated that the use of the Fv fragments offer only marginally improved stability (22).

Gene fusion proteins which can be expressed with a tag on the C- or N-terminus can be used as an epitope for detection and immunopurification of expressed proteins. The tag can be substantially removed either chemically or enzymatically. This approach has the advantage of being generic, requiring only one immunoaffinity matrix to isolate a range of different proteins expressed. This type of system has been marketed as the IBI Flag Biosystem (Immunex Corp./Kodak) for expression in *E. coli*, yeast, etc. As the system is designed for purification, the affinity can be engineered to minimize background binding and to enable mild elution (e.g. chelation with EDTA). In the Flag system an engineered epitope is added as a gene fusion poduct:

NH_3–Asp Tyr Lys Asp Asp Asp Asp Lys–protein
{—-antibody recognition sequence—-}

This is most commonly added to the N-terminus as a gene fusion and cleaved post-isolation by enterokinase which recognizes the sequence of the last five amino acids, cleaving C-terminal to the Lys residue:

{enterokinase cleavage site}
↓
–Asp Asp Asp Asp Lys–protein

9: Immunoaffinity chromatography

Protein A and protein G are proteins derived from the cell wall of the bacteria *Staphylococcus* and *Streptococcus* respectively. These proteins bind specifically to the Fc region of antibodies of the IgG type (*Figure 2*) (23). Immobilized protein A and protein G have been used extensively for the affinity purification of antibodies. A fusion protein containing a derivative of protein A, the so-called ZZ system has also been employed (24). The expressed fusion protein is readily purified using a generic IgG matrix where the immunoaffinity ligand is an antibody raised against the ZZ affinity tail of the fusion protein. The native protein is released after purification by cleavage at an Asn–Gly bond, which is sensitive to hydroxylamine.

The antibodies used for the immunoaffinity purification must be of high purity and are often themselves purified by affinity techniques such as protein A or protein G affinity chromatography (as described above), or using anti-isotype or anti-idiotype antibodies for immunopurification of the immuno-purification reagent itself. This may appear to be a circular argument but general antibodies raised against the constant region of an antibody from a particular species are often readily available commerically. Polyclonal antibodies by virtue of their source often require more clean-up before use.

Care must be observed in consideration of the molecules used as immunogens. Examples exist of immunogenic proteins isolated from electrophoresis gels in the presence of sodium dodecyl sulfate (SDS). Subsequent use of antibodies raised as the immunoaffinity matrix led to the need for SDS in the binding buffer in order to facilitate binding (25). This is presumably due to the immunogen being in effect a denatured protein, requiring denaturation of the target in order to expose the epitope. Immunogens should be as pure as possible as molecules differ greatly in their immunogenicity and minor contaminants may elicit major antibody responses.

3. Applications of immunoaffinity separations

In addition to the use of affinity column chromatography, variations have been devised. In its simplest form, immunoprecipitation, antigens are precipitated in solution using soluble antibody often in the form of antiserum (26). This approach has disadvantages as a specific ratio of antibody to antigen is required to avoid solubilization by excess antigen. This problem can be eliminated by using antibody immobilized to a solid support which can be added to the crude mixture, incubated for a brief time prior to centrifugation to recover the matrices and any bound material. In immuno co-precipitation the antibody may be added in its soluble form to bind to its target (27). The complex can then be selectively recovered by use of an affinity matrix such as protein A and protein G which recognizes the antibody.

The advent of mAbs has led to a growth in the technique for the separation of membrane receptor proteins (1, 13). This subject has been reviewed in a previous volume in this series (28). In autoimmune diseases the antibodies are often raised against cell membrane receptors and these can be used for isolation of the receptors. This technique has been used to demonstrate that patients suffering with the autoimmune disease myasthenia gravis show auto immunoreactivity against components of the acetylcholine receptor (29).

Anti-idiotypic antibodies are effectively the second in a chain of antibodies which display specificity for an antigen binding site of a previously generated antibody. As such they represent a mimic of the antigen of the first antibody. These can therefore be used as mimics of enzyme inhibitors, receptor substrates, etc. This may allow the use of antibodies for the isolation of the respective enzyme, receptors, etc.

Cells can also be purified by immunoaffinity purification. Lymphocytes and other cells have been purified utilizing cell surface markers as the antigens. This method has however seen limited use due to the popularity of fluorescence-activated cell sorting (FACS) and the observed 'stickiness' of cells leading to problems with non-specific binding (27). It has also proven difficult to optimize the elution conditions whilst retaining cell integrity.

Immunoaffinity chromatography has been successfully used for drug analysis (7, 30–32), and potential therapeutic proteins including the isolation of the lymphokine human leucocyte interferon gamma (IFN-γ), where evidence was shown of discrimination between active and inactive IFN-γ (33). Other proteins include recombinant tissue plasminogen activator from serum containing tissue culture medium (34), human plasma Factor X (35), and human urine erythropoietin (25).

4. Matrices

The range of solid phase supports for affinity chromatography is now extremely large (*Table 2*). Early work exploited cellulose-based matrices (2) whilst recent developments have involved ligand immobilization to dextran, polyacrylamide, plastic beads, and silica. However, the attributes of matrices which determine their suitability for activation and immobilization of the antibody are universal. These are:

(a) Uniformly sized particles with a macroporous structure facilitating a high area-to-volume ratio and without size exclusion effects even for large protein molecules.
(b) Hydrophilic and inert material with low non-specific absorption.
(c) Readily derivatized with a stable linkage between matrix and ligand.
(d) Rigidity to withstand high flow rates.
(e) Stability to the extremes of pH and dissociating agents used in the separation.

Table 2. Commercially available affinity matrix supports

Matrix	Type	pH stability[a,b]	Particle[a] (μm)	Supplier
Agarose	Cross-linked agarose beads, e.g. Sepharose, Affi-Gel, Reactigel	pH 3–10/12	50/100/300	Pharmacia, Bio-Rad, Pierce
Silica	Often derivitized for stability, e.g. Progel TSK	pH < 8	5/10/50/100	Toso Haas, J. T. Baker
Control pore glass	Glass particles	pH < 8	50/150	Bioprocessing
Azalactone	N-methylene-bisacrylamide and vinyldimethylazalactone	pH 1–13	50/80	3M/Pierce[c] Merck[d]
TSK Toyopearl	Glycidylmethacrylate, pentaerythritol dimethacrylate, and polyethylene glycol, e.g. AffipakACT	pH 2–12	5/45	Toso Haas
Eupergit	Methacrylamide and N,N'-methylene-bismethylacrylamide, e.g. Eupergit (Oxirane) C	pH 1–12	1/30	Röhm Pharma
Poros	Polystyrene/divinyl benzene	pH 1–14	10/20/50	Perseptive Biosystems
HyperD	Polystyrene matrix composite with hydrogel	pH 1–13	10/20/35/60	Biosepra

[a] Individual matrices from suppliers may vary.
[b] Stability of base matrix material only indicated.
[c] Marketed as Emphaze™ Biosupport.
[d] Marketed as Fractogel®EMD Aziactone Tentacle.

The most commonly used matrices are based around spherical agarose beads which can be bought commercially and chemically cross-linked to improve chemical stability and rigidity. Acrylamide has also been used; however, it is often supplied as a co-polymer with agarose to improve its physical stability. Controlled pore glass has been used to improve physical stability and thereby improve flow rates. However, as with all silica matrices, the chemical instability particularly at alkaline pHs (pH > 8) limits their use to lower pHs. Some attempts have been made to coat silica with agarose or epoxy-type resin to gain the advantages of both high rigidity and chemical stability.

Very little has been reported on the comparison of various matrices as each depends on the nature of the target, the starting material, and many other factors for its performance. However one study comparing various matrices for their performance with immobilized anti-β-galactosidase mAb for purification of β-galactosidase, indicated no apparent difference between matrices in terms of observed affinity constants suggesting the antibody–antigen interaction remained unaffected despite differing methods of antibody activation. Notably, the coupling efficiencies and specific activity differed markedly, resulting in differing capacities for the target (36). As with all affinity purifications the technique is based on selective adsorption and desorption and not continuous partitioning. As such size exclusion is not desired and the commercially available matrices should show no size exclusion effects. If large proteins or complexes are to be purified then the pore size of the matrix may have to be addressed.

For HPIAC where ridigity is paramount, immobilization techniques have been devised for glass particles, controlled pore glass, and derivitized silica. For HPIAC the most commonly used matrix is silica. The silica is modified to produce derivitizable silanol groups which can then be activated with reagents containing functions such as epoxy and thiol groups to which the protein can be immobilized (see Section 5).

Also available are affinity membranes made most commonly from regenerated cellulose (e.g. Actidisc from FMC Bioproducts, Memsep from Millipore) (37) and hollow fibres (Sepracor) (38). Alternative approaches also include liquid phase partitioning using activated polyethylene glycol, also known as immunoaffinity partitioning (39).

5. Activation and immobilization

The relative expense of mAbs production has resulted in a need to establish matrices which will retain a high capacity for the target over many cycles of reuse and without significant mAb leakage.Various functional groups are introduced onto matrices for the immobilization of ligands.

The types of functional groups exploited for the immobilization of immunoglobulins are directed towards the amino acid side chain residues (amines, sulfydryls, carboxylic acids), the termini (amino and carboxylic acid), and sugars

9: Immunoaffinity chromatography

of the oligosaccharide chains. In addition, matrices exist which are tailored for derivitized proteins such as biotinylated immunoglobulins. The major reactive groups are listed in *Table 3*. The first example of immunoaffinity chromatography utilized diazotized aminobenzyl cellulose to immobilize an antigen for subsequent antibody purification (40). The usual way for immobilization is via primary amino groups using agents such as cyanogen bromide (CNBr) or hydroxysuccinimide. These methods have the advantage of being well established and relatively easy to perform. Use of CNBr activation is complicated by the toxicity of CNBr. This may be overcome by the use of commerically available pre-activated matrices.

In theory, an immobilized IgG antibody should be capable of binding 2 mol antigen per mol antibody although a binding efficiency of around 10% is more likely to be the norm. Explanations for this poor perfomance include incorrect orientation of the antibody such that the Fab region has been utilized in the matrix coupling reaction. In the methodology described above immobilization is random with different points of the molecule forming the linkage with the solid matrix (*Figure 3*). Purely on the basis of probability, a proportion of the antibodies will be bound via the Fab region leading to functional inactivation and poor column capacity. Multipoint attachment of the mAb can also induce steric hindrance with respect to antigen binding. At high levels of activation the antibodies themselves can sterically hinder binding. Several methods have been devised to overcome these issues by utilizing site-directed immobilization. Most are based on the principle of immobilizing the immunoglobulin at the Fc region resulting in a more efficient use of the ligand.

Several matrices are available pre-activated for such approaches. Pierce supply a range of activated agarose supports marketed as ImmunoPure® immobilization kits. These kits are supplied for linkage via sulfydryl groups

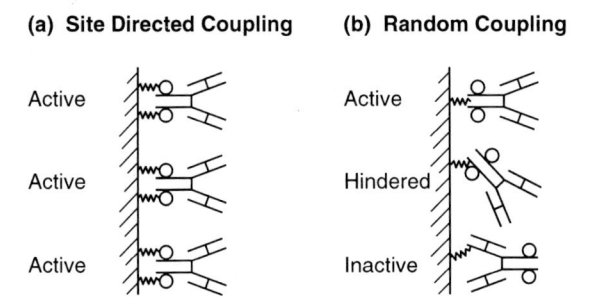

Figure 3. Random versus site-directed immobilization. (a) In site-directed coupling immunoglobulin is immobilized via functional groups such as the oligosaccharide moieties in the Fc region ensuring that the Fab region is available for antigen binding without chemical modification or steric hindrance. (b) In non-site-directed coupling immobilization leads to possible inactive or partially active forms of the immunoglobulin.

Table 3. Reactive groups for derivitization

Functional group on matrix	Structure	Target moiety on ligand
Cyanogen bromide		–NH$_2$
Epoxide/oxarine		–NH$_2$, –OH, –SH, sugars
Carbonyldiimidazole		–NH$_2$
N-hydroxysuccinimide		–NH$_2$
Vinyl sulfone		–NH$_2$, –SH, –OH
Thiol		–SH

(immobilized iodoacetyl; Sulfolink®), via the carbohydrate moieties (immobilized hydrazide; Carbolink®), and primary amines (immobilized succinimide; Aminolink®). Immobilization via the carbohydrate facilitates site-directed immobilization as does the use of the sulfydryl matrix in the presence of the reducing agent 2-mercaptoethylamine. This enables linkage of half-antibodies at the hinge region. Each kit is supplied with all the pre-packed columns, buffers, and accessories required to produce an immunoaffinity adsorbent.

5.1 Cyanogen bromide

Cyanogen bromide activation has been used most extensively due to its almost universal applicability to agarose, polysaccharide, and hydroxyl group containing synthetic polymers. They are also useful for coupling small molecules as well as large macromoleucles such as antibodies. The immobilization pro-

Table 3. Continued

Functional group on matrix	Structure	Target moiety on ligand
Tosyl chloride		$-NH_2, -SH$
Tresyl chloride		$-NH_2, -SH$
Azlactone		$-NH_2, -SH, -OH$
Hydrazine		Sugars
Avidin	Protein[a]	Biotinylated proteins
Protein A/protein G region	Protein[b]	Immunoglobulin via Fc

[a] Avidin is a tetrameric protein from egg white with a subunit M_r of 15 000.
[b] Protein A (M_r 42 000) is a cell wall protein from *S. aureus*. Protein G is a cell wall protein from *Streptococci* (M_r 35 000).

cedures are simple, reproducible, and quantitative. An important factor is the ability of the immobilization to operate under mild conditions with a range of pHs which are possible, an important consideration for immunoglobulin immobilization. CNBr-based immobilization does, however, have its disadvantages. The instability of the isourea bond leads to a low but constant leakage of ligand under operating conditions due to nucleophilic attack. The isourea moiety also acts as a weak anion exchange matrix at neutral pH. In addition to ensuring protein purity to reduce non-specific adsorption, when using CNBr matrices it is necessary to consider potential DNA due to the ability of CNBr activated matrices to immobilize nucleic acids. Despite these disadvantages, CNBr remains a popular method of immobilization.

Activation of agarose gels by CNBr is based on the conversion of the hydroxy groups of the agarose gel to a cyanate ester group (*Figure 4*). This 'activated matrix' is then capable of reacting with the ε-amino groups (lysines) of immunoglobulins and other proteins.

CNBr-activated matrices that are commercially available are highly activated and demonstrate high coupling efficiencies. However recovery of antibody

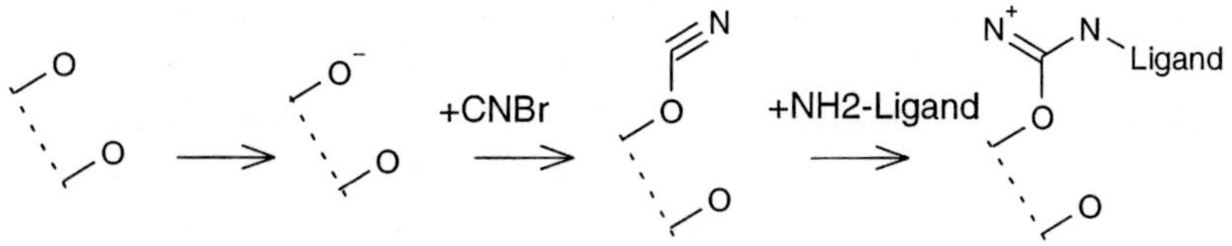

Figure 4. Activation and immobilization via cyanogen bromide.

activity is often low. Conversely, use of too low a concentration of activator may result in poor operating stability of the immunoadsorbent.

Protocol 1. Activation of agarose with CNBr and immobilization of antibody

Reagents
- Agarose gel: usually supplied as a 50% (v/v) suspension in aqueous alcohol, e.g. Sepharose (Pharmacia Biotech), Reactigel (Pierce), Affi-Gel (Bio-Rad)[a]
- Antibody solution in coupling buffer
- Solid cyanogen bromide (CNBr)
- Dimethyl formamide
- Activation buffer: 3 M potassium phosphate pH 11.6
- Coupling buffer: 0.1 M sodium bicarbonate pH 9
- Wash buffer: 0.25 M sodium bicarbonate pH 9
- Blocking buffer: 50 mM ethanolamine pH 8
- Phosphate-buffered saline containing 0.1% (w/v) sodium azide (PBSA)
- Borate buffer: 0.1 M sodium borate pH 8, 0.5 M NaCl
- Acetate buffer: 0.1 M sodium acetate pH 4, 0.5 M NaCl
- 10 M NaOH

A. *Activation*

Caution: CNBr is extremely toxic! This procedure must always be performed in a suitable fume-hood. All glassware and solutions should be treated with 10 M NaOH to destroy CNBr prior to disposal.

1. Wash the agarose gel in a Buchner funnel with 10 vol. activation buffer.[b,c]
2. Resuspend gel in 1 vol. activation buffer.
3. Transfer gel to a beaker and chill on ice with gentle stirring. Avoid vigorous stirring as this may damage the beads.
4. In a fume-hood, dissolve 5–30 mg CNBr/ml gel to 400 mg/ml in dimethyl formamide, and add dropwise to the stirring gel.[d]
5. After 10 min of stirring on ice, wash the gel thoroughly with 6 vol. ice-cold distilled water, followed by 2 vol. cold coupling buffer.[e]

B. *Immobilization*

1. Centrifuge the solution of antibody at 1000 g in the coupling buffer at a concentration of 0.5–2 mg/ml.

2. Allow the activated gel to reach 4°C, and add to a solution of antibody in coupling buffer at an antibody concentration of 5–30 mg/ml agarose gel.[e]
3. Gently agitate the gel suspension on a blood wheel overnight at 4°C.[f]
4. Wash the gel with 250 mM sodium bicarbonate pH 9 in a Buchner funnel and equilibrate in blocking buffer.
5. Incubate the gel in blocking buffer for 30 min at approx. 20°C to eliminate the remaining activated sites.
6. Wash the gel alternately with borate buffer and acetate buffer.
7. Wash the gel several times in PBSA and store in a refrigerator.

[a] Some matrices are available pre-activated with CNBr.
[b] 1 vol. is considered to be the settled volume of the gel to be prepared.
[c] It is important to ensure that the gel is not allowed to dry out any time during the activation and immobilization process.
[d] The cyanogen bromide should be prepared immediately prior to use.
[e] If unstable at pH 9, the pH of coupling can be reduced to pH 8 in borate buffer. Loss of stability can manifest itself in several forms including loss of solubility (precipitation) or loss of integrity as judged by electrophoresis.
[f] The coupling can be allowed to proceed for 2 h at room temperature. This may be preferred if the antibody chosen for immobilization precipitates at low temperature.

5.2 Immobilization with *N*-hydroxysuccinimide

N-hydroxysuccinimide (NHS) is an alternative to CNBr for the immobilization of antibodies (and other proteins) via their primary amine groups. The linkage generated by NHS is extremely stable under normal operating and storage conditions of immunoadsorbents. Coupling is relatively fast and efficient with a high degree of selectivity for primary amines, although sulfydryls will also react. As with CNBr activated matrices the supports can either be activated in the laboratory or bought as activated matrices: Affi-Gel 10 and Affi-Gel 15 (Bio-Rad), NHS activated Superose and HiTrap NHS (Pharmacia). The Affi-Gel matrices are available as loose column packing, the Pharmacia matrices are also available in pre-packed columns with methodology for immobilization *in situ*. Differences in the commercially available matrices reflect the spacer used. Affi-Gel 10 contains a non-charged ten atom spacer and is used for the immobilization of proteins with pIs of 6.5–11. Affi-Gel 15 has a positively charged 15 atom spacer arm and is useful for proteins with a pI below 6.5. An advantage of Affi-Gel is that coupling can occur over a wide pH range (pH 3–10). However, efficiency will vary markedly with protein immobilized and pH used (consult manufacturer's recommendations).

Activated NHS matrices can be generated by the reaction of the hydroxyl containing matrices at either the matrix surface using the bifunctional reagent disuccinimidyl carbonate (DCS) or attached via a spacer arm, using disuccinimidyl succinate (DMS). The use of the DCS method without the spacer arm

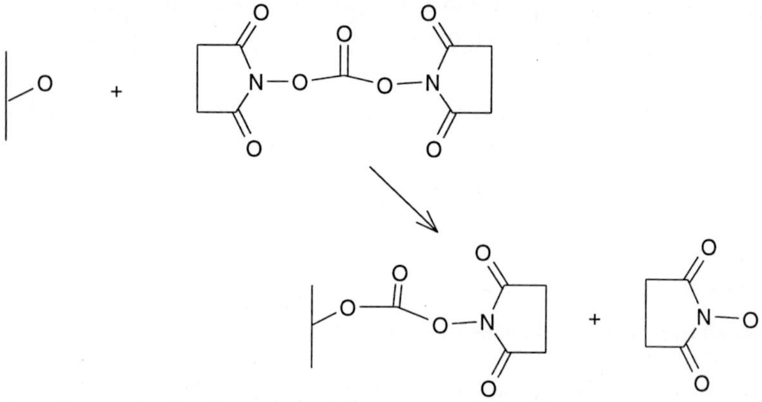

Figure 5. Activation of matrices with disuccinimidyl succinate (DMS).

(*Figure 5*) allows the unused matrix to revert by hydrolysis to the native support but will generate free carboxyl groups leading to some cation exchange.

One advantage of the NHS activated supports is their ability to immobilize ligands in either aqueous or non-aqueous solutions, although immobilization of antibodies usually proceeds in aqueous buffer solutions such as phosphate or borate (competing amines such as Tris or glycine should be avoided). For immobilization an uncharged amine is required and efficient immobilization is usually performed at pH 7.5–9, since at elevated pHs the hydrolysis of the activated matrix will accelerate leading to poor ligand immobilization (*Figure 6*). Affi-Gel 10 and Affi-Gel 15 are supplied in organic phases (isopropanol) to stabilize the matrices. As soon as exchange into aqueous buffer takes place the activated groups will begin to hydrolyse. The half-life of hydrolysis at pH 8

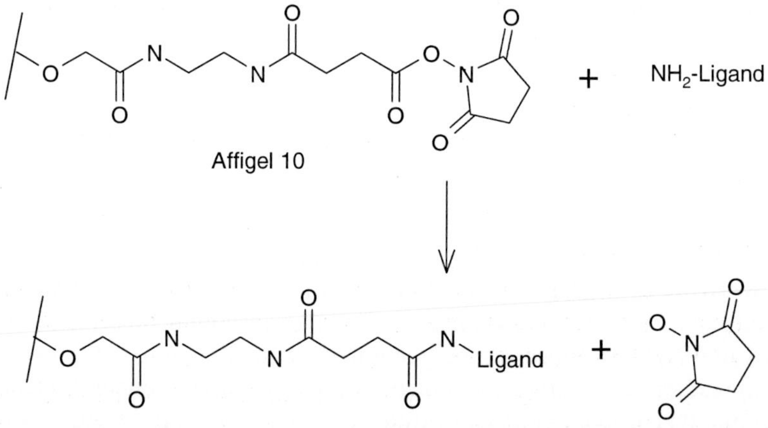

Figure 6. Immobilization of a ligand via an *N*-hydroxysuccinimide linkage.

9: Immunoaffinity chromatography

is approximately 20 min at 4 °C. This requires careful but rapid buffer transfer and ligand introduction in order to maximize the reproducibility of the immobilization.

Protocol 2. Immobilization of immunoglobulins via N-hydroxysuccinimide

Reagents
- Agarose gel: non-activated, e.g. Sepharose CL-6B (Pharmacia Biotech), or pre-activated, e.g. Affi-Gel 10 and 15 (Bio-Rad)
- Antibody solution in coupling buffer at approx. 1–5 mg/ml gel
- N,N'-disuccinimidyl carbonate (DSC)
- Dry acetone
- Dry pyridine
- Coupling buffer: 0.1 M sodium phosphate pH 7.5
- Blocking buffer: 50 mM ethanolamine pH 8
- Storage buffer: phosphate-buffered saline containing 0.1% (w/v) sodium azide (PBSA)

A. Activation

1. Wash the agarose gel in a Buchner funnel with 10 vol.[a] deionized water under low vacuum.

2. Wash sequentially with 4–10 vol. 30% (v/v) aqueous acetone, 50% (v/v) aqueous acetone, 70% (v/v) aqueous acetone, and finally dry acetone.[b]

3. Resuspend the gel in dry acetone and make the suspension a 50% gel slurry containing 80 g/litre (0.3 M) DSC.

4. Whilst mixing, in a fume-hood add one gel volume of dry pyridine containing 7.5% (v/v) anhydrous triethylamine dropwise over 30–60 min. Continue mixing for a further 60 min.

5. Wash the gel sequentially with 4–10 vol. dry acetone and 4–10 vol. dry isopropanol.

6. Store the activated gel at 4 °C as a 50% (v/v) slurry in isopropanol.

B. Immobilization

The amount of gel prepared will depend on the amount of immunoglobulin available and the scale of operation desired. The recommended level of immobilization is 4–5 mg antibody/ml gel.

1. Dissolve, desalt, or dialyse the antibody to be immobilized in the coupling buffer and cool to 4 °C.[c]

2. Wash the 50% (v/v) suspension of activated gel/Affi-Gel 10 with 10 vol. ice-cold deionized water in a sintered glass funnel to remove the storage buffer (isopropanol).[b,d]

3. Resuspend the gel to a 50% (v/v) slurry in ice-cold coupling buffer and transfer the slurry to a flask containing the immunoglobulin in coupling buffer.

249

Protocol 2. Continued

4. Stir the flask gently and allow to mix for 4–16 h at 4°C.
5. Coupling can be monitored by following the $A_{280\ nm}$ of the coupling solution. As antibody binds the absorbacne at $A_{280\ nm}$ falls.
6. Wash the gel and resuspend in 1 vol. blocking buffer and allow to mix for 1 h.
7. Transfer the slurry to storage buffer using the Buchner funnel and store at 4°C prior to use.

[a] 1 vol. is considered to be the gravity settled volume of the gel to be prepared.
[b] It is important to ensure that the gel is not allowed to dry out any time during the activation and immobilization process.
[c] In addition to phosphate buffers coupling buffers can be based on Mops, sodium bicarbonate, and sodium borate. Buffers containing primary amines such as Tris or glycine should be avoided.
[d] The water wash is necessary to avoid buffer salt precipitation in the presence of isopropanol.

5.3 Avidin immobilization of biotinylated immunoglobulins

Activation can be achieved with commerically available avidin or streptavidin agarose. The immunoglobulin to be immobilized can be modified by biotinylation. The commonest mechanism includes reaction with *N*-hydroxysuccinimde-D-biotin in a carbonate buffer pH 9 (*Figure 7*).

Once modified the immunoglobulins will bind to avidin with a binding affinity K_d of 10^{-15} M or less. This affords generation of a highly stable immunoaffinity matrix using commercially available reagents. As with CNBr activation, the largest disadvantage of this technique is the random way in which the ligand is immobilized leading to possible inactivation.

5.4 Site-directed immobilization with protein A/protein G

Protein A agarose is produced commercially and is commonly used as an extremely efficient, if relatively expensive, method of affinity purifying

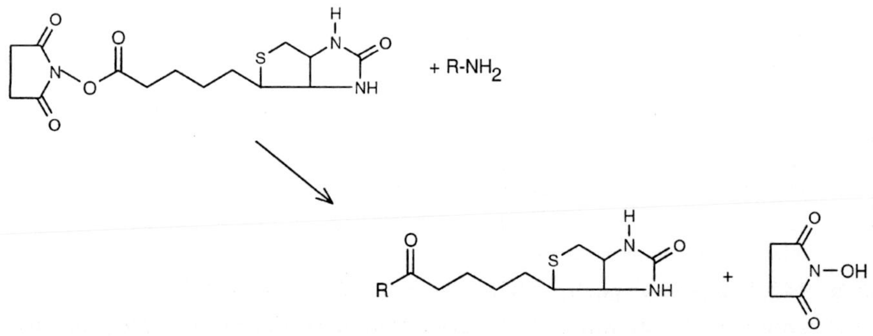

Figure 7. *N*-hydroxysuccinimide-D-biotin.

immunoglobulins (see Section 2). This method can be used to capture the immunoglobulin by the matrix in a biologically active form. The protein A–immunoglobulin complex can then be stabilized by covalent cross-linking with agents such glutaraldehyde (41) or more recently dimethyl pimelimidate (42). This method has been successfully applied to the purification of membrane proteins (43). Protein G is an alternative to protein A, capable of binding immunoglobulins from species and subclasses where protein A is not effective, e.g. mouse immunoglobulins (44–46).

Protocol 3. Immobilization and cross-linking of immunoglobulins using protein A[a]

Reagents
- Protein A agarose, e.g. protein A–Sepharose Fastflow (Pharmacia)
- Binding buffer: phosphate-buffered saline (PBS)
- Antibody solution in PBS
- Wash buffer: 200 mM triethanolamine pH 8.3
- Coupling solution: 200 mM triethanolamine pH 8.3 containing 30 mM dimethylpimelimidate dihydrochloride
- Blocking buffer: 200 mM ethanolamine pH 8
- Storage solution: PBS containing 0.1% (w/v) sodium azide

Method
1. Wash the protein A agarose using a Buchner funnel with 4–10 vol. deionized water followed by 4–10 vol. binding buffer.[b]
2. Resuspend in binding buffer and make to a 50% (v/v) slurry, and then add 10 mg antibody/ml resin.
3. Incubate by mixing for 30 min at room temperature.
4. Wash the gel with binding buffer to remove the unbound protein.
5. Wash the gel with 4–10 vol. wash buffer.
6. Resupend the gel to a 50% (v/v) slurry in coupling solution and mix by gentle rotation for 45 min at room temperature.
7. The pH of the reaction should be monitored and readjusted to pH 8.2–8.3 if necessary by dropwise addition of 1 M NaOH or 1 M HCl.
8. Stop the coupling reaction by addition of an equal volume of blocking buffer and mix the suspension for a further 10 min.
9. Wash the gel with 4–10 vol. storage solution and store in a refrigerator.

[a] Protein G may be used as an alternative.
[b] 1 vol. is considered to be the gravity settled volume of the gel.

5.5 Site-directed immobilization with hydrazide

Mammalian immunoglobulins commonly possess N-linked oligosaccharide groups in the Fc region of the molecule. In IgGs there is a conserved site for

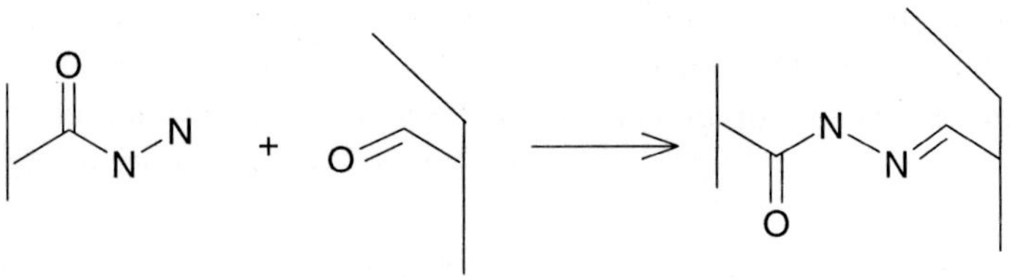

Figure 8. Immobilization of immunoglobulins using hydrazide.

glycosylation at Asn 297. As this functional group is largely restricted to the Fc region immobilization via the oligosaccharide moieties presents an efficient method of immobilization. The most common method for achieving this is to use a hydrazide activated matrix which will couple with the previously oxidized oligosaccharide (47, 48). Selectivity can be obtained via the oxidation of the vicinal diol groups present on the sugars. These are oxidized either chemically with sodium metaperiodate or enzymatically using galactose oxidase. The resultant aldehyde can then react to form a relatively stable linkage with the hydrazide (*Figure 8*).

A point to consider is that hydrazide will react with the primary amino groups of the protein backbone, although this can be prevented by maintaining the coupling reaction at pH 4.5–5.5. At such pHs the amino groups will be protonated and unreactive. Use of galactose oxidase is a useful method of selectively oxidizing the sugar moieties, however, the galactose residues may need to be first exposed by pre-treatment with neuraminidase to remove the terminal sialic acids common to glycoproteins. With this method oxidation of other residues on the protein backbone is always a possible complication (47) and whilst this method has produced stoichiometric binding of antigen to the immobilized antibody close to the theoretical of 2:1 (antigen:antibody), in some cases carbohydrate in the Fab region has led to poor activity of the immobilized matrix (48, 49).

Protocol 4. Site-directed immobilization using hydrazinolysis

Equipment and reagents
- Desalting column, e.g. Sephadex G25 column, PD10 (Pharmacia)
- Hydrazide gel
- Coupling buffer: 0.1 M sodium phosphate pH 7
- Immunoglobulin (IgG) at 1–10 mg/ml and dialysed against a suitable buffer, preferably coupling buffer
- Solid sodium periodate
- 1 M NaCl

Method
1. Add 5 mg sodium periodate/ml antibody solution. Mix gently at room temperature for 30 min.

9: Immunoaffinity chromatography

2. Desalt the oxidized antibody using a Sephadex G25 column pre-equilibrated in coupling buffer. The antibody peak is pooled.
3. Wash the hydrazide gel with ten column volumes of coupling buffer.
4. Add the desalted antibody to the gel and incubate at room temperature for 6 h.
5. Wash the column with ten column volumes of coupling buffer, followed by ten column volumes of 1 M NaCl, and ten column volumes of deionized water.
6. The gel can be equilibrated and stored in a bacteriostat such as 20% (v/v) ethanol.

Comparison between more traditional methods such as CNBr and hydrazide immobilization is difficult. However, it is generally reported that whilst the antigen binding efficiency of the hydrazine method is higher, the actual coupling efficiency is significantly lower probably due to the inaccessibility of the oligosaccharide residues (50).

An interesting development is the technique by which biotin is added to the Fc region of an immunoglobulin by use of hydrazine biotin (51, 52) (*Figure 9*). This enables site-directed immobilization by use of the avidin biotin method (see Section 5.3).

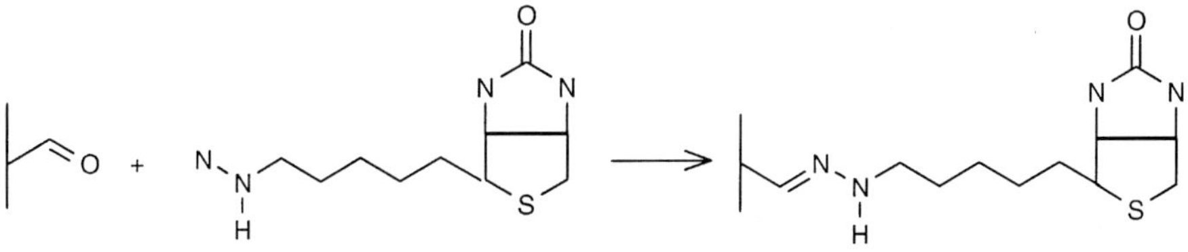

Figure 9. Biotinylation of immunoglobulins at the oligosaccharide by hydrazine biotin.

6. Determining coupling efficiency

Before using the prepared affinity matrix it is prudent to ensure that ligand immobilization has succeeded. This can be done by determination of the ligand concentration in terms of micromoles (or mg) ligand per ml of gel. By reference to the starting level of immunoglobulin the coupling efficiency can be determined. Several methods are available for achieving ligand density. The simplest is the 'indirect method' calculated by subtraction of the unbound antibody present in the solution after coupling from that present prior to coupling. Care must be taken in the method employed for determining the antibody concentration as common methods such as measurements of $A_{280\,nm}$ and the Lowry assay can give erroneous results. The UV absorbance at $A_{280\,nm}$ can be complicated by the presence of other $A_{280\,nm}$ absorbing agents such as the surfactant Triton X-100, dithiotreitol (DTT), or high levels of Tris. The Lowry assay and A_{280} can be inappropriate for NHS-activated matrices as the NHS interferes with the assay at neutral or basic pH. The solution should therefore be made acidic by the addition of 10 mM HCl prior to analysis. Amino acid analysis is a more accurate, if expensive option for determining protein concentration. As amino acid analysis is used following extended hydrolysis this method can be used to determine the protein attached to the matrix. Other methods for determining coupling efficiency include fluorescence and radioimmunoassay.

A simple method of determining coupling efficiency is described below:

(a) Determine the A_{280} (A_c) of the immunoglobulin and the volume (V_c) of coupling buffer used for the suspension of the antibody in ml.

$$\text{Total } A_{280} (T_c) = A_c \times V_c$$

(b) Determine the A_{280} (A_u) of the immunoglobulin and the volume (V_u) of coupling buffer used for the suspension of the antibody in ml.

$$\text{Total } A_{280} (T_u) = A_u \times V_u$$

$$\text{Coupling efficiency} = \frac{T_c - T_u}{T_c} \times 100\%$$

If the observed coupling efficiency exceeds 80% then the coupling may be considered complete. Otherwise an extension to the coupling time should be considered.

In order to test the functional efficiency (capacity) of the column it is important to reflect its ability to bind and subsequently release the target antigen. This is of course a function of time and how many times the matrices has been used but also reflects the nature of the immobilization of the immunoglobulin to the matrix. The capacity of a matrix activated with antibody can be determined by use of purified antigen if available. The material can be added batchwise and equilibrated before washing and elution to establish the 'equilibrium capacity' as often defined by matrix suppliers. Alternatively breakthrough kinetics can be employed, monitoring the column effluent for the target. Traditionally the column is loaded until the effluent target concentration reaches 5–10% of the concentration of the antigen applied. This method establishes the 'dynamic capacity' of the column and is by its nature more representative of the working capacity for normal column-based operations.

7. Equipment and operation

7.1 Immunoaffinity purification

Immunoaffinity columns are usually packed into glass columns for low pressure chromatography or stainless steel columns for HPIAC. The equipment required includes a pump, detector, recorder, and if possible a fraction

9: Immunoaffinity chromatography

collector. HPLC systems can be purchased from several companies including Hewlett Packard, Waters, and Perkin Elmer. Complete low pressure systems are readily available from several suppliers including Pharmacia, Bio-Rad, Amicon, Perseptive Biosystems, and Sepracor. Alternatives include stirred tanks which allow the isolation from crude cell homogenates.

Whichever method is selected the reagents used must be of the highest grade possible and all solutions must be sterile filtered prior to application to the column. These precautions may significantly prolong the lifetime of the matrix and prevent column fouling.

The type of chromatography employed and scale of operation will depend on the aim of the study. If the aim is to identify novel antibody binding targets in cell extracts then it is worth considering running the chromatography using the actual antibody and comparing the output with that from a control column preferably containing an unrelated antibody from the same species and subclass. This will allow the elimination of proteins binding via non-specific adsorption. A column containing deliberately inactivated antibody may serve the same purpose.

7.2 Sample preparation

Samples are often supplied as crude extracts which may contain cell debris, and other insoluble matter. This necessitates some form of pre-treatment of the sample to prevent column fouling and/or unwanted non-specific binding. Suitable methods include centrifugation and/or filtration through a 0.4 μm nitocellulose filter (or finer).

7.3 Binding

An important part of successful immunopurification is the choice of buffer solutions used to elicit binding of the target protein, and equally important, the conditions used to facilitate elution of the bound target. The buffer solution chosen is largely dependent upon the sample to be processed. The most commonly used binding buffers are pH 7–8, e.g. phosphate or borate, and often contain 0.1–0.5 M NaCl and/or a small amount of surfactant to reduce non-specific binding. The actual conditions will depend on the affinity constant (K_a) and the nature of the impurities present. High levels of NaCl may induce binding of impurities by promoting hydrophobic interactions.

Surfactants include the non-ionic detergents Triton X-100, Tween, and Lubrol PX. Lubrol has the advantage of being non-absorbing at 280 nm, a common wavelength for monitoring chromatography. More stringent ionic detergents such as sodium deoxycholate may be needed for solubilization of membrane proteins. When using surfactants, consideration should be given to the ultimate purpose of the purified reagent as their removal is often not a trivial exercise. Analysis of the immunoadsorbent characteristics, for example by SDS–PAGE, as an analytical technique may aid optimization of binding and elution conditions (52).

7.4 Flow rates

The sample is introduced on to the matrix in a manner limited by the mechanical ridigity of the support and the diffusion of the material into the pores of the matrix. For optimal flow rates it is always advisable to consult the manufacturer's recommendations. A flow of approximately 30–50 cm/h is suitable for low pressure chromatography and 300–500 cm/h for HPIAC.

7.5 Pre-elution washing

Following loading, the column is washed with the binding buffer to remove any unbound material. The amount of non-specific binding will depend on the original binding conditions. Use of 0.5 M NaCl in the initial buffer will result in relatively low impurity binding and reduced requirement for washing. If hydrophobic interaction of impurities is suspected, low ionic strength washing may be effective. Extended washing may remove some non-specific binding material but a balance between yield and purity has to be obtained. In general washing should be continued until a stable UV baseline is achieved.

7.6 Elution

Effective elution requires rapid desorption of concentrated functional product. The selection of the correct elution conditions is as important a part of the protocol design as any other aspect and probably the most likely to require tailoring to the particular target molecule (53). The elution conditions should facilitate release of the bound substance in a manner which retains the functional and structural integrity of both the target and the immobilized ligand. The most common way of releasing the target protein is to alter the mobile phase (*Table 4*). The most elegant and potentially selective way to facilitate release from affinity matrices is by use of a counterligand to compete for binding. This is not normally available in the case of immunoaffinity separations and so general dissociating conditions must be introduced to break the antigen–antibody binding. Often low or high pH buffers are introduced although labile proteins may be denatured by extremes of pH and the antibody may lose antigen binding capacity. This is particularly true for murine mAbs at higher pHs. Screening methods have been developed for determination of optimum elution conditions (54–57). Methods developed include modelling of immunoadsorbents on polystyrene microplates in an ELISA format (55). Other methods include use of radioloabelled antigens for sensitive assay (56) and analytical chromatography (57).

Chaotropes affect elution by altering the structure of the water associated with the interaction and by disrupting the hydrophobic interactions which may contribute to the overall binding reaction. As protein folding can be largely effected by intermolecular and intramolecular hydrophobic interactions these agents can also lead to protein denaturation. It is therefore necessary to

9: Immunoaffinity chromatography

Table 4. Mobile phases for elution from immunoaffinity supports

Class of eluent	Principle
0.1 M glycine–NaOH pH 10	High pH
50 mM diethylamine pH 11.5	
1 M NH₄OH	
0.1 M glycine–HCl pH 2	Low pH
20 mM HCl	
0.1 M sodium citrate pH 2.5	
1 M propionic acid	
50% (v/v) ethylene glycol	Organic solvent
Dimethyl sulfoxide (DMSO)	
Acetonitrile	
10% (v/v) dioxane	
0.1 M Tris–HCl pH 8 plus 2 M NaCl	High ionic strength
Deionized water	Low ionic strength
1 M ammonium thiocyanate	Chaotropes
3 M potassium chloride	
5 M potassium iodide	
4 M magnesium chloride	
6 M guanidine–HCl pH 3	Denaturant
6–8 M urea	
1% (w/v) sodium dodecyl sulfate (SDS)	Surfactant
1–10 mM EDTA or EGTA	Metal ion chelator
Sodium citrate	

remove the chaotropic agent promptly after elution using dialysis or desalting column chromatography. For this reason selection of the correct chaotrope at an appropriate concentration is critical. Chaotropes can be listed in order of stringency, as in the Hofmeister (also known as the lyotropic) series:

$$NH_4^+ < K^+ < Na^+ < Cs^+ < Li^+ < Mg^{2+} < Ca^{2+} < Ba^{2+} < PO_4^{2-} < SO_4^{2-} < CH_3COO^- < Cl^- < Br^- < NO_3^- < ClO_4^- < I^- < SCN^-.$$

Denaturants such as guanidine, urea, and ethylene glycol act by a combination of modifying protein folding (mild denaturation) and chaotropic activity. Some antibody–antigen interactions are cation-dependent and use of chelators such as EDTA can elicit elution. It should be noted that when EDTA elution is used it should be operated at pHs above neutrality to maintain solubility.

Use of differing elution conditions in series or application of a gradient of increasing concentration of the eluting agent can allow resolution of species bound to the matrix. In particular use of pH extremes prior to chaotropic agents has been shown to improve recovery from immunoadsorbents (57). Elution conditions include lowered pH (50 mM sodium acetate pH 4) and/or dissociating conditions such as 3 M potassium thiocyanate. 5 M magnesium chloride has also been used, but this should be avoided if phosphate is used as the binding buffer as insoluble magnesium phosphate will be formed.

Protocol 5. Immunoaffinity purification of target protein from crude extract

This method describes a general procedure for operation of a typical matrix.

Equipment and reagents
- Empty chromatography column of appropriate dimensions
- Suitable immunoaffinity matrix
- Suitable binding buffer: PBS
- Crude antigen containing solution/extract
- Elution buffer: 0.1 M sodium acetate pH 4
- Wash buffer: 0.01 mM sodium phosphate pH 7.4
- Regeneration buffer: 0.1 M sodium actetate pH 4 containing 8 M urea
- Storage buffer: PBS containing 0.1% (w/v) sodium azide

Method
1. Pack the column in accordance with the manufacturer's instructions.
2. Wash the column with deionized water and equilibrate in binding buffer.[a]
3. Equilibrate the starting material in the binding buffer via dialysis, desalting, or ultrafiltration, and centrifuge to remove insoluble material immediately prior to loading onto the column.[b]
4. Load the starting material onto the column at 30 cm/h and monitor the effluent by UV absorbance.[c]
5. Once loaded, wash the unbound material clear with binding buffer until the UV trace returns to baseline.
6. Wash the column with two column volumes of low ionic strength wash buffer (e.g. binding buffer minus the sodium chloride).
7. If possible, reverse the direction of the flow through the column.[d]
8. Elute the bound material with elution buffer and collect into a suitable container. The eluted protein is neutralized or dialysed as appropriate to remove the dissociating agent.
9. Adjust the eluent to neutral pH.[e]
10. Wash the column sequentially with four to ten column volumes deionized water, regeneration buffer, and finally storage buffer.

[a] If leakage of ligand is considered critical then the matrix may be subjected to at least one blank cycle using the relevant binding and elution conditions prior to the final isolation.
[b] The requirement for clarification will depend on the origin and status of the sample however 100 000 g for 1 h at 4°C will remove insoluble matter.
[c] Depending on the purity of the starting material and the titre of the target molecule it may be possible to monitor for column saturation.
[d] In order to prevent excessive exposure of the target to elution buffer the column can be reversed at this stage to alter the direction of flow.
[e] If a chaotrope is used for elution then the eluent should be desalted/dialysed promptly after elution.

7.7 Matrix stability and ligand leakage

As with all protein-based affinity matrices, immunoaffinity matrices are prone to proteolytic degradation either on exposure to the crude cell extract/medium or by microbial growth on storage. It is therefore often wise to include a cocktail of protease inhibitors such as phenylmethylsufonyl fluoride (PMSF) and leupeptin in the loading material prior to chromatography. All buffers should ideally by filtered through a 0.45 μm filter or finer.

Although some ligand leakage is inevitable, the loss of binding capacity with reuse and time is believed to be primarily the result of antibody inactivation (58). Often the capacity falls away rapidly on first use but stabilizes after several uses. This is believed to result from release of non-covalently and weakly bound ligand (13).

Stability and reuse are dependent on a number of parameters, such as the nature of the support, the immunoglobulin chosen, the target, and the elution conditions (59). A more suprising observation is the effect the level of activation used. Reports exist that matrices activated at 14 mg antibody/ml are less stable to reuse than those at 1 mg antibody/ml (57).

Storage conditions are best based on the nature of the matrix and the purpose of the end-product. Bacteriostats such as 0.05% (w/v) azide or 0.02% (v/v) chlorhexidine (hibitane) are useful storage agents as is 20% (v/v) ethanol.

7.8 Analysis of purified proteins

Proteins purified by immunoaffinity chromatography are usually analysed for purity by traditional methods such as polyacrylamide gel electrophoresis. The identification of the correct target and possibly fragments can be facilitated by Western blotting techniques. Purity may also be determined by other high resolution chromatographic techniques such as reverse-phase HPLC. If a protein has a known biological function then activity may be related to the total protein as determined by methods such as those described by Lowry (60). This enables the specific activity to be expressed in terms of activity units/mg protein. By using the total protein as a baseline the enrichment can be calculated without need for correction due to recovery. Where quantitative analytical data is required then peak area normalization is often employed using chromatographic integration. The absolute figures are related to calibration data using known standards.

References

1. Phillips, T.M. (1989). *Adv. Chromatogr.*, **29**, 133.
2. Campbell, D.H. and Weliky, N. (1967). *Methods Immunol. Immunochem.*, **1**, 365.
3. Kristiansen, T. (1976). *Scand. J. Immunol.*, **3**, 19.

4. Santucci, A., Soldani, P., Lozzi, L., Rustici, M., Bracci, L., Petreni, S., *et al.* (1988). *J. Immunol. Methods*, **114**, 181.
5. Kenney, A., Goulding, L., and Hill, C. (1988). In *Methods in molecular biology* (ed. J. Walker), Vol. 2, p. 99. Humana Press.
6. Katz, S.E. and Brady, M.S. (1990). *Anal. Chem.*, **73**, 557.
7. Katz, S.E. and Siewierski, M. (1992). *J. Chromatogr.*, **624**, 403.
8. VanGinkel, L.A. (1991). *J. Chromatogr.*, **564**, 363.
9. Jack, G.W. and Wade, H.E. (1987). *TIBTECH*, **5**, 91.
10. Balint, J.P. (1990). *Immunol. Invest.*, **19**, 81.
11. Bjercke, R.J. and Langone, J.J. (1989). *Biochem. Biophys. Res. Commun.*, **162**, 1085.
12. Yarmush, M.L., Weiss, A.M., Antonsen, K.P., Odde, J., and Yarmush, D.M. (1992). *Biotech. Adv.*, **10**, 413.
13. Josic, D., Hofmann, W., Haberman, R., Becker, A., and Reutter, W. (1987). *J. Chromatogr.*, **397**, 39.
14. Dunbar, B.S. and Schwoebel, E.D. (1990). In *Methods in enzymology* (ed. M.P. Deutscher), Vol. 182, p. 663. Academic Press, London.
15. Kohler, G. and Milstein, C. (1975). *Nature*, **256**, 495.
16. Desai, M. (1990). *J. Chem. Tech. Biotechnol.*, **48**, 105.
17. Winzor, D.J. and Dejersey, J. (1989). *J. Chromatogr.*, **492**, 377.
18. Bonde, M., Frokier, H., and Pepper, D.S. (1991). *J. Biochem. Biophys. Methods*, **23**, 73.
19. Friguet, B., Chaffotte, A.F., Djavadi-Ohaniance, L., and Goldberg, M.E. (1985). *J. Immunol. Methods*, **77**, 305.
20. Berry, M.J., Davies, J., Smith, C.G., and Smith, I. (1991). *J. Chromatogr.*, **587**, 161.
21. Pluckthun, A. (1990). *Nature*, **347**, 497.
22. Berry, M.J. and Pierce, J.J. (1993). *J. Chromatogr.*, **629**, 161.
23. Eliasson, M., Andersson, R., Olsson, A., Wigzell, H., and Uhlen, M. (1989). *J. Biol. Chem.*, **142**, 575.
24. Moks, T., Abrahamsen, L., and Osterlof, B. (1987). *Bio/Technology*, **5**, 379.
25. Yanagawa, S., Hirade, K., Ohnota, H., Saaki, R., Chiba, H., Ueda, M., *et al.* (1984). *J. Biol. Chem.*, **259**, 2707.
26. Thurston, C.F. and Henley, L.F. (1988). In *Methods in molecular biology* (ed. J. Walker), Vol. 2, p. 149. Humana Press.
27. Parkhouse, R.M.E. (1984). *Br. Med. Bull.*, **40**, 297.
28. Bailyes, E.M., Ricardson, P.J., and Luzio, J.P. (1987). In *Biological membranes: a practical approach* (ed. J.B.C. Finday and W.H. Evans), p. 73. IRL Press, Oxford.
29. Lindstrom, J. (1979). *Adv. Immunol.*, **27**, 1.
30. Bagnati, R., Gastelli, M.G., Airoldi, L., Oriundi, M.P., Ubaldi, A., and Fanelli, R. (1990). *J. Chromatogr.*, **527**, 267.
31. Van de Water, C. and Haagsma, N. (1991). *J. Chromatogr.*, **566**, 173.
32. Haasnot, W., Schilt, R., Hamers, A.R.M., Huf, F.A., Farjam, A., Frei, R.W., *et al.* (1989). *J. Chromatogr.*, **489**, 157.
33. Le, J., Barrowclough, B., and Vilcek, J. (1984). *J. Immunol. Methods*, **69**, 61.
34. Reagen, M.E., Robb, M., Bornstein, I., and Niday, E.G. (1985). *Thromb. Res.*, **40**, 1.
35. Church, W.R. and Mann, K.G. (1985). *Thromb. Res.*, **38**, 417.
36. Fowell, S.L. and Chase, H.A. (1986). *J. Biotechnol.*, **4**, 355.

9: Immunoaffinity chromatography

37. Nachman, M., Azad, A.R., and Bailon, P. (1992). *Biotechnol. Bioeng.*, **40**, 564.
38. Holton, O.D. and Vicalvi, J.J. (1991). *BioTechniques*, **11**, 662.
39. Elling, L. and Kula, M.-R. (1991). *Biotech. Appl. Biochem.*, **13**, 354.
40. Campbell, D. (1951). *Proc. Natl. Acad. Sci. USA*, **37**, 575.
41. Gyka, G., Ghetie, V., and Sjoquist, J. (1983). *J. Immunol. Methods*, **57**, 227.
42. Jack, G.W. (1994). *Mol. Biotech.*, **1**, 59.
43. Schneider, C., Newman, R.A., Sutherland, D.R., Asser, U., and Greaves, M.F. (1982). *J. Biol. Chem.*, **257**, 10766.
44. Perry, M. and Kirby, H. (1990). In *Protein purification: a practical approach* (ed. E.L.V. Harris and S. Angal), p. 149. IRL Press, Oxford.
45. Langone, J.J. (1982). *Adv. Immunol.*, **32**, 157.
46. Akerstrom, B. and Bjorck, L. (1986). *J. Biol. Chem.*, **261**, 10240.
47. O'Shanessey, D.J. and Quarles, R.H. (1987). *J. Immunol. Methods*, **99**, 153.
48. Turkova, J., Vohnik, S., Helusova, S., Benes, M.J., and Ticha, M. (1992). *J. Chromatogr.*, **597**, 19.
49. Orthner, C.L., Highsmith, F.A., Tharakan, J., Madurawe, R.D., Morcol, T., and Velander, W.H. (1991). *J. Chromatogr.*, **558**, 55.
50. Hermansen, G.T., Krishnia Mallia, A., and Smith, P.K. (1992). *Immobilised affinity ligand techniques*. Academic Press, London.
51. O'Shannessy, D.J. and Quarles, R.H. (1985). *J. Appl. Biochem.*, **7**, 347.
52. Janatova, J. and Gobel, R.J. (1984). *Biochem. J.*, **221**, 113.
53. Yarmush, M.L., Antonsen, K.P., Sundaram, S., and Yarmush, D.M. (1992). *Biotechnol. Prog.*, **8**, 168.
54. Bonde, M., Frokier, H., and Pepper, D.S. (1991). *J. Biochem. Biophys. Methods*, **23**, 73.
55. Weiss, M. and Eisenstein, Z. (1987). *J. Liq. Chromatogr.*, **10**, 2815.
56. Hornsey, V.S., Griffin, B.D., Pepper, D.S., Micklem, L.R., and Prowse, C.V. (1987). *Thromb. Haemostasis*, **57**, 102.
57. Fornstedt, N. (1990). *FEBS Lett.*, **177**, 195.
58. Sica, V., Puca, G.A., Molinari, A.M., Buonaguro, F.M., and Bresciani, F. (1980). *Biochemistry*, **19**, 83.
59. Antonsen, K.P., Colton, C.K., and Yarmush, M.L. (1991). *Biotechnol. Prog.*, **7**, 159.
60. Stoscheck, C.M. (1990). In *Methods in enzymology* (ed. M.P. Deutscher), Vol. 182, p. 50. Academic Press, London.

10

Strategies and methods for purification of Ca^{2+}-binding proteins

SANDRA L. FITZPATRICK and DAVID M. WAISMAN

1. Introduction

Ca^{2+} has been demonstrated to be the key second messenger in a variety of stimulus–response systems. These Ca^{2+} regulated processes include muscle contraction, secretory events such as exocytosis, cell cycle regulation, cellular differentiation, and gene transcription. The Ca^{2+}-binding proteins play a fundamental role in the Ca^{2+} signalling pathway: they transduce changes in intracellular Ca^{2+} into cellular activation. For example, Ca^{2+}-dependent modulatory proteins such as calmodulin or annexin II tetramer undergo conformational changes in response to increases in intracellular Ca^{2+}. The interaction of the Ca^{2+}-binding protein with Ca^{2+} results in the generation of a conformational change in the protein and when in this active conformation these proteins can bind to and affect a variety of intracellular proteins or enzymes. Alternatively, Ca^{2+}-binding proteins directly participate in cellular signal transduction by containing in the structure of the Ca^{2+}-binding protein both Ca^{2+}-binding domains and catalytic domains. For example, certain isozymes of protein kinase C contain Ca^{2+}/phospholipid-binding domains and a catalytic domain. The interaction of Ca^{2+} and phospholipid with the Ca^{2+} and phospholipid-binding domain results in activation of the catalytic domain. Other Ca^{2+}-binding proteins such as calsequestrin or calreticulin may participate in the regulation of the Ca^{2+} signal by acting as Ca^{2+} sinks. These proteins therefore function to regulate intracellular Ca^{2+} signals by influencing Ca^{2+} fluxes through the plasma membrane or the release of Ca^{2+} from intracellular Ca^{2+} storage sites. For a recent review see ref. 1.

The Ca^{2+}-binding proteins can be classified according to similarities in their Ca^{2+}-binding domains. Essentially, four Ca^{2+}-binding domains have been identified. The first, the EF hand, is shared by a large family of Ca^{2+}-binding proteins that includes calmodulin and the S100 proteins (reviewed in ref. 1). Typically, members of this family of Ca^{2+}-binding proteins may contain two, three, or four copies of the EF domain which usually bind Ca^{2+} with micromolar dissociation constants. The second family of Ca^{2+}-binding proteins, called the annexins, consist of about 13 proteins and contain four or eight copies (annexin VI only) of a Ca^{2+}- and phospholipid-binding domain called the endonexin fold. Typically, the annexins bind Ca^{2+} with millimolar dissociation constants, but in the presence of anionic phospholipids the dissociation constant decreases to micromolar Ca^{2+} concentrations (reviewed in ref. 2). The third family of Ca^{2+}-binding proteins include a family of proteins which contain the C2 domain. Members of this family include several isozymes of protein kinase C, several of the phospholipases, and the secretory granule protein synaptotagmin. These Ca^{2+}-binding proteins bind Ca^{2+}/phospholipid with micromolar dissociation constants and in several cases Ca^{2+} binding results in the translocation of the protein to the plasma membrane. The fourth family of Ca^{2+}-binding proteins include the proteins such as calsequestrin and calreticulin. These proteins typically bind Ca^{2+} with low affinity and high capacity (reviewed in ref. 3).

Isolation of individual proteins from a tissue can be a formidable task. One must use the various characteristics of the protein in question to purify it within a reasonable time to prevent protein degradation. We have developed two basic approaches to purify Ca^{2+}-binding proteins. The first approach utilizes a characteristic property of a family of Ca^{2+}-binding proteins to allow purification of these proteins. The annexin family of Ca^{2+}-binding proteins bind various cellular components in a Ca^{2+}-dependent manner (4–7). Some of these components include cellular membranes, and cytoskeletal components such as F-actin (reviewed in ref. 2).

The rationale behind the procedure for the purification of the annexins is to release proteins from the tissue with EGTA and then utilize the ability of the annexins to bind in the presence of Ca^{2+} to membranous and cytoskeletal components (the 27 000 g pellet from *Protocol 1*, step 5 serves as the source of this material). Cytosolic proteins are then removed by washing the resulting pellet with Ca^{2+}-containing buffer. The final pellet is then incubated with buffer containing EGTA which chelates the Ca^{2+} and releases the proteins bound in a calcium-dependent manner. The annexin proteins eluted by this buffer are further purified using several column chromatography procedures (8, 9).

Recently, it has been demonstrated that the annexins also bind to heparin (10). The binding of the annexins to heparin is Ca^{2+}-independent at low salt concentration and Ca^{2+}-dependent at high salt concentration. This observation has allowed the utilization of the binding of the annexins to a heparin affinity column as a useful purification step that not only produces an excellent enrichment of the partially purified proteins but also results in the purification of functionally active proteins. We have used the ability of annexin II monomer and tetramer to bind to heparin as an alternative method of purification and concentration. Many other proteins bind to heparin, but few do so

10: Strategies and methods for purification of binding proteins

in both Ca^{2+}-dependent and independent manners. By utilizing cycles of Ca^{2+}-independent and Ca^{2+}-dependent binding to the heparin affinity column, annexin II tetramer and annexin II monomer can be isolated with little proteolytic degradation.

Our laboratory has also utilized a second approach to the identification and purification of Ca^{2+}-binding proteins (11–15). This procedure uses the insoluble Ca^{2+}-binding resin, Chelex-100, to screen homogenates and column eluents for proteins that bind Ca^{2+} (16, 17). We have used this assay to purify and characterize the major Ca^{2+}-binding proteins of several bovine tissues. The assay is set-up so that initially about 95–99% of the $^{45}Ca^{2+}$ is distributed in the particulate phase (bound to the Chelex-100). Once a soluble test sample is added to the Chelex-100–$^{45}Ca^{2+}$ mixture, the presence of Ca^{2+}-binding proteins in the test sample will result in the redistribution of $^{45}Ca^{2+}$ from the Chelex-100 in the particulate phase to the soluble phase. This will result in an increase in the counts detected in the soluble phase. This assay has the advantage that the samples need not be run on SDS–PAGE or other denaturing methods before detection of Ca^{2+}-binding activity. Several proteins that have been detected by this method include calmodulin, enolase, and calreticulin. Since calmodulin is unaffected by heat treatment, heating the samples to 100°C will inactivate other proteins and allow the identification of calmodulin activity.

The Chelex-100 assay does have some limitations. For instance, if a Ca^{2+}-binding protein is present at low concentration in a test sample or if the Ca^{2+}-binding protein has a low affinity, then the amount of $^{45}Ca^{2+}$ redistributed to the soluble phase would not be distinguishable from background counts. In contrast, if the prospective Ca^{2+}-binding protein binds Ca^{2+} very tightly, it might not rapidly exchange its Ca^{2+} for $^{45}Ca^{2+}$ and so would not be detected. The assay is also not quantitative. Chelex-100 has multiple classes of Ca^{2+}-binding sites and so the assay cannot be used to estimate dissociation constants or the number of binding sites on the prospective Ca^{2+}-binding protein (16, 17).

The Chelex-100 assay can be used with crude or partially purified samples. Partial purification can be accomplished by the chromatography of 100 000 g supernatants from tissue homogenates on a variety of column types, for instance DEAE–Sephacel or hydroxylapatite. The eluent of the chromatography column is then utilized in the Chelex-100 assay and the peak of Ca^{2+}-binding activity is pooled and subjected to further purification. Therefore, the Chelex-100 assay detects Ca^{2+}-binding proteins in test samples purely on the basis of their ability to bind Ca^{2+} and does not require any prior knowledge of the biochemical properties of these proteins.

2. Purification of the annexin Ca^{2+}-binding proteins

Purification of these proteins is exemplified by their preparation from bovine lung. We have found it easiest to use frozen tissue, since this eliminates many

Table 1. Molecular weights and species distribution of the annexins

Annexin	Molecular weight (kDa)[a]	Species distribution[b]
I	35–40	Bos taurus, Cavia cutleri, Columba livia, Geodia cydonium, H. sapiens, M. musculus, R. norvegicus
II	34–39	B. taurus, H. sapiens, M. musculus, R. norvegicus, G. galus, X. laevis
III	36	B. taurus, H. sapiens, R. norvegicus
IV	28–33	B. taurus, H. sapiens, R. norvegicus, C. familiaris, Oryctolagus cuniculus, Sus scrofa
V	36	B. taurus, H. sapiens, M. musculus, R. norvegicus, G. galus
VI	67–73	B. taurus, H. sapiens, M. musculus, R. norvegicus, G. galus
VII	56–57	B. taurus, H. sapiens, M. musculus, R. macaque, X. laevis
VIII	37	H. sapiens, B. taurus
IX	33	D. melanogaster
X	36	D. melanogaster
XI	54	B. taurus, H. sapiens, Oryctolagus cuniculus, R. norvegicus
XII	35	Hydra vulgaris
XIII	3	H. sapiens, C. familiaris

[a] See ref. 18.
[b] See ref. 19.

problems of timing and availability. We have found no difference in using frozen over fresh tissue. The bovine lung is obtained from a local slaughterhouse and placed on ice as soon as possible. The lung is rinsed to remove blood. The major bronchi and vasculature are removed and the lung is frozen in 1 kg amounts in plastic freezer bags at −70°C until needed. The day before the beginning of the purification procedure the lung is transferred to a −20°C freezer and removed the morning of the annexin purification procedure. Pel-Freez Biologicals is a source of organs of various species and has distributors worldwide.

The annexins are found in many tissues and organisms. See *Table 1* for a known species distribution of the annexins. Forms of the annexins are found in plants, protists, and yeast.

Bovine liver is used when the main objective is purification of annexin VI, since this tissue does not contain any annexin V. This makes the purification simpler as many tissues such as bovine lung contain both annexin V and VI, and annexin V is sometimes difficult to resolve from annexin VI.

Once the complete annexin fraction has been dialysed, a series of column chromatography steps are required to separate and purify the individual annexin proteins. Once these columns have been run and the proteins identified, the protein fractions should be pooled, dialysed, and then frozen if it is not possible to process them further. To avoid oxidation 1 mM DTT (final concentration) should be added before freezing. We have found that many of the annexins prefer some salt present to aid in conformational stability. If it is

10: Strategies and methods for purification of binding proteins

necessary to freeze proteins, add 50 mM NaCl before doing so. Depending on which protein purification has priority, the order of purification may change to suit your laboratory's needs.

Protocol 1. Isolation of the complete annexin fraction

Equipment and reagents

- Chopping implements: hammer, 1 inch wide chisel, and a chopping board
- Meat grinder (industrial grade, motorized) and blender (4 litre capacity, Waring)
- 2 litre graduated cylinders, large bore funnels, and cheesecloth to filter the initial centrifugation
- Overhead stirring device
- 500 ml centrifuge bottles and rotor: low speed centrifuge, e.g. a Beckman JA-10 rotor and a Beckman J2-21 centrifuge
- 250 ml centrifuge bottles and rotor: low speed centrifuge and rotor, e.g. Beckman JA-14 rotor and Beckman J2-21 centrifuge
- Clinical rotator (Fisher) or equivalent
- Washing buffer (PW buffer): 20 mM imidazole pH 7.3, 100 mM KCl, 1 mM CaCl$_2$, 1 mM DTT at 4°C (approx. 4 litres of this buffer are required for complete washing of pellets)
- 50 ml centrifuge tubes, rotor: ultracentrifuge, e.g. a Ti-45 rotor and L8-70 ultracentrifuge
- 1 kg bovine lung, frozen at −70°C
- Tissue extraction buffer (TE buffer): 20 mM imidazole pH 7.3, 150 mM NaCl, 0.5 mM EGTA, 1 mM dithiothreitol (DTT), 1 mM diisopropylfluorophosphate (DIFP), 100 mg/litre soybean trypsin inhibitor (STI), 0.5 mM phenylmethylsulfonyl fluoride (PMSF) at 4°C (4 litres of buffer are required per kilogram of tissue)
- EGTA elution buffer (PE buffer): 20 mM imidazole pH 7.3, 100 mM KCl, 5 mM EGTA, 1 mM DTT at 4°C (approx. 500 ml of this buffer are required)
- Dialysis buffer (AD buffer): 20 mM imidazole pH 7.3, 50 mM NaCl, 0.1 mM EGTA, 1 mM DTT at 4°C (6–8 litres of this buffer are required)

Method

1. The night before initiating the annexin purification procedure, move the frozen lung into a −20°C freezer. In the morning, let the frozen lung thaw for approx. 20–30 min at room temperature. Take the lung and the chopping implements into a 4°C cold room. Chop the tissue into pieces 2 cm on a side and grind the tissue with a meat grinder to provide a hamburger-like texture. The flow chart for the complete annexin purification is presented in *Figure 1*.

2. Add TE buffer to the blender and add DTT and the protease inhibitors (DIFP, STI, PMSF). Add the tissue and blend until smooth. Three rounds of blending on low speed for 2 min should be sufficient. Pour into a 4 litre beaker. If there is unblended material, return to the blender and continue until unhomogenized material is gone.

3. Centrifuge at 2900 *g* in the 500 ml bottles for 15 min at 4°C. This is 4000 r.p.m. using a Beckman JA-10 rotor. Pour the supernatant through the cheesecloth to remove any floating material, then pour into a 4 litre beaker and discard the pellets.

4. Add 4 mM iodoacetic acid (2.97 g/4 litres). Let stir in the cold room for 10 min with an overhead stirrer. Adjust the supernatant to 10 mM CaCl$_2$ and allow to stir for 30 min.[a]

Protocol 1. *Continued*

5. Centrifuge at 27 000 g in the 250 ml bottles for 40 min at 4°C. This is 13 500 r.p.m. using a Beckman JA-14 rotor. Keep the pellets and discard the supernatant. The pellet is fairly sturdy but caution is advised in pouring off the supernatant. It is most efficient to use two or three rotors when processing the material.
6. Resuspend the pellet in PW buffer. Add approx. 100 ml to each bottle and use a Polytron (Brinkman Instruments) on setting 4 to resuspend the pellet. The material will go into solution fairly quickly.
7. Fill the bottles completely with more of PW buffer then centrifuge them at 27 000 g for 40 min. This is 13 000 r.p.m. in a JA-14 rotor.
8. Repeat the washing of the pellet twice more. The pellet should be a creamy colour with little trace of reddish material in the supernatant.[b]
9. Once the pellets are completely washed, use approx. 400 ml PE buffer to resuspend all of the pellets using the Polytron. Use another 100–150 ml to wash the bottles, combine all of this material into two 500 ml bottles, and rotate them on a Fisher Clinical Rotator (Fisher) for 10 min to allow the EGTA to release all of the bound proteins. Load into Ti-45 tubes or equivalent and centrifuge at 100 000 g for 1 h. This is 40 000 r.p.m. in a Beckman Ti-45 rotor.
10. Combine the supernatants and dialyse against 6–8 litres of AD buffer overnight. This pooled fraction is referred to as the complete annexin fraction. This pooled complete annexin fraction contains annexins I-VI (*Figure 2*).

[a] This allows the annexin proteins to bind to cytoskeletal components in a Ca^{2+}-dependent manner.
[b] After the second centrifugation, pellets can be consolidated from two or three bottles into one. Do not attempt to consolidate the pellets into a very small volume as it becomes more difficult to ensure that trapped proteins are efficiently removed. Subjecting the pellet to a third wash cycle will eliminate a multitude of contaminating proteins in future column chromatography steps.

The DEAE–Sepharose and HAP columns are prepared before the annexin purification procedure begins, since some time is needed to properly pack and equilibrate them.

Protocol 2. Initial separations

Equipment and reagents

- FPLC (Pharmacia Biotech or equivalent) with dual pump system and fraction collector
- Peristaltic pumps and Tygon tubing (Pharmacia Biotech)
- Collecting tubes (13 × 100 mm)
- SDS–PAGE gel apparatus (mini gels from Bio-Rad, Idea Scientific, or Pharmacia Biotech)

10: Strategies and methods for purification of binding proteins

- Dialysis equipment: (for DEAE) 20 mM Tris pH 7.5, 1 mM DTT; (for HAP) 10 mM KPi pH 7, dialysis tubing (12000–14000 M_r cut-off), and two 4 litre beakers
- 200 ml DEAE–Sepharose (Pharmacia) in a 5 × 30 cm column (Pharmacia Biotech)
- LS buffer (for DEAE column): 20 mM imidazole pH 7.3, 50 mM NaCl, 1 mM DTT (approx. 1.5 litres)
- 100 ml hydroxylapatite (HAP) (Bio-Rad) in a 2.6 × 40 cm column (Pharmacia Biotech)
- HS buffer (for DEAE column): 20 mM imidazole pH 7.3, 1 M NaCl, 1 mM DTT (approx. 1.5 litres)
- LP buffer (for HAP column): 10 mM KPi pH 7, 1 mM DTT (approx 1 litre)
- HP buffer (for HAP column): 1 M KPi pH 7, 1 mM DTT (approx. 1 litre)

Method

1. On day one of the annexin purification procedure, prepare the DEAE and HAP columns and equilibrate with LS and LP buffer respectively. All chromatography is performed at 4°C.

2. Remove the dialysed complete annexin fraction (see *Protocol 1*, step 10) the next morning and check for extreme turbidity. If the material is very cloudy, centrifuge at 100000 *g* for 30 min in a Beckman Ti-45 rotor at 4°C.

3. Add 1 mM DTT to the dialysate and place in a 1 litre Erlenmeyer flask. Pump this material onto the DEAE–Sepharose column at 2 ml/min (*Figure 3*). Once protein is detected in the flow-through of the DEAE–Sepharose column, start a second pump, which loads the flow-through material onto the HAP column.[a]

4. When the dialysate has finished loading, wash the DEAE column with LS buffer. Continue collecting and applying the flow-through to the HAP column until the protein concentration returns to baseline. One can either use a UV detector or Coomassie Blue protein assay (Bio-Rad) to determine the protein concentration of the flow-through material.

5. Connect the DEAE–cellulose column to its gradient source and commence running the gradient. For the size of column described above, use 1 litre of each LS and HS buffers. Run the gradient at 2 ml/min and collect 10 ml fractions. A simple two chamber gradient maker is all that is required, since the proteins are eluted before the 0.5 M point in the gradient. A 200 ml wash with HS buffer will ensure that all proteins have been eluted. Wash with a further 200 ml of LS buffer and the column will be ready to use again. This column can be used seven or eight times before lipids and other particulates block its ability to bind proteins. A typical chromatography profile is presented in *Figure 4*.

6. The HAP column is washed until protein levels drop as with the DEAE column and then it is hooked into an FPLC system. The annexin II tetramer is eluted at high levels (600 mM) of phosphate and a simple gradient maker will not give a sufficient gradient. The usual gradient is from 10 mM to 700 mM KPi, with a 100 ml 1 M KPi wash to ensure all protein is eluted. Fractions of 5 ml are collected with a total volume of

Protocol 2. *Continued*

   500 ml for the gradient. A typical chromatography profile is presented in *Figure 5*.

7. SDS–PAGE gels at 12.5% acrylamide are run the next day. The traces from chart recorders are used to determine where to begin sampling. Usually samples are taken every three or four fractions. By examining the SDS–polyacrylamide gels, the exact location of the proteins of interest may be determined. See *Figures 4* and *5* for the DEAE and HAP profiles. The fractions containing the proteins are pooled and dialysed to remove excess salts. For the DEAE samples, use 20 mM Tris pH 7.5, 1 mM DTT and for the HAP, dialyse against 10 mM KPi, 1 mM DTT. If your protein tends to come out of solution at low salt concentrations, maintain at least 50 mM salt in the dialysis buffers.[b]

[a] The rationale for this double pumping step is to decrease the total time necessary to load both columns. If one waited until the DEAE was completely loaded, then began loading the HAP, the HAP would be ready to run its gradient around 1 a.m. or later. See *Figure 3* for a diagram of the set-up.
[b] It is possible to add 50 mM NaCl to precipitated protein and have some of the protein return to solution, but it is better to keep some salt present in all dialysis buffers and avoid this problem.

   The proteins are fairly well separated at this point, especially the annexin I, annexin II monomer, and the annexin II tetramer from the HAP column. These are pooled, dialysed, and then individually applied to a Fast S column to remove any trace proteins. The HAP column is used to separate the annexin V and VI and the matrix is discarded afterwards. We have found that if the HAP is used over three times, the resolution begins to suffer and it becomes difficult to interpret the results. This is likely due to lipids becoming tightly bound to the HAP matrix. It is easier to replace the HAP matrix for every preparation than to risk having to redo the entire annexin purification procedure.

   Annexin II monomer is highly unstable and will experience partial proteolysis even in the absence of detectable contaminants on SDS–PAGE. If this protein is desired, maintain the DTT levels in all buffers at 1–2 mM and freeze the fractions at –70 °C if it is not to be processed for more than one day. It will degrade even in the freezer, so check for proteolysis before using.

   The Fast S column is run with MES buffer at pH 6. The pH should be adjusted as soon as possible once the protein has been eluted, especially if the column is being run overnight. The easiest way to do this is to add 0.5 ml of 1 M Tris at pH 7.5 to the collection tubes. This is necessary to ensure that the proteins are in the pH 6 buffer for a limited amount of time as the quality of protein may be adversely affected by prolonged exposure to the lower pH.

   The annexins V and VI are applied to the HAP column after dialysis of the pooled fractions from the DEAE column. A shallow gradient is necessary and

10: Strategies and methods for purification of binding proteins

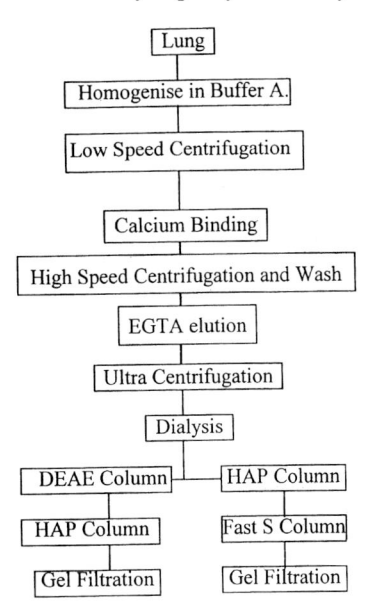

```
                    Lung
                     │
          Homogenise in Buffer A
                     │
          Low Speed Centrifugation
                     │
              Calcium Binding
                     │
      High Speed Centrifugation and Wash
                     │
                EGTA elution
                     │
             Ultra Centrifugation
                     │
                  Dialysis
                     │
         ┌───────────┴───────────┐
    DEAE Column              HAP Column
         │                       │
    HAP Column              Fast S Column
         │                       │
    Gel Filtration          Gel Filtration
```

Figure 1. Flow chart of the annexin purification procedure. This figure shows a schematic of the annexin purification procedure. *Protocol 1* consists of the steps leading to the isolation of the complete annexin fraction (see step 10). This contains annexins I–VI with only minor contamination. *Protocols 2* and *3* show chromatography steps necessary for purification of each of the seven annexins. The DEAE column separates the complete annexin fraction into three main groups: annexins I, II monomer, and II tetramer flow-through, and are applied to the HAP column. Annexins III, IV, and a 22 kDa protein are found in the low salt peak (fractions 20–40, *Figure 4*) whilst annexin V and VI are in the high salt peak (fractions 53–66, *Figure 4*). The HAP column (*Figure 5*) separates the annexin I (fractions 66–72), annexin II monomer (fractions 75–84), and annexin II tetramer (fractions 87–102). The HAP column (*Figure 7*) is also used to separate the annexin V (fractions 4–17) and annexin VI (fractions 30–34).

even then, sometimes these proteins do not separate well. If only annexin VI is desired, bovine liver should be used as the starting material, since this organ has little or no annexin V to complicate the purification. Annexins III and IV are applied to the Fast Q which results in their separation.

All of these proteins are concentrated to 3–4 mg/ml and run on gel permeation chromatography as a final step before storage. This results in the equilibration of the purified annexins in storage buffer. Many of these proteins have a salt dependency for stability and may come out of solution if the protein concentration is raised above 2 mg/ml and the salt is lower than 50 mM NaCl. Their function may also be affected by storage in minimal salt. If proteins are desalted using a PD10 column (Pharmacia), use them immediately and do not freeze. It is recommended that pooled fractions be frozen at −70°C before dialysis if they cannot be processed within two days. Each column run takes approximately one-half day, assuming that all goes well.

Sandra L. Fitzpatrick and David M. Waisman

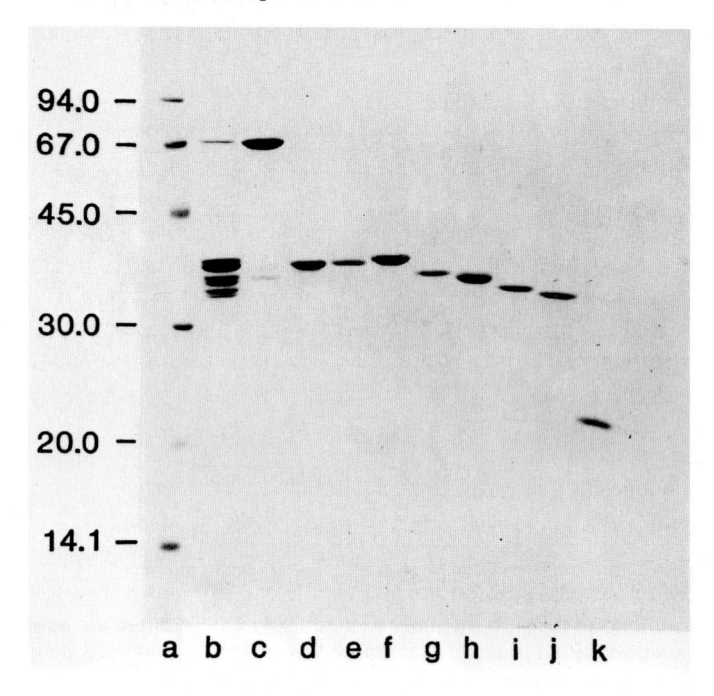

Figure 2. SDS–PAGE of purified annexins. Lanes contained: (a) molecular weight markers, (b) complete annexin fraction from *Protocol 1*, step 10, (c) annexin VI, (d) annexin II monomer, (e) annexin II tetramer, (f) annexin I, (g) annexin V, (h) annexin III, (i) unknown annexin, (j) annexin IV, (k) CAP-II.

Protocol 3. Secondary separations

All chromatography takes place at 4°C.

Equipment and reagents

- FPLC (Pharmacia Biotech) or equivalent with fraction collector
- 12 × 75 mm collection tubes
- SDS–PAGE gel apparatus (mini gels from Bio-Rad, Idea Scientific, or Pharmacia Biotech)
- Dialysis equipment: (for DEAE, Fast S, and Fast Q) 20 mM Tris pH 7.5, 1 mM DTT; (for HAP) 10 mM KPi pH 7, dialysis tubing (12 000–14 000 M_r cut-off), and 4 litre beakers
- Fast S and Fast Q columns (Pharmacia Biotech)
- HiPrep Sephacryl S-100 high resolution gel permeation chromatography column (Pharmacia Biotech 17–1194–01) (2.6 × 160 cm)
- HAP column from *Protocol 2*
- LP buffer (for HAP column): 10 mM KPi pH 7, 1 mM DTT (approx. 1 litre)
- HP buffer (for HAP column): 1 M KPi pH 7, 1 mM DTT (approx. 1 litre)
- MLS buffer (for Fast S): 50 mM MES pH 6, 1 mM DTT (approx. 1 litre)
- MHS buffer (for Fast S): 50 mM MES pH 6, 1 M NaCl, 1 mM DTT (approx. 1 litre)
- TLS buffer (for Fast Q): 20 mM Tris pH 7.3, 1 mM DTT (approx. 1 litre)
- THS buffer (for Fast Q): 20 mM Tris pH 7.3, 1 M NaCl, 1 mM DTT (approx. 1 litre)
- Gel permeation buffer: 40 mM Tris pH 7.3, 150 mM NaCl, 0.5 mM EGTA, 1 mM DTT (approx. 2 litres)

10: Strategies and methods for purification of binding proteins

A. Fast S purification of annexin I, annexin II monomer, and annexin II tetramer

1. Lower the pH of the sample to approx. pH 6 with solid MES and check the pH with pH indicator strips (Baker-pHIX from J. T. Baker). A pH meter is not recommended because the protein can interfere with its operation.
2. Use a peristaltic pump to apply the sample to the Fast S column and wash the column with 100 ml MLS buffer. Hook the column into the FPLC and run a 0–800 mM gradient in 400 ml with a 100 ml 1 M NaCl wash. Collect 4 ml fractions. In the collection tubes, place 0.5 ml 1 M Tris pH 7.5. This will raise the pH from 6 up to approx. 7.5. Equilibrate the Fast S column with MLS buffer and apply the next sample. A typical chromatography profile is presented in *Figure 6*.
3. Analyse fractions by SDS–PAGE gels at 12.5% (w/v) acrylamide. Typically, 5 µl of sample is added to 10 µl 2 × SDS disruption buffer and all 15 µl is placed in a well.
4. Pool and concentrate the pure proteins for gel permeation chromatography in Amicon concentration cells to between 3–4 mg/ml. Protein concentration is determined by Coomassie Blue reagent (Bio-Rad) or by extinction coefficient at 280 nm.

B. Fast Q purification of annexin III and annexin IV

1. Use a peristaltic pump to apply the sample to the Fast Q column and wash the column with 100 ml TLS buffer. Hook the column into the FPLC and run a 0–400 mM gradient in 400 ml with a 100 ml 1 M NaCl wash. Collect 4 ml fractions.
2. Analyse fractions and run on gel permeation chromatography as in part A.

C. HAP purification of annexin V and annexin VI

1. Use a peristaltic pump to apply the sample to the HAP column and wash the column with 100 ml LP buffer. Hook the column into the FPLC and run a 0–800 mM gradient in 400 ml with a 100 ml HP wash. A large superloop can be used if the volume of the sample is under 50 ml. Typically there is 80–100 ml in this fraction. It is not recommended that the FPLC pumps be used to apply the sample directly as the protein may damage the pump seals.
2. Run gel, concentrate, and apply to gel permeation chromatography as in part A. A typical chromatography profile is presented in *Figure 7*.

D. Gel permeation chromatography

1. Concentrate the isolated protein to 3–4 mg/ml. It is not necessary to dialyse as this column equilibrates the protein in the gel permeation buffer.

Protocol 3. Continued

2. Run the gel permeation column at 0.5 ml/min. Annexin II tetramer elutes at approx. 100 ml. Annexin VI elutes at approx. 105 ml. The other annexins (I, II monomer, III, IV, V) elute at approx. 130 ml.

3. Run SDS–PAGE gels to confirm purity and as a cross-check on the chromatography trace. Pool the peaks, concentrate to 1–2 mg/ml using the Amicon concentration cells, and determine protein concentration by Coomassie Blue reagent or extinction coefficient at 280 nm. Dispense in 1 ml aliquots and store at –70 °C.

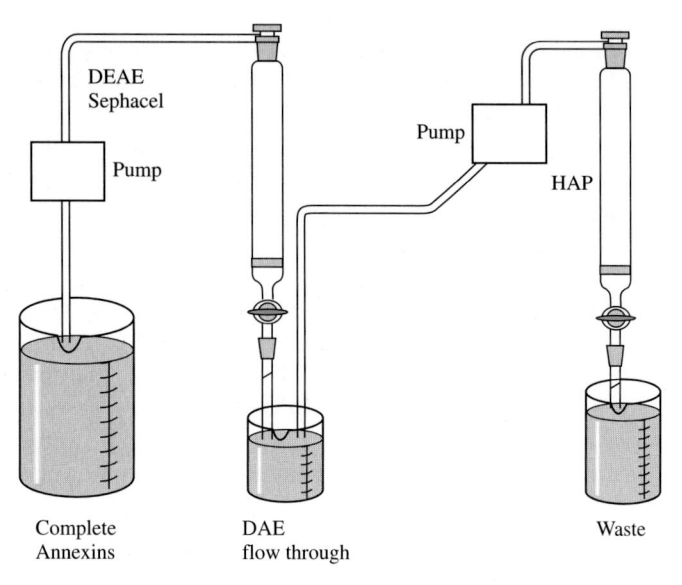

Figure 3. DEAE–Sephacel and hydroxylapatite chromatography of the complete annexin fraction. The complete annexin fraction is pooled, dialysed, and centrifuged to remove particulates. The clarified fraction is collected in the first container and pumped at 2 ml/min onto the DEAE–Sephacel column previously equilibrated with LS buffer (see *Protocol 2*, step 3). When 200 ml have accumulated in the flow-through container, the second pump is started which results in the loading of the DEAE–Sephacel flow-through onto the hydroxylapatite (HAP) column.

3. The heparin column

The annexins bind to heparin in the presence of Ca^{2+} and these proteins can be eluted with buffers containing both Ca^{2+} and 500 mM NaCl (HSC buffer) or with an EGTA buffer containing 150 mM NaCl (MSE buffer). It is possible to use a heparin affinity column to isolate the annexins by utilizing either of these methods. Typically this purification step is used before the final gel

10: Strategies and methods for purification of binding proteins

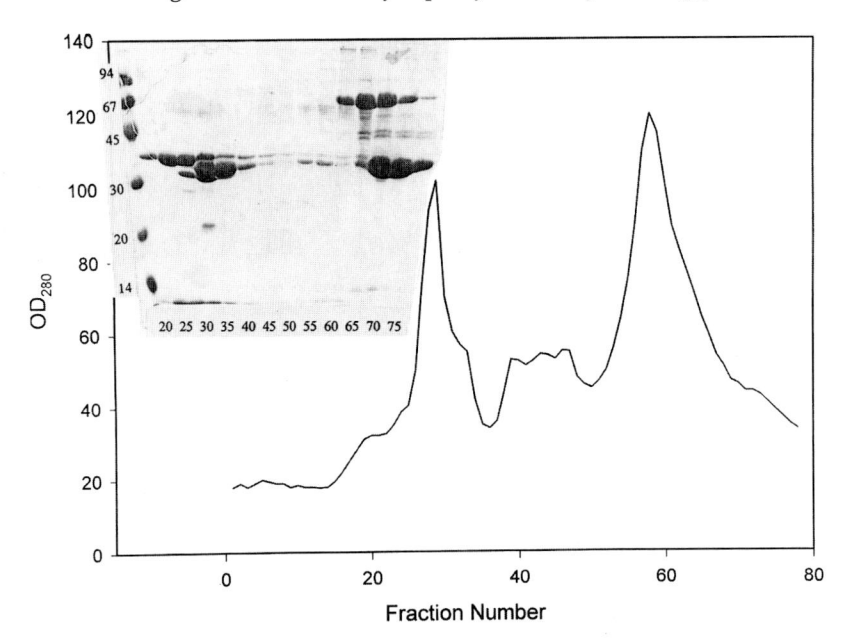

Figure 4. DEAE–Sephacel column profile. The complete annexin fraction was applied to the DEAE–Sephacel column and eluted with a linear salt gradient as described in *Protocol 2*, steps 2–5. Samples of the fractions were subjected to SDS–PAGE. Annexin III–VI were identified by their molecular mass and these proteins were pooled and subjected to further purification. Annexins III, IV, and a 22 kDa protein are found in the low salt peak (fractions 20–40) and annexin V and VI are found in the high salt peak (fractions 53–66).

permeation chromatography step. This purification step ensures that all purified annexins have retained Ca^{2+}-binding activity and also has the benefit of concentrating the purified proteins.

The two methods of elution of the annexins from the heparin affinity column involve equilibration of the column in a buffer containing Ca^{2+} and 50 mM NaCl (LSC buffer). Elution is conducted with either a EGTA-containing buffer (MSE buffer) or a Ca^{2+} buffer containing either 150 mM NaCl (MSC buffer) or 500 mM NaCl (HSC buffer). The MSC buffer is used as a wash step, since most other contaminating proteins will elute at that concentration of salt. This method has drawbacks when used with a crude sample because many proteolytic enzymes are activated by Ca^{2+} and so the annexins may be proteolysed. If a crude sample is used, the MSE buffer elution would be preferable, since it keeps the protein exposed to the Ca^{2+} for the shortest amount of time. All chromatography takes place at 4°C.

Most heparin columns used have a small bed volume, typically less than 5 ml. This small volume makes it difficult to run a gradient. If a gradient is desired, a step gradient, where a series of buffers are prepared with increasing

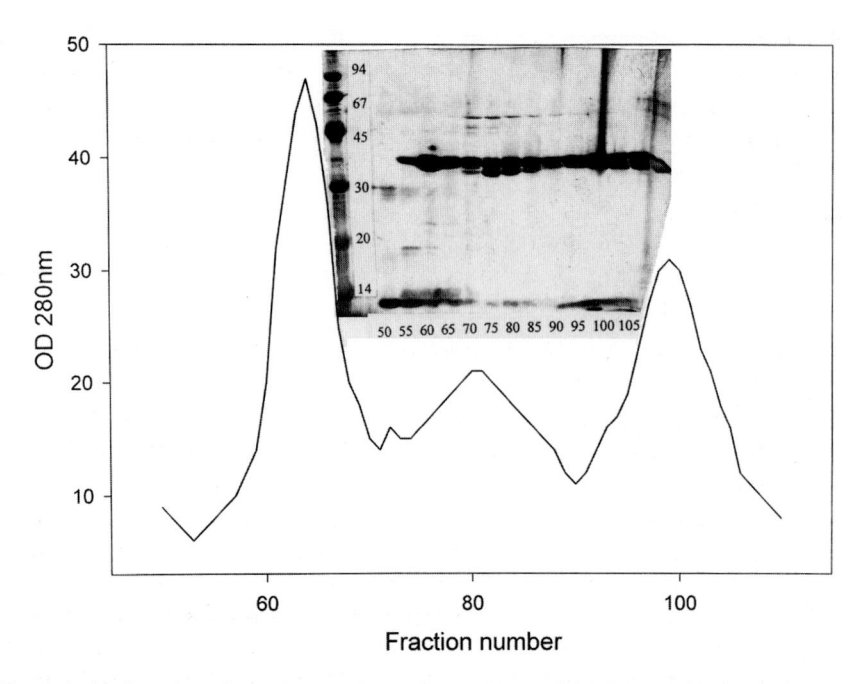

[Figure: OD 280nm vs Fraction number with SDS-PAGE gel inset showing molecular weight markers 94, 67, 45, 30, 20, 14 kDa and fractions 50–105]

Figure 5. Hydroxylapatite column chromatography profile. The DEAE–Sephacel flow-through, which contains annexin I, annexin II, and annexin II tetramer was directly applied to a column of hydroxylapatite as described in *Protocol 2*, step 3 and 6. The column was developed with a linear phosphate gradient and aliquots of the fractions were subjected to SDS–PAGE. Annexin I (fractions 66–72), annexin II monomer (fractions 75–84), and annexin II tetramer (fractions 87–102) were identified by molecular mass, pooled, and subjected to further chromatography on Fast S.

concentrations of either NaCl or EGTA, is the most efficient way of purification of the annexins. The small volume of buffer precludes using most fraction collectors with on-line $A_{280\,nm}$ absorbance and it has been found that the most efficient method is to collect directly into tubes and then assay for protein. The Gilson system has a short tubing length and has been used successfully.

Protocol 4. Binding to heparin and elution

Equipment and reagents

- Heparin–Sepharose CL-6B (Pharmacia Biotech, 17-0467-01)
- 5 ml capacity columns, 1 × 5 cm diameter (Bio-Rad Laboratories, Pharmacia Biotech)
- Tubing, collecting tubes, fraction collector (Gilson)
- 200 ml LSC buffer: 30 mM Hepes pH 7.4, 50 mM NaCl, 1 mM CaCl$_2$, 1 mM DTT
- 200 ml MSE buffer: 30 mM Hepes pH 7.4, 150 mM NaCl, 5 mM EGTA, 1 mM DTT
- 200 ml MSC buffer: 30 mM Hepes pH 7.4, 150 mM NaCl, 1 mM CaCl$_2$, 1 mM DTT
- 200 ml HSE buffer: 30 mM Hepes pH 7.4, 500 mM NaCl, 5 mM EGTA, 1 mM DTT
- 200 ml HSC buffer: 30 mM Hepes pH 7.4, 500 mM NaCl, 1 mM CaCl$_2$, 1 mM DTT
- Protein detection by Coomassie Blue reagent (Bio-Rad)

10: Strategies and methods for purification of binding proteins

Method

1. Hydrate heparin–Sepharose resin with LSC buffer. Approx. 2 g resin will make a 4 ml column. Degassing the resin is not necessary since the column is not under pressure. Run 20 ml LSC buffer through the column to equilibrate the resin.
2. Equilibrate the sample by adding 1 M $CaCl_2$ to give 1 mM final Ca^{2+} concentration. The salt concentration of the sample should be less than 100 mM salt or the proteins will not bind tightly. Dialyse or dilute the sample to lower the salt concentration.
3. Apply the sample to the 4 ml column and collect the flow-through. Flow is usually controlled by gravity.
4. Wash the column with LSC buffer until no protein is detected by Coomassie Blue reagent. Collect 1 ml fractions using the fraction collector.
5. If the sample was a crude mixture, a 10 ml wash with MSE buffer will elute the annexins. It is then possible to equilibrate this pooled sample with LSC buffer and repeat the procedure as discussed for purification of pure proteins. To elute the annexins from a reasonably pure sample, wash the heparin affinity column with MSC buffer and elute the annexins with the HSC buffer. Collect 1 ml fractions. Approx. 10–15 ml buffer is necessary to completely elute the annexins.
6. Before reusing the column, wash with 40 ml HSE buffer to ensure that all protein has been eluted from the column and then equilibrate the column with LSC buffer. Store at 4°C. The SDS–PAGE analysis of the purified annexins is presented in *Figure 2*.

4. The Chelex-100 competitive Ca^{2+}-binding assay

Aliquots of column fractions are mixed with a known amount of Chelex-100 (suspended in five volumes of 20 mM Tris pH 7.5) and $^{45}Ca^{2+}$ in a small 3 ml tube with a stopper. The samples are rotated for 30 min at 4°C and then centrifuged at 5000 g for 5 min to pellet the Chelex-100. 0.1 ml of the supernatant is counted and the c.p.m. in the supernatant is plotted. A schematic of the assay is depicted in *Figure 8*. A typical profile that utilizes the Chelex-100 assay to detect the major bovine liver Ca^{2+}-binding proteins is presented in *Figure 9*.

If a column such as DEAE–cellulose has been used to partially purify the original sample, it is necessary to dialyse the samples to remove excess salt which will interfere with the assay. Rather than dialysing each fraction individually in dialysis tubing, we use the Microdialysis System from BRL.

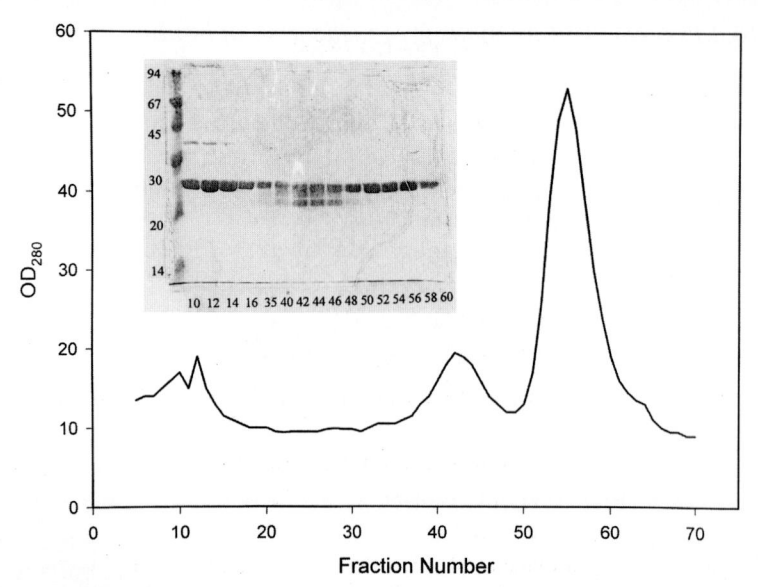

Figure 6. Fast S (Pharmacia) column chromatography profile. Partially purified annexin I, annexin II, and annexin II tetramer were pooled after chromatography on hydroxylapatite, dialysed, and each partially purified annexin was applied to a column of Fast S as described in *Protocol 3*, steps 1–3. The proteins were eluted with a linear salt gradient. Annexin I is eluted at very low salt, annexin II monomer typically elutes at about 0.4 M NaCl (fractions 35–45), and the annexin II tetramer typically elutes at 0.6 M NaCl (fractions 48–60). Aliquots of fractions were subjected to SDS–PAGE. As shown in the inset, each of the annexins is homogeneous after this purification step.

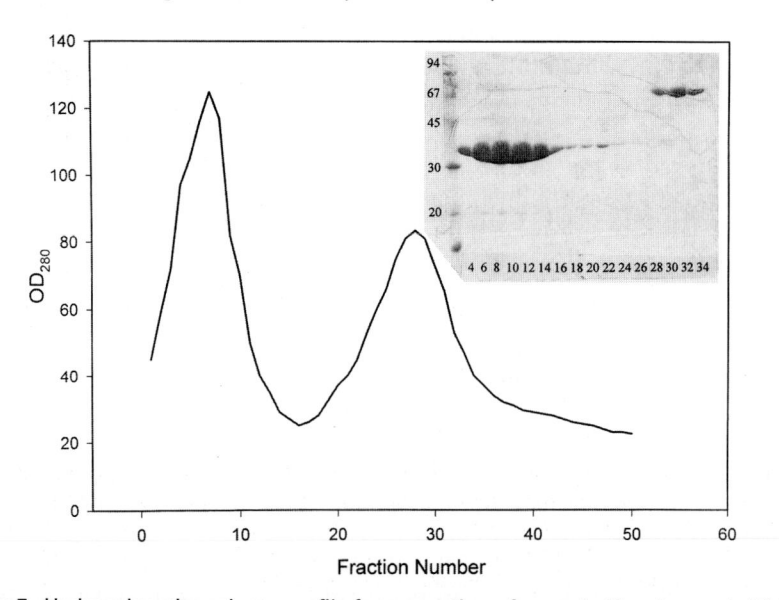

Figure 7. Hydroxylapatite column profile for separation of annexin V and annexin VI. The peak of proteins that eluted from the DEAE–Sephacel column pooled, dialysed, and applied to a hydroxylapatite column as described in *Protocol 3*. The proteins were eluted with a linear phosphate gradient. Samples of the fractions were subjected to SDS–PAGE and annexin V (fractions 4–17) and annexin VI (fractions 30–34) were pooled.

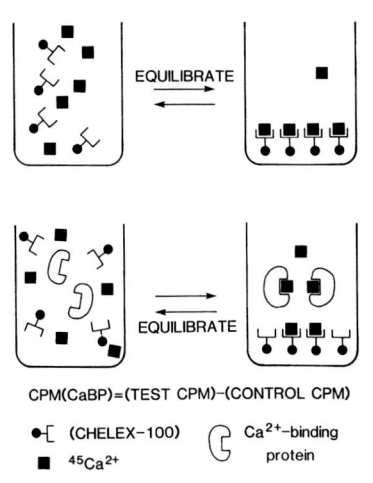

Figure 8. Schematic of the Chelex-100 assay. This shows the two stages in the Chelex-100 assay. First is the standardization of the $^{45}Ca^{2+}$ as described in *Protocol 5*, step 2. 95% of the available $^{45}Ca^{2+}$ should be bound by the Chelex-100. The next stage is the addition of a suspected Ca^{2+}-binding protein as described in *Protocol 5*, steps 3–6. By comparing the c.p.m. with protein present, or absent, the amount of $^{45}Ca^{2+}$ bound by the protein can be determined.

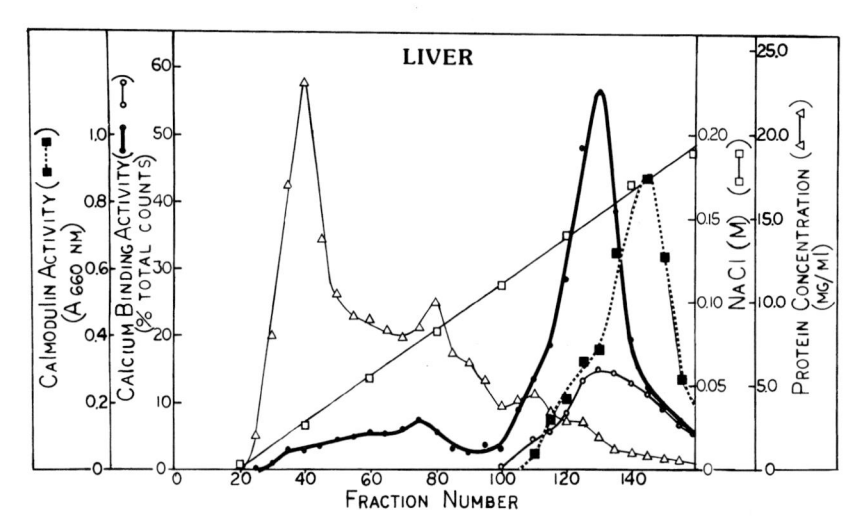

Figure 9. Chromatography of liver 100 000 g supernatant on DEAE–cellulose. The 100 000 g supernatant of bovine liver was diluted tenfold into a buffer containing 10 mM Tris pH 7.5 (final volume of 8 litres) and 400 ml DEAE–cellulose (1:1 suspension in 10 mM Tris pH 7.5) was added batchwise. The resultant slurry was stirred for 60 min at 4°C, filtered through a sintered glass funnel, and the DEAE–cellulose was poured into a 5 cm × 30 cm column. The column was washed extensively with 10 mM Tris pH 7.5 and then with a linear salt gradient from 0–0.3 M NaCl in 10 mM Tris pH 7.5. $^{45}Ca^{2+}$ bound is expressed before (closed circles) and after (open circles) incubation of aliquots at 100°C for 2 min. The heat treatment inactivates many proteins, but under these conditions calmodulin is not inactivated. This treatment allows specific calmodulin activity to be determined.

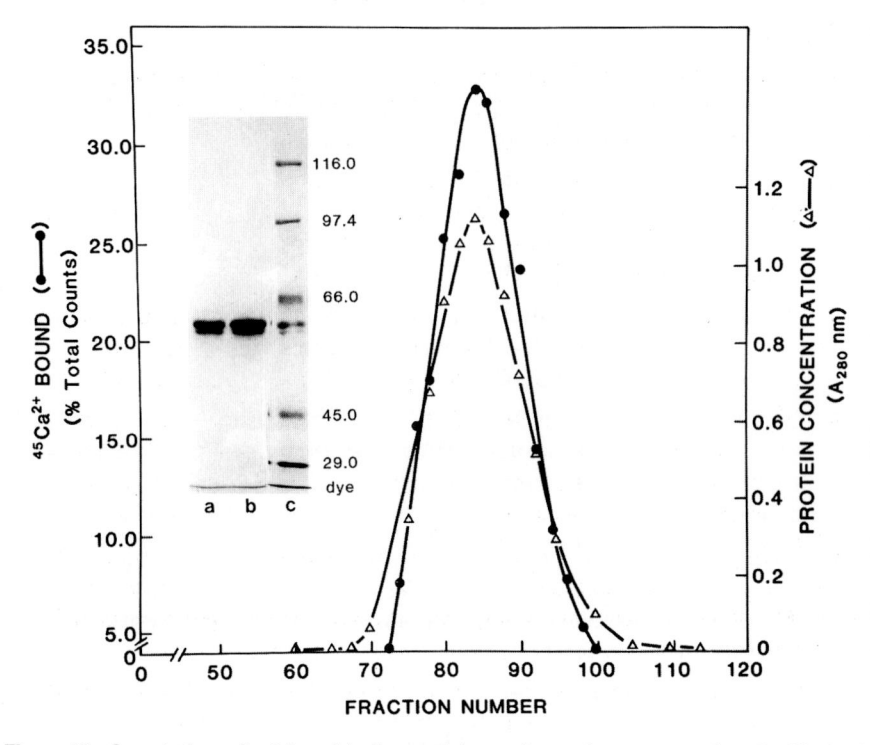

Figure 10. Correlation of calcium-binding activity and protein concentration. Purified calreticulin (15) was subjected to chromatography on Sephadex G-100 and protein concentration ($A_{280\ nm}$) and calcium-binding activity determined. A molecular weight of 51 400 was estimated by this procedure. Insert: 10% polyacrylamide gel electrophoresis in the presence of sodium dodecyl sulfate for the two forms of calreticulin (a) and (b). (c) Calreticulin was electrophoresed in the presence of β-galactosidase (116 kDa), phosphorylase b (97.4 kDa), bovine serum albumin (66 kDa), ovalbumin (45 kDa), and carbonic anhydrase (29 kDa).

One version can dialyse 28 samples at once. Dialysis is performed with 0.5 ml samples and run overnight in a cold room. After dialysis, pure preparations of Ca^{2+}-binding proteins show an excellent correlation between Ca^{2+}-binding activity and protein concentration (*Figure 10*).

Protocol 5. Chelex-100 assay

Equipment and reagents

- Chelex-100 (Bio-Rad) in buffer A (minus 400 mesh)
- $^{45}Ca^{2+}$ (Amersham)
- Scintillation counter (LKB RackBeta 1217 or equivalent), racks, vials, and scintillation fluid (Amersham)
- Clinical rotator (Fisher)
- Microdialysis System (Bethesda Research Laboratories) Model No. 1200 MD, peristaltic pump, and tubing
- Buffer A: 100 mM Tris pH 7.5, 50 mM NaCl

10: Strategies and methods for purification of binding proteins

Method

1. The Chelex-100 is washed twice with buffer A and suspended in buffer A at a 1:10 ratio (Chelex-100:buffer). Dilute the $^{45}Ca^{2+}$ into buffer A to give about 100 000 c.p.m./ml.
2. Standardize the amount of Chelex-100 necessary to bind 95% of the available $^{45}Ca^{2+}$. This is accomplished by varying the amount of Chelex-100 added to a mixture of buffer A and $^{45}Ca^{2+}$. Typically about 0.025 ml Chelex-100 is required.
3. To a 1.5 ml microcentrifuge tube, add buffer A, $^{45}Ca^{2+}$ (1/10 dilution), and test sample (typically 0.1 ml) or test sample buffer (control). Add 25 µl of rapidly stirring Chelex-100. Final volume should be 1 ml.
4. The mixture is placed in a clinical rotator for 20 min. After this, the tubes are centrifuged at 1500 *g* for 2 min to pellet the Chelex-100. 0.1 ml samples of the supernatant are taken and counted in the scintillation counter.
5. A control with no protein is always included. Dialysis of samples is recommended since excess salt can interfere with the assay and samples obtained by elution by a salt gradient will have increased counts in the supernatant as the salt concentration in the sample increases.
6. The results of the assay are expressed as $^{45}Ca^{2+}$ bound where $^{45}Ca^{2+}$ bound = (c.p.m. in test − c.p.m. in control) × 100/total c.p.m.

References

1. Niki, I., Yokokura, H., Sudo, T., Kato, M., and Hidaka, H. (1996). *J. Biochem. (Tokyo)*, **120**, 685.
2. Waisman, D. M. (1995). *Mol. Cell Biochem.*, **149/150**, 301.
3. Bleackley, R. C., Atkinson, E. A., Burns, K., and Michalak, M. (1995). *Curr. Top. Microbiol. Immunol.*, **198**, 145.
4. Mathew, J. K., Krolak, J. M., and Dedman, J. R. (1986). *J. Cell Biochem.*, **32**, 223.
5. Moore, P. B., Kraus-Friedmann, N., and Dedman, J. R. (1984). *J. Cell Sci.*, **72**, 121.
6. Creutz, C. E., Dowling, L. G., Sando, J. J., Villar-Palasi, C., Whipple, J. H., and Zaks, W. J. (1983). *J. Biol. Chem.*, **258**, 14664.
7. Creutz, C. E. (1981). *Biochem. Biophys. Res. Commun.*, **103**, 1395.
8. Ikebuchi, N. W. and Waisman, D. M. (1990). In *Stimulus response coupling: the role of intracellular calcium-binding proteins* (ed. V. L. Smith and J. R. Dedman), pp. 357–81. CRC Press, Boca Raton, Florida.
9. Khanna, N. C., Helwig, E. D., Ikebuchi, N. W., Fitzpatrick, S., Bajwa, R., and Waisman, D. M. (1990). *Biochemistry*, **29**, 4852.
10. Hubaishy, I., Jones, P. G., Bjorge, J., Bellagamba, C., Fitzpatrick, S., Fujita, D. J., *et al.* (1995). *Biochemistry*, **34**, 14527.
11. Tokuda, M., Khanna, N. C., and Waisman, D. M. (1987). In *Methods in enzymology* (ed. A. R. Means and P. M. Conn), Vol. 139, p. 68.

12. Tokuda, M., Khanna, N. C., and Waisman, D. M. (1987). *Cell Calcium*, **8**, 229.
13. Waisman, D. M., Khanna, N. C., and Tokuda, M. (1986). *Biochem. Biophys. Res. Commun.*, **139**, 596.
14. Waisman, D. M., Tokuda, M., Morys, S. J., Buckland, L. T., and Clark, T. (1985). *Biochem. Biophys. Res. Commun.*, **128**, 1138.
15. Waisman, D. M., Salimath, B. P., and Anderson, M. J. (1985). *J. Biol. Chem.*, **260**, 1652.
16. Waisman, D. M. and Rasmussen, H. (1983). *Cell Calcium*, **4**, 89.
17. Johnstone, S. A. and Waisman, D. M. (1990). In *Stimulus response coupling: the role of intracellular calcium-binding proteins* (ed. V. L. Smith and J. R. Dedman), pp. 21–37. CRC Press, Boca Raton, Florida.
18. Niki, I., Yokokura, H., Sudo, T., Kato, M., and Hidaka, H. (1996). *J. Biochem.*, **120**, 685.
19. Morgan, R. O. and Fernandez, M.-P. (1995). *Mol. Biol. Evol.*, **12**, 967.

11

Purification of DNA-binding proteins

HIROSHI HANDA, YUKI YAMAGUCHI, and TADASHI WADA

1. General procedure for DNA-binding proteins

Sequence-specific DNA-binding proteins are involved in many biological processes, such as DNA repair, recombination, replication, and transcription. To characterize these processes, it is quite important to purify the products that participate in them to homogeneity. The purified proteins can then be used to analyse their biochemical properties, to raise antibodies against them, and to isolate the genes encoding them.

In this chapter, we describe general procedures involved in the purification of DNA-binding proteins. It includes the preparation of cell nuclear extracts, the use of heparin–Sepharose, and the preparation and use of DNA affinity chromatography.

2. Preparation of nuclear extracts

The first step in studying DNA-binding proteins is the preparation of cell extracts from eukaryotic cells and tissues. The following procedure was originally developed for studies of RNA and protein synthesis and has been used extensively in studies of *in vitro* transcription and RNA splicing (1). The preparation of HeLa cell nuclear extracts is described here. To summarize: cells are collected from spinner cultures by centrifugation, washed in cold phosphate-buffered saline, and suspended in hypotonic buffer to swell. After homogenization in a glass–glass Dounce homogenizer, nuclei are pelleted by centrifugation, and the cytoplasmic fraction is removed. The crude nuclear pellet is resuspended in a high salt buffer and subjected to the Dounce homogenizer again. By stirring gently, soluble proteins are released from the nuclei and recovered in the supernatant by centrifugation. Finally, the extract is dialysed and any precipitated protein is removed by centrifugation. The nuclear extract may be used immediately or stored at –80 °C.

The procedure detailed in *Protocol 1* allows one to prepare both a cytoplasmic fraction (known as S100) and a nuclear extract, however this chapter only describes the preparation of nuclear extracts.

2.1 Cell culture

HeLa cells should be grown in spinner bottles to a density of 5–10 × 10^8 cells/litre.

2.2 Preparation of nuclear extracts

Protocol 1. Preparation of nuclear extracts

Equipment and reagents

- Continuous flow refrigerated centrifuge, e.g. Tomy RS 20IV (Tomy Seiko Co. Ltd. or equivalent) and rotor, e.g. No. 8N
- Beckman low speed and high speed centrifuge, plus rotors, e.g. CS-6KR and Avanti (Beckman or equivalent)
- 750 ml polypropylene, 50 ml polypropylene, and 50 ml polycarbonate centrifuge tubes
- Glass Dounce homogenizer (type B)
- Cellulose dialysis tubing
- Magnetic stirrer
- 10 ml disposable pipette
- HeLa cells from spinner cultures
- 10 × phosphate-buffered saline (PBS) (per litre): 80 g NaCl, 2 g KCl, 11.5 g $Na_2PO_4 \cdot 7H_2O$, 2 g KH_2PO_4
- Hypotonic buffer (buffer A): 10 mM Hepes–NaOH pH 7.9, 1.5 mM $MgCl_2$, 10 mM KCl—add 0.5 mM dithiothreitol (DTT) and 0.5 mM phenylmethylsulfonyl fluoride (PMSF) immediately before use
- High salt buffer (buffer C): 20 mM Hepes–NaOH pH 7.9, 25% (v/v) glycerol, 0.42 M KCl, 1.5 mM $MgCl_2$, 0.2 mM ethylenediaminetetraacetic acid (EDTA)—add 0.5 mM DTT and 0.5 mM PMSF immediately before use; store at 4°C
- Dialysis buffer (buffer D): 20 mM Hepes–NaOH pH 7.9, 20% (v/v) glycerol, 0.1 M KCl, 0.2 mM EDTA—add 0.5 mM DTT and 0.5 mM PMSF immediately before use; store at 4°C
- Liquid nitrogen

A. Isolation of the nuclei

1. Collect cells by a continuous flow centrifuge (1000 *g*) at 4°C using the Tomy refrigerated centrifuge using a flow rate of about 400 ml/min. After centrifugation, transfer the rotor contents into 750 ml centrifuge tubes and centrifuge for 10 min at 1000 *g* at 4°C in the Beckman CS-6KR centrifuge. Decant the supernatants and discard. Resuspend the cell pellet in ice-cold PBS and centrifuge again.

2. Resuspend the cells in ice-cold PBS and transfer them into 50 ml polypropylene centrifuge tubes. Centrifuge for 5 min at 2000 r.p.m. (1000 *g*) at 4°C in the Beckman CS-6KR centrifuge. Measure the packed cell volume (PCV).

3. Resuspend the cells in five times the PCV of buffer A and let stand on ice for 10 min to swell. Centrifuge the cells for 5 min at 2000 r.p.m. (1000 *g*) at 4°C in the Beckman CS-6KR centrifuge. Decant the supernatants and discard.

11: Purification of DNA-binding proteins

4. Resuspend the cells in two times the PCV of buffer A and lyse them by ten strokes in a glass Dounce homogenizer using a type B pestle.[a] The homogenizer should be chilled before use.
5. Centrifuge cells for 10 min at 3000 r.p.m. (1000 g) at 4°C in a Beckman Avanti 30 centrifuge. Transfer the supernatant (the cytoplasmic fraction) to a 50 ml polypropylene tube using a disposable pipette. Quick-freeze this fraction in liquid nitrogen and store at –80°C.
6. Recentrifuge the pellet for 20 min at 12 000 r.p.m. (13 600 g) at 4°C in a Beckman Avanti 30 centrifuge and discard the supernatant using a disposable pipette.

B. Extraction of the nuclei

1. Resuspend the pellet in one times the PCV of buffer C and homogenize with 12 strokes in a glass Dounce homogenizer using a type B pestle. Transfer homogenate to a 200 ml beaker and mix gently for 30 min at 4°C with a magnetic stirrer.[b]
2. Centrifuge the nuclear extracts for 30 min at 12 000 r.p.m. (13 600 g) at 4°C in a Beckman Avanti 30 centrifuge and transfer the supernatant to a dialysis tube.
3. Dialyse against 1 litre buffer D for 2.5 h at 4°C, change the dialysis buffer, and dialyse for 2.5 h more.
4. To remove any precipitates, centrifuge the dialysate for 30 min at 12 000 r.p.m. (13 600 g) and at 4°C in a Beckman Avanti 30 centrifuge. The supernatant may be used immediately, or frozen quickly in liquid nitrogen and stored at –80°C.

[a] After ten strokes, check for cell lysis under a microscope. If less than 90% of the cells are lysed, continue to homogenize with several more strokes, and then check cell lysis again.
[b] Important: gentle mixing is important to avoid nucleic acid contamination.

The quality of nuclear extracts can be estimated by *in vitro* transcription or gel mobility shift assays (1–3). The protein concentration of the extract should be about 10 mg/ml. 4–5 µl of the extract is generally used for *in vitro* transcriptional analysis and 1 µl is usually sufficient for detection of a DNA: protein band in gel mobility shift analysis. In order to ensure active extracts, healthy, mid-log phase cells should be prepared. This method also works well on other cultured cells.

Nuclear extracts are a good starting material for the purification of DNA-binding proteins. Usually the extracts are fractionated by conventional chromatography in order to remove protease and nuclease in the extracts. For this purpose, in the next section heparin–Sepharose affinity chromatography is introduced.

3. Heparin–Sepharose

Now almost all of the methodologies for purification of DNA-binding proteins use affinity chromatography, since it allows much time to be saved by avoiding complicated strategies. The acid polysaccharaide, heparin, has repeating units of six sugar residues, each consisting of an alternating sequence of sulfate derivatives of N-acetyl-D-glucosamine and D-induronate. Heparin resembles nucleic acid because of its linear polyanionic structure and accordingly, heparin–Sepharose is often employed in the purification of DNA-binding proteins. Many of the bacterial restriction endonucleases (4, 5), DNA ligase (5), avian reverse transcriptase (6), have been purified by heparin–Sepharose chromatography, suggesting that heparin–Sepharose should be a good candidate material for purification of DNA-binding proteins. The following procedure was originally used for fractionation of HeLa cell nuclear extracts to analyse transcription factors of RNA polymerase II (7), some of which are sequence-specific DNA-binding proteins.

3.1 Protein fractionation by heparin–Sepharose

Protocol 2. Heparin–Sepharose

Equipment and reagents

- Heparin–Sepharose CL-6B (Pharmacia Biotech)
- 1 × 10 cm or 1.6 × 20 cm column (Pharmacia C10/10 or C16/20 or equivalent)
- Peristaltic pump, UV monitor, and fraction collector
- HeLa cell nuclear extract (see *Protocol 1*)
- Starting buffer: 20 mM Hepes–NaOH pH 7.9, 20% (v/v) glycerol, 0.1 M KCl, 12.5 mM MgCl$_2$, 0.5 mM EDTA—add 1 mM DTT and 0.5 mM PMSF immediately before use; store at 4°C
- Elution buffer: 20 mM Hepes–NaOH pH 7.9, 20% (v/v) glycerol, 12.5 mM MgCl$_2$, 0.5 mM EDTA, plus appropriate concentration of KCl—add 1 mM DTT and 0.5 mM PMSF immediately before use; store at 4°C
- High salt buffer: 20 mM Hepes–NaOH pH 7.9, 20% (v/v) glycerol, 1 M KCl, 12.5 mM MgCl$_2$, 0.5 mM EDTA—add 1 mM DTT and 0.5 mM PMSF immediately before use; store at 4°C
- Regeneration buffer: 20 mM Hepes–NaOH pH 7.9, 2.5 M NaCl, 1 mM EDTA

A. *Packing the column*

1. If using the freeze-dried gel, the powder should be swollen for 15 min, washed with 200 ml/g starting buffer in a glass filter funnel, and then resuspended in a small amount of the starting buffer. 1 g freeze-dried gel will give about 4 ml swollen gel. During this step, which should be performed at room temperature, it is important to avoid the gel drying out.

2. Pour the swollen gel into the column gently under gravity and measure the column volume. A typical column size for fractionation of HeLa cell nuclear extracts is 40 ml.

11: Purification of DNA-binding proteins

B. *Setting up the system*
1. Set up the system as shown in *Figure 1*.
2. Equilibrate the column with ten times the column volume of starting buffer. The linear flow rate must not be over 10 ml/cm^2/h.

C. *Loading and washing the column*
1. Load the HeLa cell nuclear extracts onto the heparin–Sepharose column.[a]
2. Collect fractions[b] of an equal size to the column volume and take aliquots (1 ml) for assays. Flash-freeze the proteins in liquid N$_2$ and store at –80°C.
3. Wash the column with starting buffer until the UV monitor indicates a constant value or until at least ten times the column volume of starting buffer have been used.

D. *Elution*
1. Elute the bound protein with elution buffer.[c]
2. Collect one-half of the column volume of fractions and take aliquots (1 ml). Flash-freeze the fractions and the 1 ml aliquots in liquid N$_2$ and store at –80°C.
3. Wash the column with at least three times the column volume of high salt buffer.

E. *Regeneration*
1. Wash the column with regeneration buffer.[d]

[a] One column volume of HeLa cell nuclear extracts can be loaded for good separation.
[b] The volume of the fraction depends on the column volume. One column volume is recommended for one fraction.
[c] The optimal salt concentration of the elution buffer must be determined experimentally. To do this, a batchwise method is recommended (see *Protocol 3*).
[d] Heparin–Sepharose is stable from pH 5–10 and can be treated with urea (6 M) or guanidine hydrochloride (6 M).

Almost all sequence-specific and non-specific DNA-binding proteins will bind to heparin–Sepharose because they may interact with negatively charged structures. Therefore, it is important to determine the optimal salt concentration of the elution buffer which gives the most efficient purification of the target protein. For this purpose, a batchwise method described in Section 3.2 is useful. After heparin–Sepharose chromatography, other conventional chromatographic methods, ion exchange, gel filtration, and DNA affinity chromatography are also appropriate for further purification.

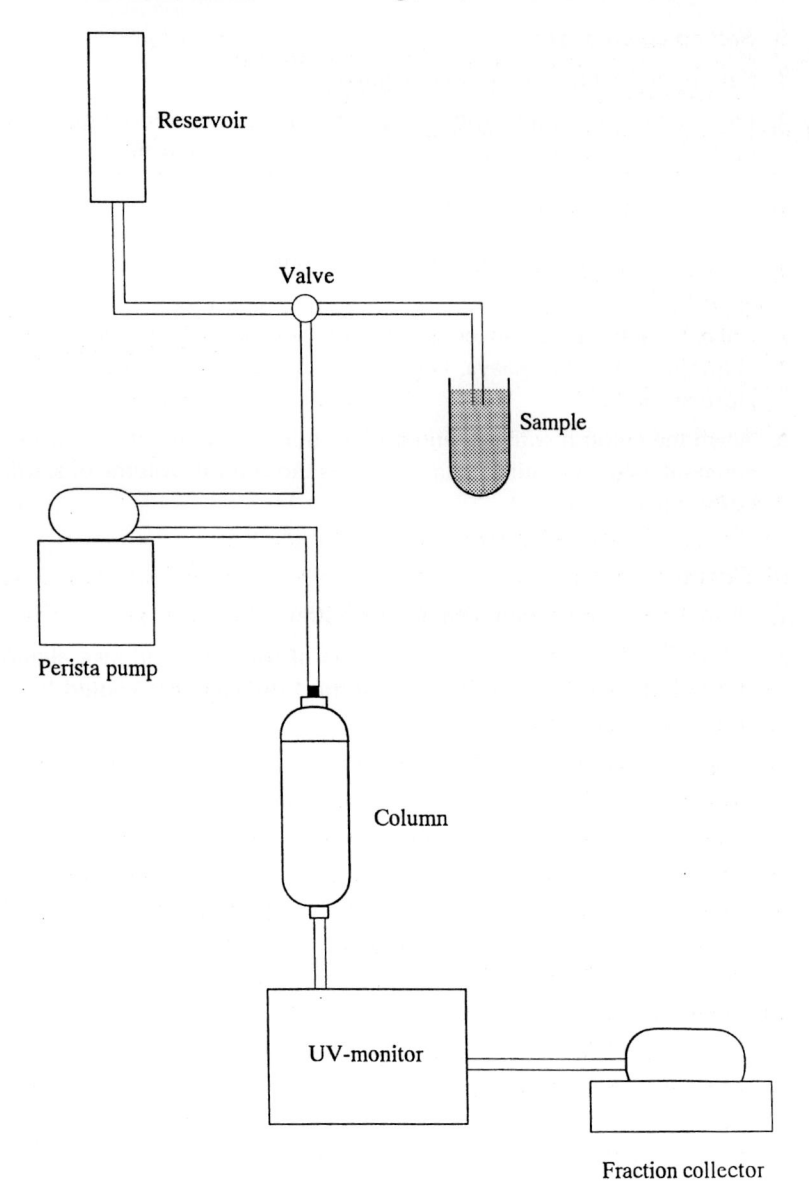

Figure 1. Schematic representation of the system for heparin–Sepharose.

11: Purification of DNA-binding proteins

3.2 Batchwise method for determination of salt concentration of elution buffer

This method is summarized in *Figure 2*.

Protocol 3. Batchwise method

Equipment and reagents
- Refrigerated microcentrifuge
- Rotating incubator, e.g. blood wheel
- Cold chamber (4°C)
- Heparin–Sepharose CL-6B (Pharmacia Biotech)
- HeLa cell nuclear extract (see *Protocol 1*)
- Buffer A: 20 mM Hepes–NaOH pH 7.9, 20% (v/v) glycerol, 12.5 mM $MgCl_2$, 0.5 mM EDTA—add 1 mM DTT and 0.5 mM PMSF immediately before use; store at 4°C
- Buffer B: buffer A plus 1 M KCl

Method
1. Prepare heparin–Sepharose (see *Protocol 2*, part A).
2. Add about 500 μl heparin–Sepharose to a 1.5 ml microcentrifuge tube.
3. Add 1 ml buffer A containing 0.1 M KCl[a] to the tube and mix the gel by tipping the tube.
4. Centrifuge the gel at 3000 r.p.m. for 5 sec in a microcentrifuge.
5. Discard the supernatant.
6. Repeat steps 3–5 at least three times.
7. Add 500 μl HeLa cell nuclear extract to the gel.
8. Incubate the sample in a rotating incubator for 5 min at 4°C.
9. Centrifuge the gel for 5 sec at 3000 r.p.m. and transfer the supernatant (this corresponds to the flow-through fraction and should be assayed) to a new 1.5 ml centrifuge tube. Freeze in liquid N_2 and store at −80°C.
10. Wash the gel three times with 1 ml buffer A containing 0.1 M KCl.[a]
11. Prepare six 1.5 ml microcentrifuge tubes.
12. Add 500 μl buffer A containing 0.1 M KCl[a] to the gel, mix, and transfer 100 μl of the gel to each 1.5 ml microcentrifuge tube.
13. Centrifuge 15 000 r.p.m. for 5 sec and discard the supernatant.
14. Add 50 μl buffer A containing 0.1 M, 0.2 M, 0.3 M, 0.4 M, 0.5 M, or 1 M KCl[a] to each tube, respectively, and mix the gel by tipping the tube.
15. Stand on ice for 5 min and centrifuge the mixtures at 15 000 r.p.m. for 5 sec.
16. Collect the supernatants in which the protein is present.
17. Assay the protein fractions using a DNA-binding assay (e.g. a gel mobility shift assay) (2, 3).

Protocol 3. Continued

18. Determine the optimal salt concentration for the elution buffer on the basis of these results.

[a] Buffer A containing KCl is prepared by mixing buffer A and B together in the appropriate proportions.

The batchwise method is useful for determination of several critical conditions during purification of DNA-binding proteins. For example, the optimal concentration of Mg^{2+} in the buffer, which affects DNA-binding activity of the desired protein, can be determined by this method. Alternatively, when a variety of non-specific competitor DNAs, including poly(dI:dC), poly(dA:dT), poly(dG:dC), and salmon sperm DNA are tested to determine their effect on DNA:protein interactions, it is recommended to employ this method before a large scale purification.

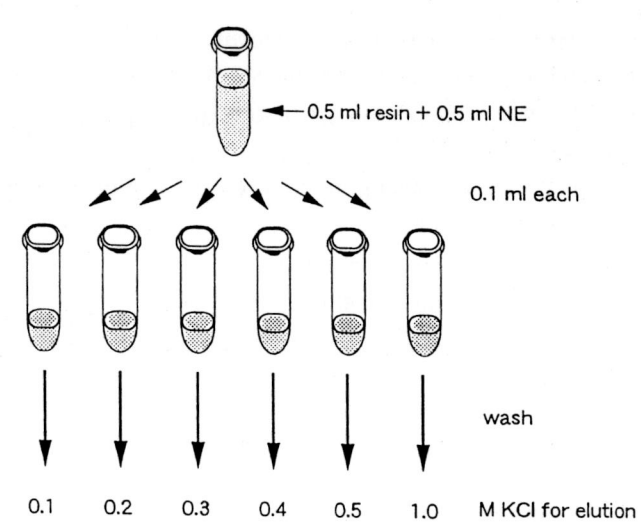

Figure 2. Batchwise method. Nuclear extracts (NE) are mixed with the appropriate resin and incubated, then the mixture is divided into six aliquots. After washing with low salt (0.1 M KCl) buffer, bound proteins are eluted with buffer containing the indicated salt concentration, i.e. 0.2–1 M KCl.

4. DNA affinity chromatography for purification of sequence-specific DNA-binding proteins

It has generally been considered difficult to purify sequence-specific DNA-binding proteins by conventional chromatography, since most of these proteins constitute less than 0.01% of the total cellular protein. However, the development of affinity chromatography procedures based on the sequence-specific DNA-binding properties of these proteins allows one to purify these

11: Purification of DNA-binding proteins

proteins several hundred-fold and with high yield (8 and refs therein). DNA affinity chromatography uses an oligonucleotide containing the specific recognition sequence(s) for the protein of interest, which is covalently or non-covalently attached to a solid support. Different types of resins and methods for coupling oligonucleodies have been employed, including CNBr-activated agarose (9, 10) or St-GMA-based latex particles (11), to which multimerized synthetic oligonucleotides are covalently attached. The following section describes how to design and prepare oligonucleotides used for affinity chromatography, how to prepare the DNA affinity resins, and how to determine the chromatographic conditions to use with these resins.

4.1 Design and preparation of oligonucleotides

To design oligonucleotides for sequence-specific separations, the DNA-binding site for the protein to be purified must first be determined. Footprinting techniques using DNase I or other enzymatic or chemical reagents are suitable for this purpose. No strict limitation on the length of the oligonucletide exists: 14-mers to 61-mers have been used successfully. It is recommended that oligonucleotides that cover the region protected in the footprinting experiment plus an additional few flanking bases are synthesized.

Multimerized synthetic oligonucleotides are the preferred affinity ligand. This maximizes the purification efficiency by increasing the number of active sites and avoids potential steric hindrance of the protein:DNA interactions, which might occur when short, monomeric oligonucleotides are directly attached to the resin. In some cases, however, monomeric oligonucleotides are also used successfully.

If multimerized oligonucleotides are to be used, they should be designed to create a single-stranded overhang (4- to 6-mers) as shown in *Figure 3*. This

```
5'-GGGG-BINDING SITE-3'
   3'-BINDING SITE-CCCC-5'
```

↓ phosphorylate the 5'-ends with T4 polynucleotide kinase

↓ multimerize with T4 DNA ligase

```
5'-GGGG-BINDING SITE-(GGGG-BINDING SITE)n-3'
   3'-BINDING SITE-(CCCC-BINDING SITE)n-CCCC-5'
```

Figure 3. Schematic representation of the oligonucleotides to be synthesized for multimerization. The oligonucleotides should be designed with a single-stranded overhang (top), since they are multimerized efficiently and unidirectionally (bottom).

allows the annealed oligonucletides to be ligated efficiently. The overhang may also help the subsequent coupling reaction if an amine coupling method is employed, as is used in CNBr-activated agarose or latex particles. In these methods, DNA is coupled to the resin via primary amine groups and the bases of single-stranded DNA are more reactive than those of double-stranded DNA. A detailed procedure for the preparation of multimerized oligonucleotides is presented in *Protocol 4*.

4.2 Preparation of DNA oligomers

Protocol 4. Preparation of DNA oligomers

Equipment and reagents

- Centrifugal evaporator
- Water-bath at 37°C, water-bath at 100°C, and incubator at 4°C
- Spectrophotometer
- Table-top microcentrifuge
- Electrophoresis equipment for polyacrylamide gels
- Nick column G50 (Pharmacia Biotech)
- Milli Q water (or equivalent)
- Synthesized DNA oligomers (see *Figure 3*)
- 100 mM ATP
- 0.5 M EDTA

- T4 polynucleotide kinase
- T4 DNA ligase
- T4 DNA polymerase
- [γ-^{32}P]ATP
- Phenol:chloroform
- 10 × ligation buffer: 660 mM Tris–HCl pH 7.4, 100 mM MgCl$_2$, 10 mM spermidine, 150 mM DTT, 2 mg/ml bovine serum albumin (BSA)
- 10 × exobuffer: 330 mM Tris–acetate pH 7.9, 660 mM KCH$_3$CO$_2$, 100 mM Mg(CH$_3$CO$_2$)$_2$, 5 mM DTT

A. Preparation of multimerized oligonucleotides

1. Synthesize two complementary oligonucleotides containing the binding site for the protein of interest.
2. Resuspend the synthetic oligonucleotides in Milli Q water at the final concentration of 0.5 mg/ml.
3. Anneal the oligonucleotides by heating a mixture containing 100 μg of each oligonucleotide to 100°C for 10 min, followed by slowly cooling to room temperature.
4. Divide the DNA solution (400 μl) into five microcentrifuge tubes.
5. Add 20 μl 10 × ligation buffer, 2 μl 100 mM ATP, 96 μl Milli Q water, and 2 μl T4 polynucleotide kinase (15 U/μl), and incubate the samples for 1 h at 37°C. Addition of 1 μCi [γ-^{32}P]ATP to this reaction helps calculation of the rate at which the DNA is fixed to resin.
6. Ligate the oligonucleotides by adding 10 Weiss units[a] T4 DNA ligase to each reaction and incubating for 2 h at 15°C or overnight at 4°C.
7. Monitor the length of the ligated DNA by 12% (w/v) native polyacrylamide gel electrophoresis. The average length should be around 10-mers. If monomers are not ligated as expected, repeat steps 5 and 6 after phenol:chloroform extraction and ethanol precipitation.

11: Purification of DNA-binding proteins

8. Purify the ligated DNA by phenol:chloroform extraction followed by ethanol precipitation.

B. *Preparation of 5'-protruding ends for efficient immobilization (optional)*
If you skip part B (optional), go directly *Protocol 4*, part C.
By analysis of protruding ends of the DNA being immobilized, the DNA fragment with about 15 nucleotides at the 5'-protruding ends gave the most efficient purification of the transcription factor ATF/E4TF3; however, longer protruding ends reduced the purification efficiency (11). On the basis of this result, a method for preparing DNA oligomers containing suitable numbers of single-stranded nucleotides at the 5' end is described here.

1. Resuspend the precipitated DNA in 200 μl Milli Q water.
2. Add 40 μl 10 × exobuffer, 120 μl Milli Q water, and 40 μl T4 DNA polymerase (5 U/μl).
3. Incubate at 37°C for 30–45 sec to digest the DNA by the 3'–5' exonuclease activity of T4 DNA polymerase.
4. Add 4 μl 0.5 M EDTA and 400 μl phenol:chloroform concomitantly and vortex for 1 min.
5. Centrifuge at 15 000 r.p.m. for 5 min, transfer the aqueous phase to a new 1.5 ml microcentrifuge tube, followed by ethanol precipitation.

C. *Purification of oligonucleotides*

1. Resuspend the DNA in 40 μl Milli Q water and purify it using a Nick column[b] (or its equivalent).
2. Measure the concentration of the DNA at 260 nm wavelength using a spectrophotometer (an OD of 1 corresponds to 50 μg/ml).[c]
3. Use the DNA solution immediately for the coupling reaction.

[a] One Weiss unit is defined as the amount of enzyme that catalyses the exchange of 1 nmole of PPi for 20 min at 37°C in an ATP–PPi substitution reaction (12).
[b] A Nick column is used to remove reagents containing amino groups, such as nucleotides, that can prevent efficient coupling of the DNA to the resins. For the same reasons, Milli Q water should be used for separation of the DNA from other materials.
[c] If the concentration of the DNA is less than 0.2 μg/μl, precipitate it with ethanol and appropriate salt, and adjust the concentration to at least 0.2 μg/μl with Milli Q water.

4.3 Preparation of DNA affinity resins

Several resins for immobilizing the DNA, including agarose (13), Teflon-based beads (14), or latex beads (11, 15, 16) have been used and there are a variety of methods for attaching DNA fragments to various supports. The most popular and general method uses CNBr-activated Sepharose (8) which is available commercially and thus avoids using CNBr, which is toxic. After reacting the DNA with CNBr-activated Sepharose, the unreacted CNBr is inactivated with ethanolamine. The DNA affinity Sepharose can then be used for purification of sequence-specific DNA-binding proteins. The first protocol (*Protocol 5*) describes how to couple the DNA to CNBr-activated Sepharose and purify sequence-specific DNA-binding proteins by chromatography on DNA affinity Sepharose, whilst *Protocol 6* describes a method using latex beads. This enables direct purification of sequence-specific DNA-binding proteins from crude cell extracts within a few hours (11, 16). The latex beads have several useful advantages. First, they are extremely small (average diameter of 0.2 μm), and therefore provide a larger surface area for total DNA coupling than DNA Sepharose. Secondly, the beads have a hydrophilic surface, which minimizes non-specific protein binding. Thirdly, the beads are free to move in the binding buffer so that DNA-binding proteins can bind to DNA on the surface of the beads without any steric interference. Finally, the batch-wise method employed in this protocol reduces contamination with non-specific DNA-binding proteins because the method minimizes the volume of the beads. More than ten sequence-specific DNA-binding proteins have been purified by the method.

Briefly, the protein fractions are mixed well with non-specific competitor DNA, such as poly(dI:dC) or single-stranded DNA and are then added to the DNA affinity resin. After washing with buffer containing the appropriate concentration of salt, the bound protein of interest is eluted with buffer containing a high salt concentration. The DNA affinity resin can then be regenerated for reuse. Usually, the presence of the desired protein in the eluted fraction is verified using a DNA-binding assay and the purity of the fraction is estimated by SDS–PAGE followed by silver staining to visualize the protein. The method described in *Protocol 5* relies on DNA immobilization to CNBr-activated Sepharose whilst that in *Protocol 6* uses latex beads as the matrix.

4.4 DNA affinity Sepharose

Protocol 5. Purification of DNA-binding proteins on a Sepharose matrix

Equipment and reagents

- Buchner funnel with a sintered glass filter
- Beckman CS-6KR centrifuge (or equivalent)
- Rotating incubator or blood wheel
- Poly-Prep chromatography column (Bio-Rad)
- CNBr-activated Sepharose 4B (Pharmacia Biotech)
- Multimerized DNA oligomers
- Partially purified protein fraction

- Milli Q water (or equivalent)
- 1 mM HCl
- 1 M ethanolamine–HCl pH 8
- 10 mM potassium phosphate pH 8
- 1 M KCl
- Non-specific competitor DNA: poly(dI:dC), poly(dA:dT), poly(dG:dC) (Pharmacia Biotech), or salmon sperm DNA (Sigma) can all be used

11: Purification of DNA-binding proteins

- 2.5 M NaCl
- Buffer A: 50 mM Tris–HCl pH 7.9, 20% (v/v) glycerol, 0.1 M KCl, 0.5 mM EDTA, 0.1% (v/v) Nonidet P-40 (NP-40)—add 1 mM DTT and 0.5 mM PMSF immediately before use; store at 4°C
- Buffer B: buffer A plus appropriate concentration of KCl[a]
- Column storage buffer: 10 mM Tris–HCl pH 8, 0.3 M KCl, 1 mM EDTA, 0.02% (w/v) NaN$_3$

A. Coupling of DNA oligomers to CNBr-activated Sepharose

1. Swell 1 g dry CNBr Sepharose powder in 1 mM HCl for 15 min. About 3.5 ml swollen gel should be recovered.
2. Wash the gel with 200 ml 1 mM HCl[b] in a Buchner funnel with a sintered glass filter.
3. Wash the gel with 100 ml Milli Q water and then with 100 ml 10 mM potassium phosphate pH 8.
4. Combine 100 µg of the DNA oligomers, which were prepared according to *Protocol 4*, with 1 ml CNBr-activated Sepharose in 2–3 ml 10 mM potassium phosphate pH 8 in a 15 ml polypropylene centrifuge tube. Incubate the mixture overnight at room temperature using end-over-end agitation using a rotating incubator.
5. Centrifuge the resin at 2000 r.p.m. (1000 g) for about 10 sec and discard the supernatants.
6. Add 2–3 ml 1 M ethanolamine pH 8 to the resin and incubate for 2 h at room temperature on a rotating incubator.
7. Wash the resin with 5 ml 10 mM potassium phosphate buffer and wash with 5 ml 1 M KCl.
8. Resuspend the gel in column storage buffer. The DNA affinity column may be used immediately. 1 ml CNBr-activated Sepharose will carry about 20–30 µg of the DNA.[c]

B. Purification of sequence-specific DNA-binding proteins

All of the steps should be performed at 4°C.

1. Pour 1 ml DNA affinity Sepharose into a disposable Poly-Prep chromatography column without allowing the column to drain.
2. Equilibrate the gel with 10 ml buffer A.
3. Add the partially purified protein fraction[d] which has been mixed with non-specific competitor DNA[e] to the gel and let stand for 10 min. Do not drain the column at this stage.
4. Mix the gel and the protein fraction gently using a 10 µl disposable pipette by hand and let it stand for 30 min.
5. Allow the unbound protein fraction to flow through the column by gravity.
6. Reapply the unbound protein fraction to the column and repeat steps 3–5.

Protocol 5. Continued

7. Wash the column with 10 ml buffer A.
8. Add 5 ml buffer B to elute the protein.
9. Collect 0.5 ml fractions (total of ten fractions), quick-freeze the protein in liquid N_2, and store at $-80\,°C$.
10. Wash the column with 10 ml 2.5 M NaCl to regenerate it and wash with column storage buffer.
11. Store the column, which should be capped to avoid drying, at $4\,°C$.

[a] The salt concentration of buffer B is determined experimentally. It is recommended to employ the batchwise method as described in *Protocol 3*.
[b] HCl is used to keep the pH low for protection of the activated group.
[c] Count the radioactivity of the DNA to estimate the DNA coupling efficiency.
[d] To remove proteases and nucleases, partial purification by conventional chromatography prior to affinity chromatography is recommended. Before loading the protein fractions on DNA affinity Sepharose, it is necessary to dialyse the protein fractions against buffer A.
[e] The salt concentration of the protein fraction should not be over 0.1 M KCl and the amount of competitor DNA added should be determined by DNA-binding or DNase I footprint assays.

4.5 DNA affinity latex beads

Protocol 6. Purification of DNA-binding proteins on latex beads

Equipment and reagents

- Refrigerated microcentrifuge
- Incubator or water-bath at $50\,°C$
- Latex beads[a]
- Multimerized DNA oligomers
- 10 mM potassium phosphate pH 8
- HeLa cell nuclear extracts
- Milli Q water (or equivalent)
- 2.5 M NaCl
- Storage buffer: 10 mM Tris–HCl pH 8, 0.3 M KCl, 1 mM EDTA, 0.02% (w/v) NaN_3
- 1 M ethanolamine–HCl pH 8
- Buffer A: 50 mM Tris–HCl pH 7.9, 20% (v/v) glycerol, 0.5 mM EDTA, 0.1% (v/v) NP-40—add 1 mM DTT and 0.5 mM PMSF immediately before use; store at $4\,°C$
- Buffer B: buffer A plus 1 M KCl
- Non-specific competitor DNA: poly(dI:dC), poly(dA:dT), poly(dG:dC) (Pharmacia Biotech), or salmon sperm DNA (Sigma)
- Liquid nitrogen

A. *Coupling reaction*

1. Transfer 10 mg latex beads to a 1.5 ml microcentrifuge tube.
2. Centrifuge the latex beads at 15 000 r.p.m. (18 000 *g*) for a few minutes until the supernatant becomes clear.
3. Decant and discard the supernatant, and add 400 µl 10 mM potassium phosphate pH 8.
4. Resuspend the latex beads thoroughly[b] and centrifuge at 15 000 r.p.m. (18 000 *g*) for a few minutes.

11: Purification of DNA-binding proteins

5. Repeat steps 3 and 4 twice.
6. Wash the latex beads twice with Milli Q water.
7. Measure the radioactivity of the DNA being immobilized.[c]
8. Add 100 μg of the DNA to latex beads and mix well.
9. Incubate the sample at 50°C for 3h (overnight incubation can also be used).

B. Masking

1. Centrifuge the latex beads carrying the DNA at 15 000 r.p.m. (18 000 g) for a few minutes at room temperature.
2. Transfer the supernatant containing the unbound DNA to a new 1.5 ml microcentrifuge tube for reuse.[d]
3. Wash the latex beads twice with 500 μl 2.5 M NaCl.
4. Wash the latex beads twice with 500 μl Milli Q water.
5. Resuspend the latex beads in 1 ml 1 M ethanolamine pH 8.
6. Incubate the sample for 24 h at room temperature.

C. Storage

1. Centrifuge the latex beads at 15 000 r.p.m. (18 000 g) for a few minutes.
2. Wash once with 500 μl Milli Q water.
3. Wash twice with 500 μl storage buffer.
4. Resuspend the latex beads in 200 μl storage buffer.[e] Usually 20–40% of input DNA becomes coupled to the beads.
5. The beads should be stored at 4°C.

D. Purification of the sequence-specific DNA-binding protein

This method enables purification of the protein of interest directly from nuclear extracts. This section describes how to purify proteins from HeLa cell nuclear extracts using a batchwise procedure (*Figure 5*). All of the steps should be performed at 4°C.

1. Add non-specific competitor DNA to 1 ml HeLa cell nuclear extract[f] and incubate on ice for 15 min.
2. Centrifuge the sample at 15 000 r.p.m. (18 000 g) for 15 min to remove insoluble material.
3. During step 2, transfer latex beads containing 5–10 μg of the DNA oligomers to a 1.5 ml microcentrifuge tube and wash three times with 100 μl buffer A.
4. Transfer the supernatant from step 2 to the latex beads, mix thoroughly, and let stand on ice for 30 min. The colour of the solution should be white due to a stable dispersion of the beads in the solution.

Protocol 6. *Continued*

5. Centrifuge the latex beads at 15 000 r.p.m. (18 000 g) for 5–10 min to clarify the supernatant.[g]

6. Transfer the supernatant to a new 1.5 ml microcentrifuge tube, quick-freeze in liquid N_2, and store at –80 °C.

7. Wash the latex beads three times with 200 μl buffer A containing 0.1 M KCl.[h]

8. Wash the latex beads three times with 200 μl buffer A containing 0.3 M KCl.[h,i]

9. Add 20 μl buffer B and mix well.

10. Let stand on ice for 5 min and centrifuge at 15 000 r.p.m. (18 000 g) for a few minutes.

11. Collect the supernatant. About 70% of the bound protein should be recovered in one step.

12. Repeat steps 9–11 twice.

13. Quick-freeze the protein fractions in liquid N_2 and store at –80 °C.

14. Wash the latex beads once with 200 μl 2.5 M NaCl and three times with 200 μl storage buffer to regenerate. The latex beads can be recycled at least ten times when HeLa cell nuclear extract is used as the starting material.

[a] Preparation of latex beads is described elsewhere (16).
[b] An effective method of resuspending the beads is to use a 'fraction holder' which has many hard rods on one side. The tightly-capped microcentrifuge tube is then rubbed on the rods several times (see *Figure 4*). After mixing, the latex beads are centrifuged at 15 000 r.p.m. for a few minutes.
[c] For estimation of the DNA coupling efficiency.
[d] After phenol:chloroform extraction and ethanol precipitation, the DNA oligomers can be reused for another coupling reaction.
[e] Count the radioactivity of the DNA bound to the beads to estimate the DNA coupling efficiency.
[f] There are many DNA-binding proteins in nuclear extracts. To increase the purity of the desired protein, proper use of competitor DNA is essential. There are a variety of competitor DNAs, including poly(dI:dC), poly(dA:dT), poly(dG:dC), salmon sperm DNA, and single-stranded DNA. As the correct choice of the competitor depends on the protein to be purified, the amount and type of competitor must be determined experimentally. Usually a mini-scale purification helps to determine these conditions (see Section 4.7). In the case of purification of transcription factor ATF/E4TF3 from HeLa cell nuclear extracts, 20 μg single-stranded salmon sperm DNA and 2 μg poly(dI:dC) were added to 1 ml of the extracts (8 mg total proteins) (11).
[g] The actual length of this centrifugation step depends on the buffer used and the proteins involved.
[h] Buffer A containing KCl is prepared by mixing buffer A and B in appropriate quantities.
[i] This washing step is optional. The salt concentration in the wash buffer should be determined experimentally by a mini-scale method (see Section 4.7).

11: Purification of DNA-binding proteins

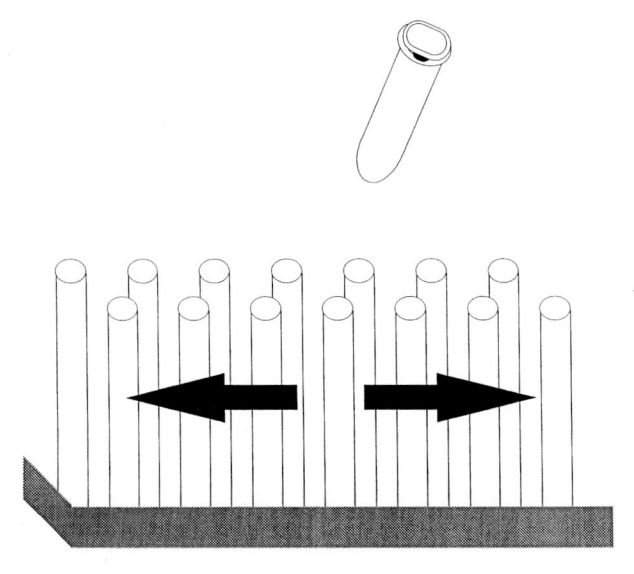

Figure 4. A method for mixing the latex beads and solution homogeneously. Use a fraction holder or equivalent which has many hard rods on one side and rub the tightly-capped microcentrifuge tube on the rods until the flocculated beads are completely suspended.

4.6 Troubleshooting

4.6.1 Low coupling efficiency of DNA

If the coupling efficiency of the DNA is low, it is recommended that before the coupling reaction, the DNA oligomers should be purified by using other types of gel filtration chromatography or polyacrylamide gel electrophoresis, because some metabolites which contain amino groups, such as nucleotides, inhibit efficient binding of the DNA oligomers to resins.

4.6.2 Low recovery of purified protein

When the affinity of the desired protein with DNA is low, crude nuclear extracts should not be used as a source of purification. Fractionation of crude nuclear extracts by conventional column chromatography should help to solve this problem and should increase the purity and recovery of the desired proteins.

4.6.3 Contamination with non-specific DNA-binding proteins

In order to avoid contamination with non-specific DNA-binding proteins, it is recommended to add non-specific DNAs to crude nuclear extracts or the protein fractions before purification. However, if the source is HeLa cell nuclear extracts, two non-specific DNA-binding proteins, ADP ribosylase with a molecular weight of 116 kDa and Ku antigens with molecular weights of 80

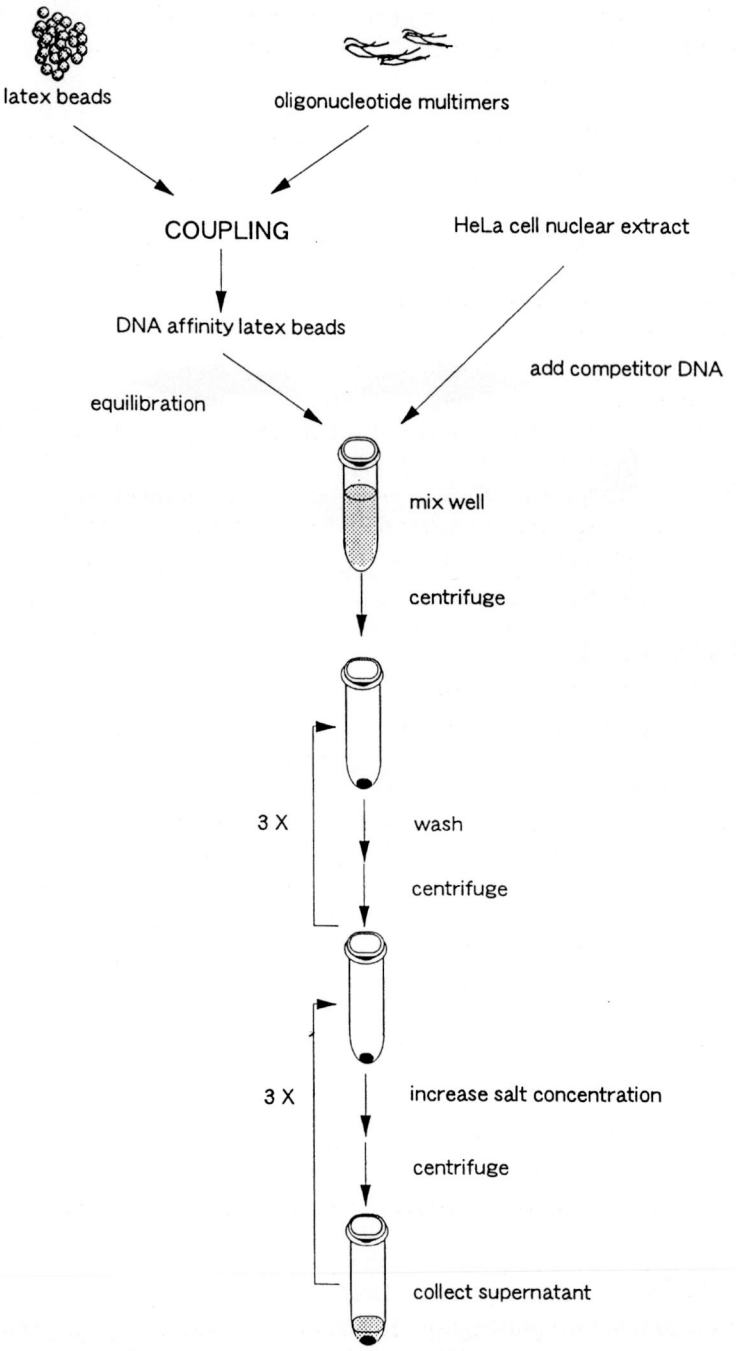

Figure 5. Schematic representation of purification of the sequence-specific DNA-binding protein by DNA affinity latex beads.

and 70 kDa are commonly co-purified with the desired proteins. A second washing buffer, containing 0.2–0.3 M NaCl or KCl, is effective in reducing the background contamination.

4.6.4 Observation of sequence-specific DNA-binding protein(s)

When multiple polypeptides are included in the purified fraction, they may bind to the target sequence specifically, since some transcription factors form a family, in which multiple proteins recognize and bind to an identical DNA sequence. For example, transcription factors ATF/E4TF3 is purified as at least eight polypeptides with different molecular weights 116, 80, 65, 60, 55, 47, 45, and 43 kDa from HeLa cell nuclear extracts by DNA affinity latex beads and each protein binds to the ATF/E4TF3 sequence (11). To test for the DNA-binding activities of each polypeptides included in the purified fraction, it is a method of choice to renature the polypeptides after fractionation by a SDS–polyacrylamide gel (11, 17). Renatured proteins can then be tested for their sequence-specific DNA-binding activity individually by DNA-binding assays. However, since some proteins may only bind to DNA as a heterodimer, a combination of these renatured proteins should also be tested in DNA-binding assays.

4.7 Comments

4.7.1 Mini-scale purification

In order to determine a critical condition, such as amount of non-specific competitor DNA or requirement of Mg^{2+} for the desired protein's activity, a mini-scale purification using a batch method should be performed. For example, to determine the amount of non-specific competitor DNA to use in an affinity chromatography experiment, prepare several tubes containing 100 µl of cell extracts and add different amounts of the competitor DNA to each tube. The extracts are then subjected to DNA affinity chromatography and the protein fractions analysed by a DNA-binding assay and SDS–PAGE followed by silver staining to visualize the protein. Usually, the larger the amount of the competitor DNA the higher the specific activity obtained but a reduction in the DNA-binding protein recovery.

4.7.2 Other methods

Other methods, including biotinylated DNA fragments immobilized on Sepharose using biotin–avidin or biotin–streptavidin interaction, are also available for purification of sequence-specific DNA-binding proteins (18).

References

1. Dignam, J. D., Lebovitz, R. M., and Roeder, R. G. (1983). *Nucleic Acids Res.*, **11**, 1475.
2. Carthew, R. W., Chodosh, L. A., and Sharp, P. A. (1985). *Cell*, **43**, 439.

3. Singh, H., Sen, R., Baltimore, D., and Sharp, P. A. (1986). *Nature*, **319**, 154.
4. Farooqui, A. A. (1980). *J. Chromatogr.*, **184**, 335.
5. Bickle, T. A., Pirrotta, V., and Imber, R. (1977). *Nucleic Acids Res.*, **4**, 2561.
6. Golomb, M., Vora, A. C., and Grandgenett, D. P. (1980). *J. Virol. Methods*, **1**, 157.
7. Watanabe, H., Imai, T., Sharp, P. A., and Handa, H. (1988). *Mol. Cell. Biol.*, **8**, 1290.
8. Kerrigan, L. A. and Kadonaga, J. T. (1995). In *Current protocols in molecular biology* (ed. F. M. Ausubel, R. Brent, R. E. Kingston, D. D. Moore, J. G. Seidman, J. A. Smith, and K. Struhl), p. 12.10.1. Wiley (Interscience), New York.
9. Axén, R., Porath, J., and Ernback, S. (1967). *Nature*, **214**, 1302.
10. Cuatrecasas, P., Wilchek, M., and Anfinsen, C. B. (1968). *Proc. Natl. Acad. Sci. USA*, **61**, 636.
11. Inomata, Y., Kawaguchi, H., Hiramoto, M., Wada, T., and Handa, H. (1992). *Anal. Biochem.*, **206**, 109.
12. Weiss, B., Jacquemin-Sablon, A., Live, T. R., Fareed, G. C., and Richardson, C. C. (1968). *J. Biol. Chem.*, **243**, 4543.
13. Arndt-Jovin, D. J., Jovin, T. M., Bähr, W., Frischauf, A.-M., and Marquardt, M. (1975). *Eur. J. Biochem.*, **54**, 411.
14. Duncan, C. H. and Cavalier, S. L. (1988). *Anal. Biochem.*, **169**, 104.
15. Kawaguchi, H., Asai, A., Ohtsuka, Y., Watanabe, H., Wada, T., and Handa, H. (1989). *Nucleic Acids Res.*, **17**, 6229.
16. Wada, T., Watanabe, H., Kawaguchi, H., and Handa, H. (1995). In *Methods in enzymology* (ed. P. K. Vogt and I. M. Verma), Vol. 254, p. 595. Academic Press, London.
17. Hager, D. A. and Burgess, R. R. (1980). *Anal. Biochem.*, **109**, 76.
18. Chodosh, L. A. (1995). In *Current protocols in molecular biology* (ed. F. M. Ausubel, R. Brent, R. E. Kingston, D. D. Moore, J. G. Seidman, J. A. Smith, and K. Struhl), p. 12.6.1. Wiley (Interscience), New York.

A1

List of suppliers

This core list of suppliers appears in all books in the Practical Approach series. If there are any relevant suppliers that you would like to add to this list for the book you are working on, please send them with your chapter.

Clay Adams, Division of Becton Dickinson and Co., Parasipany, NJ 07054, USA.
Aldrich Chemical Co.
Aldrich Chemical Co. Inc., 1001 West Saint Paul Avenue, Milwaukee, WI 53233, USA.
Aldrich Chemical Co., The Old Brickyard, New Road, Gillingham, Dorset SP8 4JL, UK.
Alltech Associates, Unit 6/7 Kellet Road Industrial Estate, Carnforth, Lancashire LA5 9XP, UK.
Amersham
Amersham International plc., Lincoln Place, Green End, Aylesbury, Buckinghamshire HP20 2TP, UK.
Amersham Corporation, 2636 South Clearbrook Drive, Arlington Heights, IL 60005, USA.
Amicon
Amicon Inc., 72 Cherry Hill Drive, Beverly, MA 01915, USA.
Amicon, Upper Mill, Stonehouse, Gloucestershire GL10 2BJ, UK.
Analytica of Branford
Analytica of Branford, 29 Business Park Drive, Branford, CT 06405, USA.
Analytica of Branford, Quad Service, Technoparc, 29 rue Edouard Jeanneret, 78300 Poissy, France.
Anderman
Anderman and Co. Ltd., 145 London Road, Kingston-Upon-Thames, Surrey KT17 7NH, UK.
Applied Biosystems
Applied Biosystems, Inc. (Division of Perkin-Elmer), 850 Lincoln Centre Drive, Foster City, CA 94404, USA.
Applied Biosystems, Kelvin Close, Birchwood Science Park N., Warrington, Cheshire WA3 7PB, UK.

List of suppliers

Applied Biosystems, Seergreen, 7 Kingsland Grange, Woolston WA1 4SR, UK.

Beckman Instruments

Beckman Instruments UK Ltd., Progress Road, Sands Industrial Estate, High Wycombe, Buckinghamshire HP12 4JL, UK.

Beckman Instruments Inc., PO Box 3100, 2500 Harbor Boulevard, Fullerton, CA 92634, USA.

Becton Dickinson

Becton Dickinson and Co., Between Towns Road, Cowley, Oxford OX4 3LY, UK.

Becton Dickinson and Co., 2 Bridgewater Lane, Lincoln Park, NJ 07035, USA.

Bethesda Research Laboratories (BRL), PO Box 6010, Rockville, MD 20850, USA.

Bio

Bio 101 Inc., c/o Stratech Scientific Ltd., 61–63 Dudley Street, Luton, Bedfordshire LU2 0HP, UK.

Bio 101 Inc., PO Box 2284, La Jolla, CA 92038–2284, USA.

Bio-Rad Laboratories

Bio-Rad Laboratories Ltd., Bio-Rad House, Maylands Avenue, Hemel Hempstead HP2 7TD, UK.

Bio-Rad Laboratories, Division Headquarters, 3300 Regatta Boulevard, Richmond, CA 94804, USA.

BioSepra UK, Clarendon House, 125 Shenley Road, Borehamwood, Hertfordshire WD6 1AG, UK.

Boehringer Mannheim

Boehringer Mannheim UK (Diagnostics and Biochemicals) Ltd., Bell Lane, Lewes, East Sussex BN17 1LG, UK.

Boehringer Mannheim Corporation, Biochemical Products, 9115 Hague Road, PO Box 504, Indianapolis, IN 46250–0414, USA.

Boehringer Mannheim Biochemica, GmbH, Sandhofer Strasse 116, Postfach 310120, D-6800 Ma 31, Germany.

British Drug Houses (BDH) Ltd., Poole, Dorset, UK.

Difco Laboratories

Difco Laboratories Ltd., PO Box 14B, Central Avenue, West Molesey, Surrey KT8 2SE, UK.

Difco Laboratories, PO Box 331058, Detroit, MI 48232–7058, USA.

Du Pont

Dupont (UK) Ltd. (Industrial Products Division), Wedgwood Way, Stevenage, Hertfordshire SG1 4Q, UK.

Du Pont Co. (Biotechnology Systems Division), PO Box 80024, Wilmington, DE 19880–002, USA.

European Collection of Animal Cell Culture, Division of Biologics, PHLS Centre for Applied Microbiology and Research, Porton Down, Salisbury, Wiltshire SP4 0JG, UK.

List of suppliers

Falcon (Falcon is a registered trademark of Becton Dickinson and Co.)
Finnigan
Finnigan Corporation, 355 River Oaks Park Way, San Jose, CA 95134, USA.
Finnigan MAT Ltd., Paradise, Hemel Hempstead, Hertfordshire HP2 4TG, UK.
Fisher Scientific Co., 711 Forbest Avenue, Pittsburgh, PA 15219–4785, USA.
Flow Laboratories, Woodcock Hill, Harefield Road, Rickmansworth, Hertfordshire WD3 1PQ, UK.
Fluka
Fluka-Chemie AG, CH-9470, Buchs, Switzerland.
Fluka Chemicals Ltd., The Old Brickyard, New Road, Gillingham, Dorset SP8 4JL, UK.
Gibco BRL
Gibco BRL (Life Technologies Ltd.), Trident House, Renfrew Road, Paisley PA3 4EF, UK.
Gibco BRL (Life Technologies Inc.), 3175 Staler Road, Grand Island, NY 14072–0068, USA.
HP, 1601 California Avenue, Palo Alto, CA 94404, USA.
Arnold R. Horwell, 73 Maygrove Road, West Hampstead, London NW6 2BP, UK.
Hybaid
Hybaid Ltd., 111–113 Waldegrave Road, Teddington, Middlesex TW11 8LL, UK.
Hybaid, National Labnet Corporation, PO Box 841, Woodbridge, NJ 07095, USA.
HyClone Laboratories, 1725 South HyClone Road, Logan, UT 84321, USA.
International Biotechnologies Inc., 25 Science Park, New Haven, Connecticut 06535, USA.
Invitrogen Corporation
Invitrogen Corporation, 3985 B Sorrenton Valley Building, San Diego, CA 92121, USA.
Invitrogen Corporation, c/o British Biotechnology Products Ltd., 4–10 The Quadrant, Barton Lane, Abingdon, Oxon OX14 3YS, UK.
J & W Scientific, Inc.
J & W Scientific Inc., 91 Blue Ravine Road, Folsom, CA 95630, USA.
Fisher Scientific, Bishop Meadow Road, Loughborough, Leicestershire LE11 ORG, UK.
Kodak: Eastman Fine Chemicals, 343 State Street, Rochester, NY, USA.
Labconco Corporation, 8811 Prospect Avenue, Kansas City, MO 64132–2696, USA.
LC Packings
LC Packings (USA) Inc., 80 Carolina Street, San Francisco, CA 94103, USA.
LC Packings, Dufourstrasse 30, 8008 Zurich, Switzerland.

List of suppliers

Life Technologies Inc., 8451 Helgerman Court, Gaithersburg, MN 20877, USA.
Merck
Merck Industries Inc., 5 Skyline Drive, Nawthorne, NY 10532, USA.
Merck, Frankfurter Strasse, 250, Postfach 4119, D-64293, Germany.
Microbioresources, 1945 Industrial Drive, Aubum, CA 95603, USA.
Micromass (formerly Fisons), Floats Road, Wytenshawe, Manchester M23 9LZ, UK.
Micro Mass UK Ltd., 3 Tudor Road, Altrincham, Cheshire WA14 5RZ, UK.
Micro-Tech, 140 S. Wolfe Road, Sunnyvale, CA 94086, USA.
Millipore
Millipore (UK) Ltd., The Boulevard, Blackmoor Lane, Watford, Hertfordshire WD1 8YW, UK.
Millipore Corp./Biosearch, PO Box 255, 80 Ashby Road, Bedford, MA 01730, USA.
New England Biolabs (NBL)
New England Biolabs (NBL), 32 Tozer Road, Beverley, MA 01915–5510, USA.
New England Biolabs (NBL), c/o CP Labs Ltd., PO Box 22, Bishops Stortford, Hertfordshire CM23 3DH, UK.
Nikon Corporation, Fuji Building, 2–3 Marunouchi 3-chome, Chiyoda-ku, Tokyo, Japan.
Pel-Freez Biologicals, PO Box 68 Rogers, Arkansas, USA.
Perkin-Elmer
Perkin-Elmer Ltd., Maxwell Road, Beaconsfield, Buckinghamshire HP9 1QA, UK.
Perkin Elmer Ltd., Post Office Lane, Beaconsfield, Buckinghamshire HP9 1QA, UK.
Perkin Elmer-Cetus (The Perkin-Elmer Corporation), 761 Main Avenue, Norwalk, CT 0689, USA.
PerSeptive Biosystems (UK) Ltd., 3 Harforde Court, Foxholes Business Park, John Tate Road, Hertford SG13 7NW, UK.
PE Sciex
PE Sciex (Division of Perkin-Elmer), 71 Four Valley Drive, Concord, ON, Canada.
PE Sciex, 850 Lincoln Center Drive, Foster City, CA 94404, USA.
Pharmacia Biotech Europe, Procordia EuroCentre, Rue de la Fuse-e 62, B-1130 Brussels, Belgium.
Pharmacia Biosystems
Pharmacia Biosystems Ltd. (Biotechnology Division), Davy Avenue, Knowlhill, Milton Keynes MK5 8PH, UK.
Pharmacia LKB Biotechnology AB, Björngatan 30, S-75182 Uppsala, Sweden.
Polymicro Technologies
Polymicro Technologies, Inc., 18019 N 25th Avenue, Phoenix, AZ 85023, USA.

List of suppliers

Polymicro Technologies, Composite Metal Service Ltd., The Chase, Hallow, Worcester WR2 6LD, UK.
Presearch Ltd., 13 Business Centre West, Avenue One, Letchworth Garden City SG6 2HB, UK.
Promega
Promega Ltd., Delta House, Enterprise Road, Chilworth Research Centre, Southampton, UK.
Promega Corporation, 2800 Woods Hollow Road, Madison, WI 53711–5399, USA.
Qiagen
Qiagen Inc., c/o Hybaid, 111–113 Waldegrave Road, Teddington, Middlesex TW11 8LL, UK.
Qiagen Inc., 9259 Eton Avenue, Chatsworth, CA 91311, USA.
Rainin Instrument Co.
Rainin Instrument Co. Inc., Mack Road, Box 4026, Woburn, MA 01888, USA.
Anachem, 20 Charles Street, Luton, Bedfordshire LU2 0EB UK.
Rohm Pharma, Westerstadt, PO Box 4347, D-6100, Darmstadt 1, FRG.
Schleicher and Schuell
Schleicher and Schuell Inc., Keene, NH 03431A, USA.
Schleicher and Schuell Inc., D-3354 Dassel, Germany.
Schleicher and Schuell Inc., c/o Andermann and Co. Ltd.
Scientific Resources
Scientific Resources Inc., PO Box 1290, Eatontown, NJ 07724 USA.
N.L.G. Analytical, 3 Edinburgh Drive, Macclesfield, Cheshire SK10 3PZ, UK.
Shandon Scientific Ltd., Chadwick Road, Astmoor, Runcorn, Cheshire WA7 1PR, UK.
Sigma Chemical Company
Sigma Chemical Company (UK), Fancy Road, Poole, Dorset BH17 7NH, UK.
Sigma Chemical Company, 3050 Spruce Street, PO Box 14508, St. Louis, MO 63178–9916, USA.
Sorvall DuPont Company, Biotechnology Division, PO Box 80022, Wilmington, DE 19880–0022, USA.
Stratagene
Stratagene Ltd., Unit 140, Cambridge Innovation Centre, Milton Road, Cambridge CB4 4FG, UK.
Stratagene Inc., 11011 North Torrey Pines Road, La Jolla, CA 92037, USA.
Supelco
Supelco, Inc., Supelco Park, Bellefonte, PA 16823, USA.
Supelco, Fancy Road, Poole, Dorset BH12 4QH, UK.
Tomy Seiko Co. Ltd., 2-2-12 Asahi-cho, Nerima-ku, Tokyo 176, Japan.
Toso Haas (UK), 7 Lonsdale Union, Cambridge CB1 6LT, UK.
United States Biochemical, PO Box 22400, Cleveland, OH 44122, USA.
VirTis, Gardiner, NY 12525, USA.
Wellcome Reagents, Langley Court, Beckenham, Kent BR3 3BS, UK.

Index

N-acetylgalactosamine (GalNAc) 119
Affi-Gel 10 135, 205, 247, 248
affinity-CE 110
4-aminoanilido–GTP 164
N^6-2-aminoethyl–NAD 158
C^8-6-aminoethyl–NADP$^+$ 159–60
angiotensin converting enzyme
    226–7
annexins 265–77, 266
    purification 267–8, 268–70
    purification scheme 271
antibody 236
antibody immobilization 242–53
    via biotin/avidin 250, 253
    via CNBr 246–7
    via hydrazide 251–3
    via NHS 249–50
    via protein A/protein G 250–1
antigen 237

binding constant 109–11, 237
biotinylation 228, 229–30, 245
blue dextran 3
borohydride 139, 142
butyl–Sepharose 10

Ca^{2+}-binding proteins 263–5
^{45}Ca^{2+}-binding 265, 277
calmodulin 263
capillary
    diameter 78
    gel electrophoresis (CGE) 102, 111
    isoelectric focusing (cIEF) 100–2
capillary electrophoresis (CE) 77–117, 103
    buffers 91, 90–3
    phosphate buffer 93
    sample concentration 94–7
    sample preparation (cIEF, CZE, MEKC) 93
    sample preparation (CGE) 94
capillary wall
    adsorbed coating 85–6
    buffer additives 86–90
    coating 80–5, 90
    modification 79–90
capillary zone
    electrophoresis (CZE) 98–9
    isotachophoresis (cITP) 99–100
carbodiimide 162, 164, 210, 212
N,N'-carbonyldiimidazole (CDI) 203, 244
    activation 204
carboxymethyl (CM) 6
chaotrope 214, 256–7

Chelex-100 277, 279, 280–1
chromatography
    column 11–12
    column packing 14–16
    column sanitation 20
    fraction collection 14
    sample loading 16–18
Cibacron Blue 154, 165, 167, 169, 173–4
CNBr 197, 244
CNBr activation 196–8, 243, 244–7
CNBr–Sepharose
    coupling to 135, 160, 197, 222
    lisinopril coupling 223, 224
    oligonucleotide coupling 295
concanavalin A (Con A) 119
coupling efficiency 193, 225–6, 253–4, 298, 299
cyanuric chloride 198–200

decamethonium bromide 89
detector
    conductivity 66–8
    dynamic range 49–50
    electrochemical 68–70
    electronic 66
    light scattering 72–5
    linearity 50–1
    multi-electrode 70–2
    noise 52–3
    noise measurement 53
    response 51
diethylaminoethyl (DEAE) 6, 274, 279
dihydrolipoamide 207
disuccinimidyl succinate (DMS) 248
5,5'-dithiobis(2-nitrobenzoic acid) (DTNB) 209
DNA
    affinity beads 296–8, 300
    affinity chromatography 290
    Sepharose 294–5
DNA-binding proteins 283
    purification 295–6, 296–8
    sequence-specific 301
dye
    content 169
    coupling 165
    designer 171–3
    leaching 170
    polymer shielding 173–5
    triazine 167–8

electro-osmotic flow (EOF) 79, 80, 83, 102
enterokinase 238

Index

epoxide 244
exclusion limit (V_e) 2, 4

FAST-Q 273
FAST-S 273, 278
fluidics 12–13
Fluorad 87
fluorescence
  derivatization 106–7
  native 106
  tag 106
fluorescence detector
  fixed wavelength 63–6, 106
  multi-wavelength 66

gel permeation (gel filtration) 1–4, 273
  sample size 4
glycan 121–3, 138, 144–5
N-glycanase (PNGase A, PNGase F) 139, 140
glycopeptide preparation 139
glycoprotein 111, 120
glutaraldehyde 202
  activation 202–3
gradient maker 19
  dimensions 20
GTP–Sepharose 164

heparin 274–7
  protein purification 276–7, 286–7
hexamethonium bromide 89
HPLC column sizes 24
hydrazine 139, 141, 160, 245
hydrophobic interaction chromatography (HIC) 9–11
hydroxylapatite 269, 270, 273, 278
N-hydroxysuccinimide (NHS) 136, 205, 244
  activation 205–6, 247–50

IgG 236
imidazole 163
immunoaffinity
  applications 235
  chromatography 233–6
  eluents 257
  matrix 241
  protease inhibitors 214, 259
  purification 258
iodoacetic acid 210
ion exchange
  chromatography 5–9
  elution salts 8
  media 6
  salt gradient 9
ion pairing 87–9

lactate dehydrogenase 177
lectin 119–51
  chromatography 142, 146
  glycoproteins 136–8
  oligosaccharides 138
  elution conditions 134
  immobilization 135
  specificity 125, 128–33

maleimide 211
mass-sensitive detector 54
mass spectroscopy (MS) 107–8, 113–14
matrix
  activated 194–5, 220
  see also antibody immobilization
  regeneration 230
methacryloxypropyltrimethoxysilane (MAPS) 81
methyl α-glucoside 143
methyl α-mannoside 143
micellar electrokinetic capillary chromatography (MEKC) 100
microcolumn
  detector 32
  flow splitting 35–7
  HPLC 23–46
  packing 25
  pre-column focusing 30–2
  pumping 33, 37
  sample loading 29–32
  solvent compressibility 40
  solvent gradient 38, 41–4
monoclonal antibody (mAb) 237, 242, 243

nuclear extract 283, 284–5
nucleotide–ligand
  base coupling 156–60
  chromatography 153–5, 166, 178–84
  matrix 155
  phosphate coupling 161
  sugar coupling 160

oligonucleotide
  design 291–2
  preparation 292–3
oligosaccharide
  N-linked 120, 124
  O-linked 124, 126–7
  processing 124
over-coupling 193

peptide mapping 108–9
phenyl–Sepharose 10
pI determination 7
polybrene 86
poly(ethylene imine) 85–6, 164, 173
polyvinylpyrrolidone 174
Procion Red 154, 165, 173
pronase 139
protein A 239–45
protein G 239–45

Index

quaternary aminoethyl (QAE) 6

reactive groups 244, 245

sample purity 108
siloxane bond 81–2
silylation 81
solid phase reductant
   see dihydrolipoamide 207
solute property detector 54
spacer arms 193, 220
spermine 89
sulfo-MBS 211
sulfopropyl (SP) 6
sulfydryl determination 209
superloop 18
surfactant 100, 237
synthetic peptide 191
   affinity chromatography 213–14

elution 214
immobilization 192–3, 201, 204, 206, 211
reduction 208–9
synthesis 191–2

thiol group 244
tosyl chloride 245
tresyl chloride 245
tubing, volume/length 17

UV detector
   fixed wavelength 13–14, 58–61, 105
   multi-wavelength 61–3

vinyl sulfone 244
viscous polymer 103–4
void volume (V_o) 2
   determination 3